PETROLEUM & PUBLIC SAFETY

MAKING THE MODERN SOUTH

David Goldfield, Series Editor

PETROLEUM & PUBLIC SAFETY

RISK MANAGEMENT IN THE GULF SOUTH 1901–2015

James B. McSwain

LOUISIANA STATE UNIVERSITY PRESS BATON ROUGE

Published by Louisiana State University Press

Manufactured in the United States of America
First printing

Designer: Barbara Neely Bourgoyne
Typeface: Chaparral Pro
Printer and binder: Sheridan Books

Cataloging-in-Publication Data are available at the Library of Congress.

ISBN 978-0-8071-6912-4 (cloth : alk. paper) — ISBN 978-0-8071-6913-1 (pdf) —
ISBN 978-0-8071-6914-8 (epub)

The paper in this book meets the guidelines for permanence and durability of the Committee on Production Guidelines for Book Longevity of the Council on Library Resources. ∞

This book is dedicated to four scholars that God used to encourage me to publish my ideas:

Professor David Goldfield, University of North Carolina, Charlotte, Editor of the *Journal of Urban History*

Professor John Boles, Rice University, Editor Emeritus of the *Journal of Southern History*

Professor Martin V. Melosi, University of Houston

Professor Joseph A. Pratt, University of Houston

Your Committee find, that the substance known by the name of Petroleum, Rock Oil, or Earth Oil, as it is generally received in its crude and unrefined state, is largely charged with volatile matter, highly inflammable in its nature, and evolved, to some extent, at the usual temperature of the atmosphere, and much more freely by an increased degree of heat.

—New York Board of Fire Insurance Companies, December 23, 1861

FUEL OIL.
Under your direction we have published rules and requirements regulating the hazard of the storage and use of fuel oil, and for the construction and installation of oil-burning equipments. These were compiled after an extended inspection of the oil-burning equipments, and in answer to continued requests from the oil fields.

—Consulting Engineers of the National Board of Fire Underwriters, April 20, 1903

The handling of risk in society is ultimately done by people. A central activity for any successful risk handling process is the exchange of risk related information between these.

—H. Veland and T. Aven, "Risk Communication in the Light of Different Risk Perspectives," *Reliability Engineering and System Safety,* February 2013

CONTENTS

PREFACE

I started this book many years ago when a colleague, Dr. Dixie Hicks of Faulkner University, asked me to look through an archival collection for several letters related to her dissertation. During this search I found a letter written by an Alabama man, who I later discovered married a woman from Mobile in the 1890s and was involved in the oil business in Corsicana, Texas. I started reading the 1901 issues of the *Mobile Daily Register* to find a story on their wedding and noticed that there was a great deal of coverage of City Council meetings in which heated arguments over the safety of storing large amounts of petroleum in the city went on for months. It was an issue brought to the forefront by the opening of the Beaumont oil field and the giant Spindletop well, oil from which was easily piped to Port Arthur and Sabine, Texas, and loaded on barges or steamers bound for New Orleans, Mobile, and the Mississippi River. From this Mobile fight between fire-insurance agents/underwriters and oil men I learned that similar conflicts raged in New Orleans, Houston, and Galveston. I spent many pleasant and productive summers in those towns searching through archives, reading microfilm, and gradually putting together the pieces, although conceptually I struggled to make sense of these developments.

I settled on the idea that everyone involved wanted to do something that held out the possibilities of gain and loss, implying risk and uncertainty—the realm of calculative probabilities or likelihood and risk management. I was also intrigued by the cities in question and their legal power to address a seemingly mundane problem involving oil tanks. I learned that some U.S. cities before 1860 had identified burning fluids (alcohol) and coal oil as hazards needing regulation. Cheap kerosene after 1860 became popular as a safe and bright illuminant but eventually elicited deep safety concerns. Many refiners sold cheap grades that were deadly because they had not boiled off the dangerous components—naphtha and benzine—which caused horrible accidents in the nineteenth and early twentieth centuries. States and cities devised oil-inspection laws for kerosene, laying by the 1890s the legal groundwork for controlling crude petroleum, fuel oil, and ultimately flammable and combustible liquids up to the present.

My conceptual framework matured after I read several important studies. William J. Novak's excellent book, *The People's Welfare: Law and Regulation in Nineteenth-Century America* (1996), introduced me to municipal police powers, a topic hidden by the tide of post–World War II litigation that expanded and consolidated federal power over states and cities; Jackson Lears's insightful work, *Something for Nothing: Luck in America* (2003), opened to me issues of human agency, chance, risk, and indeterminacy in historical context; Joy Jackson's *New Orleans in the Gilded Age: Politics and Urban Progress, 1880–1896* (1969) is a model of carefully crafted urban history; John O. King's *Joseph Stephen Cullinan: A Study of Leadership in the Texas Petroleum Industry, 1897–1937* (1970) taught me much about the men and affairs surrounding the 1901 opening of the Spindletop-Beaumont oil field; and David A. Moss confirmed for me the importance of risk in American culture in his masterful study, *When All Else Fails: Government as the Ultimate Risk Manager* (2002).

ACKNOWLEDGMENTS

Many libraries and archives provided me invaluable help. They include the Research Library, New York Public Library; Sara Hart, Insurance Association Library of Boston; Carol Ellis and the McCall Rare Book Library and Archives, University of South Alabama; Auburn University Library and Special Collections; the Southwest Branch, National Archives and Record Administration, Fort Worth, Texas; the National Archives and Record Administration, Washington, D.C.; the Rosenberg Library, Galveston, Texas; the City of Galveston; the Secretary of the City of Houston; the Clerk of the Fifth Judicial District, Harris County, Texas; the Houston Metropolitan Research Center (Houston Public Library); the University of Houston Library; the University of Texas Law School Library; the John Willard Brister Library, University of Memphis; the Hagley Museum and Library, Wilmington, Delaware; the Pennsylvania Historical Society; the Library Company of Philadelphia; the Ace Insurance Company Archives (Philadelphia); the Filson Historical Society, Louisville, Kentucky; the Alabama Department of Archives and History; the University of Michigan Library, Hathi Trust Digital Collection; the Alabama State Law Library and librarians Susan Wilkinson and Alma Surles; the City of New Orleans, Notarial Archives; the City of New Orleans, Conveyance Records; the New Orleans Public Library and its Louisiana Room stalwart Irene Wainwright; the Board of Commissioners of the Port of New Orleans (Port Authority); the Orleans Parish Levee Board; the City of Algiers, Louisiana; Tulane University Library, especially Eric Wedig and Hayden Battle; and the Auburn Public Library, notably the staff in Interlibrary Loan and the Electronic Services librarians Michael Devore and Denise Ballentine.

Many individuals contributed to this book, including David Goldfield, Martin Melosi, Joseph Pratt, Michael Thomason; Marsha Boosinger and

Tim Dodge of Auburn University Library; Marcus Kieltyka of Yakima, Washington, and formerly Central Washington University; Steve Murray, Scotty Kirkland, and Keri Hallford, ADAH; Professor Howard Kunreuther, Wharton School of Business, University of Pennsylvania; Sue Marsh of the Charles S. Morgan Library, National Fire Protection Association; the support staff of Nota Bene software; Wei Bridges, Photoshop expert, Auburn, Alabama; Dr. Marty Olliffe, archivist, Wiregrass Museum, Troy University, Dothan, Alabama; Craig Colton, Department of Geography, Louisiana State University; and Jack F. Colbert, Library of the Center for Disease Control, who kindly provided me with a valuable bibliographic record of hard-to-find conference proceedings. Mary Lee Eggart of Baton Rouge, Louisiana, an exceptionally talented artist and cartographer, expertly crafted the three maps in this book, based on my primitive descriptions and awkward templates. I thank her for this invaluable assistance.

I am also indebted to my friends Joe and Kathy Wharton, whose faithful friendship and prayers encouraged me both in Lafayette and Silver Lake, California, and for Joe's untiring efforts to read chapter drafts and improve them. Of course, I am solely responsible for any errors or mistakes in the book. I also received help from my friends John and Barbara Stewart, Memphis, Tennessee, who as retired teachers sympathized with my struggles to research and write while teaching very large classes. My friends Dr. Jerry and Joanne Haughton and Geary Irwin sustained me over the years with wise spiritual counsel and good conversation, as did the friendship of Dick, David, Greg, and Chris Irwin. Good friends and fellow-historians, Drs. Ed and Dixie Hicks, offered advice and support, giving a very tired fellow valuable rest-time on Lake Martin, Equality, Alabama, as well as occasional lessons in fixing cars and boats that produced in my case only limited results. My colleagues Dana Chandler, archivist, Tuskegee University Archives; Mary Edith Powell, Allied Health (retired), Tuskegee University; Glenn Drummond, U.S. Corp of Engineers (retired), and formerly Macon County, Alabama, archivist and county engineer; and Tim Bryant, PhD candidate, Economics Department, Auburn University, and his wife Kelly, Graphic Design Department, Auburn University, have supplied digital copies of photographs, great conversations, exploration of Native American sites, reliable knowledge of Alabama history, and true friendship. I must also thank Ellen Weiss, Tulane University School of Architecture (retired), who allowed me to stay in her home in New Orleans while she was on vacation, making it possible to spend several summers using the libraries of Tulane University and the New Orleans Public Library.

In July 1998 I received a grant from the Hagley Museum and Library that enabled me to work for several weeks at that wonderful facility and to enjoy staff assistance, including Linda Gross. In 2006 the Filson Historical Society in Louisville, Kentucky, designated me as a Fellow, allowing me to spend two weeks working in their excellent collections related to insurance. The hospitality and competence of the staffs and directors of these institutions aided me greatly. I also received leave in the fall of 2012 from Tuskegee University to work on this project, a valuable period for reading, writing, and reflection.

I note also that chapter 2 appeared in an earlier form as "Fire Hazards and Protection of Property: Municipal Regulation of the Storage and Supply of Fuel Oil in Mobile, Alabama, 1894–1910," in the *Journal of Urban History* 28 (July 2002): 599–628; and similarly chapter 3 in earlier form as "Urban Government and Environmental Policies: Regulating the Storage and Distribution of Fuel Oil in Houston, Texas, 1901–1915," *Journal of Southern History* 71, no. 2 (May 2005): 279–320.

ABBREVIATIONS

ADAH	Alabama Department of Archives and History
AFMFIC	Associated Factory Mutual Fire Insurance Companies
AIA	American Insurance Association (successor to the NBFU)
AmEx&Rev	*American Exchange and Review*
AM&IW	*American Manufacturer and Iron World*
ANSI	American National Standards Institute
API	American Petroleum Institute
ASME	American Society of Mechanical Engineers
ASTM	American Society for Testing and Materials
BA	*Boston Advertiser*
BDA	*Boston Daily Advertiser*
BH	*Boston Herald*
BEJ	*Boston Evening Journal*
BJ	*Boston Journal*
BMMFIC	Boston Manufacturers' Mutual Fire Insurance Company
BOC	Board of Commissioners, Galveston
BOT	Board of Trade, New Orleans
BTU	British Thermal Unit
CCG	*Cincinnati Commercial Gazette*
CCG&RO	*Charter of the City of Galveston . . . and the Revised Ordinances* 1893
CCM	City Council Minutes (Houston, Galveston)
CDG	*Cincinnati Daily Gazette*
CPD	*Cleveland Plain Dealer*
CTrib	*Chicago Tribune*
DET	*Daily Evening Traveller* (Boston)
DIO	*Daily Inter Ocean* (Chicago)
DMN	*Dallas Morning News*

E&MJ	*Engineering and Mining Journal*
Factory Mutuals	*The Factory Mutuals 1835–1935* (1935)
F&LC	Fire & Lighting Committee (New Orleans City Council)
FUAN	Fire Underwriters' Association of the Northwest
FUAP	Fire Underwriters' Association of the Pacific
GCC	Galveston City Council
GCC, CCM	Galveston City Council, City Council Minutes
GCC, CCR	Galveston City Council, City Council Records
GCSFR	Gulf, Colorado & Santa Fe Railway
GDN	*Galveston Daily News*
GHH	Galveston, Houston and Henderson Railway
GWC	Galveston Wharf Company
HC	*Houston Chronicle*
HDP	*Houston Daily Post*
HMRC	Houston Metropolitan Research Center
HSP	Historical Society of Pennsylvania
IDC	International Development Corporation
InsEng	*Insurance Engineering*
InsTm	*The Insurance Times*
JFI	*Journal of the Franklin Institute*
JSH	*Journal of Southern History*
L&L&G	Liverpool and London and Globe Insurance Company
LH	*Louisiana History*
LILAB	Library, Insurance Library Association of Boston
LSCOC	Lone Star and Crescent Oil Company
ManRecord	*Manufacturers' Record*
MMA	Mobile Municipal Archives
MPR	*Mobile Press Register*
MDR	*Mobile Daily Register;* and *Mobile Register*
NBFU	National Board of Fire Underwriters
NBS	National Bureau of Standards
N.C.S.	New Orleans New Council Series
NECMA	New England Cotton Manufacturers' Association
NEIEx	New England Insurance Exchange
NELA	National Electric Light Association
NFPA	National Fire Protection Association
NODI	*New Orleans Daily Item*
NODP	*New Orleans Daily Picayune*

NODS	*New Orleans Daily States*
NOPL/LD/CA	New Orleans Public Library, Louisiana Division, City Archives
NOTD	*New Orleans Times-Democrat*
NOTP	*New Orleans Times-Picayune*
NSC	National Supply Company
NYBFIC	New York Board of Fire Insurance Companies
NYBFU	New York Board of Fire Underwriters
NYDTB	*New York Daily Tribune*
NYEP	*New York Evening Post*
NYHRD	*New York Herald*
NYT	*New York Times*
NYTB	*New York Tribune*
OIJ	*Oil Investors' Journal*
PBFU	Philadelphia Board of Fire Underwriters
PBOCCG	Proceedings of the Board of Commissioners of the City of Galveston
PCDR	*Petroleum Centre Daily Record* (Petroleum Centre, Pa.)
Peckham	*Report on the Production, Technology, and Uses of Petroleum*
Phil Inq	*Philadelphia Inquirer*
PNECMA	*Proceedings of the New England Cotton Manufacturers' Association*
P.L.	Public Law
QRA	quantitative risk assessment
RE&SS	*Reliability Engineering & System Safety*
SBCCI	Southern Building Code Congress International
SCCF	*Special Com. on Cheap Fuel* (New Orleans Board of Trade)
SciAm	*Scientific American*
SDR	*Springfield* (Mass.) *Daily Republican*
SFC	*San Francisco Chronicle*
SFEB	*San Francisco Evening Bulletin*
SETA	South-Eastern Tariff Association
SOC	Standard Oil Company
SPR	Southern Pacific Railroad
SR	*Springfield* (Mass.) *Republican*
S&LC	Streets and Landings Committee (New Orleans City Council)
SSFPC	Southern Standard Fire Prevention Code
STWPR	*Stowell's Petroleum Reporter*

TC	*The Chronicle: A Journal Devoted to the Interests of Insurance* (New York)
TFPA	Texas Fire Prevention Association
TIA	*The Iron Age*
TIJ	*The Insurance Journal* (Hartford)
TIP	*The Insurance Press*
TMM	*The Mechanic's Magazine* (London)
TNECMA	*Transactions of the New England Cotton Manufacturers' Association*
TNORC	Texas & New Orleans Railroad Company
TS	*The Standard* (Boston)
TSA	Texas State Archives
UL	Underwriters' Laboratories
USGS	United States Geological Survey
USIG&M	*United States Insurance Gazette and Magazine of Useful Knowledge*

PETROLEUM & PUBLIC SAFETY

INTRODUCTION

This study examines the regulation of the storage and consumption of petroleum in the early twentieth-century Gulf south, focusing on events in Mobile, New Orleans, Galveston, and Houston. City officials managed the risks posed by petroleum hazards that came from the development of the Texas-Louisiana oil fields, which the famous Spindletop strike south of Beaumont, Texas, set in motion in January 1901.

I demonstrate that Gulf South officials, fire-insurance representatives, and the oil interests engaged in the politics of risk management in reliance upon accumulated risk knowledge. Through rhetoric and political interactions, they produced regulations for the storage and consumption of petroleum, establishing themselves as agents or social actors in a complex technical system built around petroleum use and regulation.

Serious reflection and reasoning before a city-council committee charged with drafting a preliminary oil ordinance, or before the entire council in a public forum, clarified the views of councilmen and interested parties. Exchanging viewpoints demonstrated that elementary notions of hazards, risk, and disasters were central to the arguments, discourse, and rhetoric of participants.[1]

Further, I ground this regulatory process in its historical antecedents such as fire-prevention engineering, standardization, and petroleum-hazards mitigation, as developed and expressed in the second half of the nineteenth century through statutes, ordinances, and insurance rules that were eventually refined into uniform and technically informed regulations and codes. From the unfolding of these elements, a petroleum-risk culture emerged in the United States from 1860 to 1900 in which humans purposely acted through technologies, business and government entities, as well as professional associations—guided by experience, basic science, and technical experimentation at various stages—to mitigate or eliminate petroleum hazards.

Finally, I focus on the storage-and-consumption aspect of the petroleum industry because it was a central issue for fire-insurance companies and the constant subject of state laws, municipal ordinances, and insurance rules. When fire risks elicit discussion in insurance meetings or periodicals before 1900, at issue are the hazards and risks that petroleum or its refined products pose to buildings, residences, manufacturing facilities, harbors, or sections of a city. Oil-field activities only come into view when lightning sets off conflagrations in a tank farm, hitting tanks clustered around several drilling rigs, or damaging small supply tanks installed to feed portable boilers that power drilling equipment and pumps.[2]

The question then arises why concentrate attention on Mobile, Galveston, Houston, and New Orleans? The answer is that a perennial record of natural and man-made hazards in the Gulf South pointed me to these cities as communities in which the politics of risk management of flammable liquids could be examined in a historical and geographic context, deepening and detailing the role risk played in the nineteenth- and twentieth-century South in the face of incipient industrialization and modernization.[3]

Their differences in population, financial and commercial resources, regional heritage—Houston as a Southwest town, New Orleans's cultural ties to Africa, the Caribbean, and two European empires—as well as the economic significance of their ports, maritime traffic, and railroad connections make them appropriate subjects for analysis.

Mobile and peripheral communities enjoyed growth by 1906 based on the lumber business and shipbuilding that was bolstered by the founding in 1916 of the Alabama Dry Dock and Shipbuilding Company, as well as the arrival during WWI of the Tennessee Iron and Coal Company in nearby Chickasaw, Alabama. Mobile by 1919 was firmly established as an international port of call, attracting steamship companies and shipbuilding enterprises.[4]

By 1910 industrialization in Houston had gained traction, and by 1915 the city had access to the Gulf of Mexico through the Ship Channel. The burgeoning oil industry supplied fuel for machinery and services. Finding and producing oil gradually created networks of skilled labor, capital, and markets. Constant economic interactions in Houston eventually achieved a critical level of reciprocity and demand. After World War II, economic growth sustained machine-tool shops and manufacturers; fabricators, suppliers, and distributors of oil field equipment; and marine transportation and chemical processing—all of which constituted capital-goods creation and industrialization. As a supportive community and financial bastion for the petroleum industry, Houston became a national and international center of engineering expertise, technology, finance, and management.[5]

While Galveston in 1915 had ample wharves, huge cotton transshipment operations, steamship lines, and many railroad connections, Houston eventually surpassed the Oleander City, enjoying over time huge economic benefits from the development of the East Texas oil field, its favorable position as a railroad nexus, and a successful campaign to complete the Ship Channel.[6] Houston had an ambitious commercial elite who were loyal promoters of the city while the Moody, Sealy, and Kempner families controlled Galveston through banking and insurance, as well as the Galveston Wharf Company. This Sealy enterprise ended in 1940 after decades of resentment over the grip this private entity had on docks and wharfage. Houston financed waterways access to its port. In contrast, Galveston had to pour its capital into cleaning up the mess left by the 1900 hurricane, and gave way to the port of Houston in tonnage and cargo values after WWI.[7]

New Orleans's population grew dramatically in the 1890–1910 period, from 242,000 in 1890 to 287,000 in 1900 and 339,000 in 1910. Although Mobile's population increased from 31,000 in 1890 to 38,500 in 1900 and finally 51,500 in 1910, New Orleans in 1910 was six and one-half times bigger than its Alabama neighbor, and much larger than Houston and Galveston as well.[8]

However, census information shows that workers in New Orleans in 1909 produced products valued at approximately $78.8 million (down to $69.8 million in 1914), which increased to $182.8 million in 1919. Houston, a smaller city, increased from $23 million in 1909 ($25.8 million in 1914) to $86.8 million in 1919, a bigger percentage increase than the Crescent City, and greater values than Galveston at $6.3 million (1909), $8.7 million (1914), and $14.9 million. Mobile products in 1909 were valued at $5.5 million, growing slightly to $5.8 million in 1914, and after WWI in 1919 to $27.4 million, a five-fold increase but in absolute terms not comparable with the larger Gulf South municipalities.[9]

These cities, despite their disparities, survived the labor shortages and fall off in harbor traffic of WWI. In New Orleans and Houston, petroleum played a role in energy markets. By 1918, Houston's Ship Channel proved to be an excellent location for oil refineries and transportation of petroleum.[10]

In terms of oil consumption, the figures suggest that Houston underwent industrial growth and that New Orleans was not far behind. In addition to coal usage, Houston manufacturers consumed 560,339 barrels of fuel oil in 1919; New Orleans, 480,814 barrels; Galveston less (82,436); and Mobile much less (16,172). All four had access to the Gulf of Mexico, although Galveston and Houston had the additional advantage of being close to the oil fields of East Texas, forcing officials, oilmen, and the insurance interests to take serious notice of current risk knowledge and professional expertise about petroleum storage and consumption.[11]

The four communities also have accessible records, such as long runs of important newspapers; minutes of municipal councils, commissions, and boards; court documents and maps; supplemented by proprietary indices of ordinances, legal digests of case law, and editions of fire-prevention codes and ordinances that reveal a permanent regulatory concern over flammable liquids.

I did not find similar resources in other Gulf South communities such as Pensacola or Port Arthur that had this scope or depth. Beaumont was a boom town at ground zero in the oil field and a great source of news about claims, leases, wells, drilling, production, company formations, and safety issues, but incomplete or inaccessible newspaper sources made it difficult to identify or follow protracted debates between oil and insurance men over regulations. Port Arthur was an important loading point for tank steamers that railroads and pipelines served, but there were no comprehensive sources that documented any heated debates in city government over petroleum risks.

These historical materials show that many Gulf Coast residents and visitors believed that petroleum would bring industrialization and prosperity to the area, otherwise occupied with transshipment of commodities, port activities, and general commerce. Securing its safe use was a prerequisite to the success of local and regional promotion campaigns that southern civic and business elites practiced relentlessly during the nineteenth and twentieth centuries. It fell to Gulf South leaders in 1901–2 to weigh the safety of citizens against the potential economic benefits that huge quantities of newly discovered petroleum might bring to their communities.[12]

Fire-insurance underwriters worked actively with these officials, link-ing state or municipal risk management to a growing body of information and hazard-mitigation methods based on the cumulative experiences of oil-field and refinery workers, mechanics, engineering practitioners, insurance inspectors, and special agents who regularly dealt with the hazards of flammable liquids from 1860 to 1900.[13]

For example, H. K. Lindsay, general agent in Cincinnati, Ohio, for the Lancashire Insurance Company, addressed the Insurance Library Association of Boston in February 1891, explaining in detail the distillation of petroleum, how it should be conveyed to furnaces, and the requisite safety restrictions for using petroleum as fuel. Having studied petroleum hazards for thirty years, he cautioned his audience that his remarks applied to "isolated plants," but the use of fuel oil in "compactly-built cities" was a pressing problem that needed "early solution."[14]

To this must be added a strong, nineteenth-century tradition in Europe and Great Britain of developing industrial applications of scientific findings that generated reports about the chemical properties of burning fluids, coal, and crude-petroleum distillates in solid, liquid, or gas form; descriptions of appliances and apparatuses to exploit these substances; explanations of their use in various commercial processes, as well as patent notices. Periodicals such as the *Chemical News, Chemical Gazette,* and several mining magazines and technical journals gave interested readers in the United States access to this material and to articles and papers read at scientific, technical, and academic meetings.[15]

Understanding the historical context of these developments requires knowing that in the aforementioned risk culture the hazards of petroleum and its products grew in step with mechanical and technical advances associated with industrialization. Shop- and factory-trained engineers worked alongside mechanics, machinists-proprietors, and fabricators, relying on shared ideas and useful techniques, as well as experimentation, to design and build machinery, turn out general and customized orders, maintain manufacturing processes, and create new products with a close eye on customer satisfaction. Drilling rigs, pipelines, pumps, boilers, refineries, tank cars, coal- and water-gas generators and lamps came out of this tradition and were artifacts of petroleum technology and risk.[16]

After the Civil War, shop culture continued in railroad operations, engine and metal fabrication, as well as in the design and maintenance of steam engines for naval propulsion and manufacturing. Over time, institutionally trained engineers gradually rationalized informal experimentalism, practice, and tacit knowledge—particularly in the 1885–1900 era—disseminating it through professional meetings and publications, as well as specialized technical or industry journals and magazines.[17]

Some engineers and scientists, working with people or groups concerned with fire protection, loss reduction, and hazard mitigation in mills or commercial operations, investigated specific fire hazards and conditions, gaining insights into such issues as oil-lamp safety, petroleum lubricants, automatic sprinklers, acetylene generation (illuminant), electricity, and the origins of petroleum-related catastrophes. Experiments, tests and trials, performed in laboratories or manufacturing facilities produced fire-protection techniques and devices that attracted the attention and support of insurance companies, fire patrols, and inspection bureaus.[18]

Henry Morton, for example, president of the Stevens Institute of Technology, contributed to the campaign to produce national rules for using electricity, so that insurance inspections and electrical contractors could carry out projects in any state in compliance with a set of basic regulations. He was a "consulting expert" to an NBFU subcommittee on acetylene hazards, which was directed to devise rules for a "standard acetylene generator" based on "recognized technical and scientific knowledge of the subject matter." Stevens's alumnus, engineer, business man, and obituarist Alexander C. Humphreys reported that accolades from friends, coworkers, and admirers of Morton described him as a "great scientist" and "scientific friend." Most important was the fact that at his death he had served as a long-term "scientific advisor to the New York Board of Fire Underwriters," and was also an expert consultant, educator, chemist, and accomplished engineer.[19]

By 1900, technicians, engineers, and technically astute insurance leaders had acquired a measure of expertise regarding the dynamics of petroleum fires and explosions, as well as other hazardous events related to industrialization, contributing to the development, organization, and dissemination of fire-protection standards and risk knowledge. Encapsulated in model codes and sample ordinances, these standards rationalized and codified the risk behavior of industries, businesses, the military, and governments. This was the collective work of a coalition of affiliated technical and professional societies and groups that were engaged in what came to be known as fire-protection engineering.[20]

Two prominent examples were the National Fire Protection Association (NFPA), whose members included national, regional, and local fire-insurance associations, bureaus, and inspection agencies, and its patron and founder, the National Board of Fire Underwriters (NBFU), composed of major stock, fire-insurance companies. At its formation in 1896, the NFPA declared: "The objects of this Association shall be to promote the science and improve the methods of fire protection and prevention; to obtain and circulate information on these subjects and to secure the cooperation of its members in establishing proper safeguards against loss of life and property by fire."[21]

In the interest of their member firms and associations, the NFPA and other organizations such as the Associated Factory Mutual Fire Insurance Companies (AFMFIC) formulated standards and rules applicable to a range of fire hazards, promoting approved practices, materials, and techniques that multiplied risk knowledge and fire-risk data.[22] Insurance companies made indemnification contracts contingent upon compliance with these standards. The AFMFIC rigorously studied fire hazards, advancing fire-resistant construction and pioneering the use of sprinklers in textile mills. The NFPA routinely revised its standards and procedures, making use of the determinations of technical and professional societies such as the American Society for Testing Materials (ASTM, established in 1898) or the ASME (American Society of Mechanical Engineers, 1880). To address the concerns of fire-insurance underwriters and their clients over previously unrecognized or unanticipated hazards, the NBFU crafted rules and regulations based on NFPA findings or work at the Underwriters' Laboratories (UL, 1896), and published this material for voluntary adoption as statutes and ordinances.[23]

Before proceeding I will identify several terms or concepts that will clarify what I am doing in this study. Three related concepts explain the connection between risk knowledge and expertise and the struggles of our four Gulf South cities to manage petroleum consumption. Risk, hazards, and disasters are the central elements of a conceptual framework, present in the nineteenth century but not fully developed and applied until the twentieth century, that sought to anticipate the harmful, sometimes catastrophic, consequences that might arise from human decisions, events, industrial processes, or the use of hazardous materials. In the post–World War II world this risk paradigm became robust and fundamental to a variety of disciplines and activities, from engineering, architectural design, environment and development studies, to manufacturing, finance, medicine, government policy and planning. It is enshrined in international, federal, state and local law, agency operations, and policy objectives and requirements, and studied by scholars who have created a massive body of risk literature that informs this project, making it important to understand how risk models apply to the Gulf South.[24]

We start with the observation that risk is an intrinsic component of human existence, requiring every person to make sense of hazards and the dangers they pose.[25] The term "hazard" reflects a distinction between reality and possibility. Hazards are agents, events, conditions and substances, such as careless workers, improperly constructed dams or bridges, hurricanes, vapor leaks, combustibles or flammable liquids, spills, landslides, or earthquakes that can potentially initiate or trigger a destructive sequence of events, or by virtue of intrinsic characteristics hurt or kill people and/or damage what they value. This means that objective, unfavorable effects can come about as the result of "natural events or human activities." Tornadoes, lightning, or human misjudgment initiate fires, cause explosions, or destroy a financial portfolio, giving rise to analysis of the probability of their occurrence.[26]

The term "risk" has two meanings in this study. First, there is an informal and conventional definition, a derivative of professional convenience that means "the hazard assumed by the insurer." In insurance parlance, *risk* is the liability assumed in a fire-insurance contract, in which companies promise to indemnify someone for a loss arising from the anticipated consequences of known hazards. The liability implied in *risk* arises from the "occurrence of a specified contingency," which from the perspective of an underwriter is an uncertain event, condition, or agent that could cause an insurance company to lose capital in a payout to a client. *Risk* is an assumed obligation to "compensate the insured for specific future losses."[27]

Although the range and number of hazards or "risks" have increased throughout history, the core notion of possible or predictive harm to humans or the things they value remains unchanged. This points to a second meaning for *risk* that is a formal and analytical concept, as in *risk assessment, risk analysis,* or *risk management*. It describes the identification of hazards—events or conditions, which if they occur or come about, constitute threats to humans or what they value, as well as the quantifiable determination, if possible, of the probability or likelihood of these events taking place and the consequences—deaths per annum, wages lost over a five-year period, or property damage reckoned in dollars from tornadoes over a six-month season. *Risk* is usually understood as the mathematical function or product of these numerical expressions.[28]

This in turn leads to risk management, "selecting policy alternatives" or hazard mitigation strategies outlined in insurance rules and requirements, building codes, professional protocols, fire-protection handbooks, state laws, and municipal ordinances.[29] the purpose is to eliminate potentially harmful events, reduce the likelihood of their occurrence, and minimize their consequences. Through practical acts such as building a levee or keeping potentially dangerous and flammable substances away from densely populated areas, risk management can reduce a community's vulnerability or its susceptibility to harm.[30]

Two additional terms—"standardization" and "rationalization"—also deserve a brief examination, supplemented in the chapters that follow by technical and regulatory usage to clarify their analytical value. *Standardization* means establishing uniform rules, procedures, or materials to be applied throughout a vast range of activities and methodologies such as manufacturing, testing, professional norms, or prescriptions for military justice, logistics, operational regulations, or strategies. It is especially important when the concept is applied to technical systems, fire-protection engineering, or insurance. *Rationalization* implies reasoned analysis and descriptions (as opposed to traditional claims and customs), identification and putting things in logical order. Rationalization is a fundamental task in fire-insurance activities.

RISK AND INDUSTRIALIZATION

Risk and industrialization in the nineteenth and twentieth centuries converged in conjunction with economic arrangements, technology, population growth, and the development of large technical systems such as the Pennsylvania Railroad, Standard Oil Company (SOC), or American Telephone and Telegraph.[31]

The production, processing, and consumption of hydrocarbons generated previously unmatched complexities, lethal hazards, and levels of risk owing to the technologies and energy needs related to mechanization; heavily capitalized, industrial enterprises; and the centralization of the national economy and coordination of transportation networks. Nineteenth-century insurance leaders recognized this development, arguing that through organized and concerted action underwriters should gather and preserve risk knowledge about the spiraling number of "fire risks in modern life," brought on by "modern science and inventive genius."[32]

This applies to the cities in this study. Modern integration of society's arrangements for producing and distributing goods replicated and confirmed the economic advantages of scales of efficiency, manufacturing methods, transportation, exploitation of labor, and the efficacy of mass markets that industrial capitalism fostered. Many towns and cities in nineteenth-century North America participated in this reworking of human culture and had to come to grips with its attendant hazards and risk levels.[33]

The intersection of the physical world with human activity historically shaped the risk culture of the Gulf South through hurricanes, tornadoes, storms, and flooding in low-lying communities such as Houston or Galveston, but petroleum, a known energy carrier through the Corsicana field south of Dallas, Texas, in the 1890s, was the vanguard of potential economic changes and industrial hazards that loomed on the horizon in 1901. Petroleum added a new dimension to the region's risk culture, forcing individuals and communities to come to terms with this potential industrial threat based on their perceptions of the risks it entailed.[34]

Incremental industrialization and the energy to power it forced Gulf South communities in 1901–2 to ascertain what level of petroleum risks their citizens would tolerate. Perceptions of danger, practical know-how, accepted procedures, and a slowly growing body of professionally certified facts and workable safety measures that were applicable to the petroleum industry and related commercial activities guided the search for risk thresholds and fueled debates over how to regulate the storage and use of petroleum. The new oil held out the prospect of commercial advancement but also the possibility of oil-field fires, tank explosions, pipeline accidents, waterfront conflagrations, and destruction of urban property.[35]

Organizing oil companies, intense drilling, and accelerating oil production in the region in the early twentieth century confirmed that expectations of wealth were well founded. However, this did not distract fire-insurance companies. Petroleum was a proven hazard with enormous destructive potential, and the best line of defense was to enfold it in rules and regulations and to persuade municipalities to adopt them as ordinances.[36]

The Gulf South cities of this study followed this path throughout the twentieth century, beginning in 1901–2 when disputes over managing petroleum risks played out in City Council meetings, committee hearings, and public discussions of proposed oil ordinances. City officials transformed insurance rules and later consensual, codified rules and requirements into legally enforceable regulations that gave these municipalities undisputed control over storing and handling petroleum. They achieved this by making use of the risk-management template laid down by nineteenth-century fire underwriters in Chicago, San Francisco, and the Atlantic Northeast where industries experimented with and adopted liquid fuel under conditions laid down in policy forms, special provisions, and permit conditions.[37]

Before we examine what the Gulf South cities did in relation to petroleum storage and use that constituted risk management, and before we try to understand how this regulatory activity played out in each city as a political process of public debate, advocacy, and policy development, we must first consider the nineteenth-century background of these issues.

To that end, chapter 1 examines the backdrop for analyzing the regulation of petroleum risks in the Gulf South. I describe a historically situated petroleum-risk culture that emerged by 1900 to frame and shape the politics of risk management that played out in Mobile, Houston, New Orleans, and Galveston. This risk culture provided Gulf South officials, insurance interests, and oil men with accumulated risk knowledge as well as practices and strategies for hazard mitigation.

In charting government and insurance concerns about flammable hydrocarbons, chapter 1 explores what eventually became a broad, socio-technical system, defined as all aspects of producing, storing, distributing, and consuming petroleum, including the human role in devising, operating, supervising, and regulating this system for mankind's well-being.[38] At the core of this system was the coevolution of energy-resources infrastructure, social-technical institutions, and insurance strategies and organizations, offering insight into the governance of a complicated drawn-out energy transition (biomass, coal, petroleum) that featured increasingly intense energy use and greater complexity, attributes associated with the modernization of the twentieth-century South.[39]

1

PETROLEUM AND NINETEENTH-CENTURY RISK CULTURE

The historical roots of Gulf South rules and regulations that governed the storage and use of fuel oil can be traced to a nineteenth-century risk culture, one important component of which was the discovery, distribution, and public consumption of flammable hydrocarbons, such as petroleum and its products.[1] Canadian Abraham Gesner contributed to this culture when he obtained three patents in 1854 to distill asphaltum, petroleum, and other bitumens. Using this process he redistilled crude coal-oil made from heating coal into several fractions he called kerosenes. One produced a brilliant light and proved to be a highly marketable burning fluid.[2]

During the period from 1860 to 1900, crude petroleum emerged as a cheap alternative to coal for producing kerosene.[3] To match the demand for this popular illuminant, refiners in Pittsburgh, New York, Philadelphia, and Cleveland expanded production, benefiting after 1870 from process efficiencies, extension of the rail system into the oil fields, and reliable bulk-rail transportation.[4]

Controlling the hazards of crude petroleum, widely consumed kerosene, and other flammable liquids was an important task in nineteenth-century risk culture that provided Gulf South cities in 1901–2 a valuable hazard-mitigation, risk-management tradition. States, municipalities, and insurance groups took part in a century-long struggle to control petroleum hazards. The federal government participated, banning in 1866 the storing of combustible materials such as cotton, hay, or straw on the decks of passenger steamers unless covered by canvas or other material, and restricting "coal oil or crude petroleum" to decks or in "open holds" away from fire and furnaces.[5]

Congress added more safety provisions for steam-powered vessels in 1871, stipulating that refined petroleum could be onboard such ships if it had an ignition point of no less than 110 degrees Fahrenheit, and if there was no other practical way of transporting it. The act empowered steamboat inspectors working out of the Treasury Department to prescribe additional regulations in such cases. The secretary of the treasury could also permit vessel owners to burn "petroleum or other mineral oils" for propulsion and devise and enforce regulations for its safe use. Flammable liquids and explosives such as gunpowder, "coal oil, crude or refined petroleum," and the lighter products of the first refining of petroleum, such as benzine, could be carried as freight if shipped in secure, labeled packages, boxes, or casks.[6]

Acting in step with the evolution of lighting and heating technology, as well as petroleum production, refining, and transportation, federal and state governments as well as municipalities formed a risk-management tradition shaped by safety and liability concerns for passenger vessels and insured property in which lamps, fixtures, and gas machines burned turpentine, camphene mixtures of turpentine and alcohol, as well as naphtha, gasoline, and kerosene.[7]

Early in the nineteenth century, state and city officials laid the groundwork for this tradition, as was evident in Massachusetts, where camphene and burning fluids became alternatives in the 1830s and 1840s to whale oil or lard. Burning turpentine in lamps was not satisfactory, but redistilling it produced a volatile fraction known as camphene. Adulterating turpentine or camphene with alcohol made brighter but more volatile illuminants known as burning fluids that called for careful regulation.[8]

In the 1850–70 period the commonwealth and the city of Boston in the face of numerous accidents and fires passed statutes and ordinances that regulated the manufacturing, refining, storage, and sale of camphene, burning fluids, crude petroleum or its products.[9] They limited the amount of flammable liquids that could be kept in buildings; banned storage and refining under sidewalks, in dwellings or alleys, and along wharves; imposed fire and flash tests conducted by inspectors; and required a license to handle flammable liquids in any building above the cellar.[10]

Flammable liquids provoked serious political controversies and investigations in Philadelphia. A catastrophic fire in February 1865 ignited a coal-oil yard, destroyed five squares of buildings, and killed many residents. In response the state legislature banned large-scale storage of crude and refined petroleum, benzine, and naphtha in "built-up districts" in the city. Four years later the *Evening Telegraph* published an account of its investigation into the widespread sale of adulterated kerosene, a Philadelphia market that supposedly had been reined in by an 1868 state law establishing a city oil inspector, only to see it subverted by political corruption and bribery.[11]

In 1872–73 the Franklin Institute, long concerned with exploiting science for practical purposes, joined the battle over controlling flammable liquids in Philadelphia. William H. Wahl, secretary of the institute, read a short paper before its Optical Section, March 1872, in which he sketched out the development of "artificial illumination," taking note of the enrichment or carburetion of water gas with hydrocarbons, production of gas from coal distillation, and refinement of crude petroleum to obtain light fractions such as naphtha, benzine, and gasoline.[12]

The next year, 1873, the institute commissioned a two-part committee report on the dangers of these illuminants. The first part introduced statistics for 1871 that identified petroleum as the "most prolific cause of conflagrations" in the city. It also described the ineffective legal efforts to test coal oil and kerosene in Pennsylvania, and cautioned readers that cheap illuminating oils were hazards comparable to gunpowder and should be never be used near open flames.[13]

In the second part William Wahl identified nine petroleum distillates, including naphtha and kerosene, recounted their specific gravities and boiling points, and explained how their flash and ignition points identified various mixtures of these substances by which one could detect dangerous, adulterated kerosene that flashed at room temperatures. The public should use petroleum illuminants with extraordinary care, Wahl exhorted readers, because without proper precautions petroleum fractions and poor-quality kerosene caused "terrible accidents," death and destruction of property.[14]

While properly constructed lamps and high-quality kerosene from coal or crude petroleum mitigated these hazards, Wahl noted that random tests of oil samples from Philadelphia dealers showed that many of the most dangerous were advertised as "safe oils." He also charged that "carburetting" illuminating gas—passing it through the vapors of the light products of petroleum to make it burn brightly—was very dangerous as was the vaporization of naphtha, gasoline, or kerosene in air-gas machines for light, both uses objectionable because they employed "incendiary petroleum oils." Deeply impressed by Wahl's scientifically informed mitigation material, the NBFU printed ten thousand copies of the report and distributed them to all board members and eight thousand general, special, and local agents.[15]

In 1874, Pennsylvania set down mandatory oil-inspection procedures, to which the legislature added extensive rules for pipeline operations in 1878 and 1883 that were revised in 1888 and still in place in 1900. Acts passed in 1865 and 1871 were part of a comprehensive 1887 statute that restricted the refining, manufacturing, or storage of crude petroleum, benzine, or naphtha to specific areas in Philadelphia. The restrictions did not apply to locations where existing large-scale refining and storage had the space for "sufficient excavations or embankments" to control overflows "in case of fire or accident."[16]

Controlling the hazards posed by petroleum and its products was also a contentious issue in New York City in the 1860–1900 era.[17] As early as 1813 the state authorized the city to restrict the keeping and storage of dangerous substances including pitch, tar, turpentine and its fractional distillates.[18] On the eve of the Civil War the city fire marshal reported a tragic record of fires and injuries from camphene and burning fluids in the city. Three years later the fire marshal cautioned officials about the dangers that coal-oil lamps posed to women and girls. He also asked for an ordinance "prohibiting the storage of Petroleum in large quantities within the city fire limits." In 1866 he documented damaging and lethal kerosene fires and explosions.[19]

As early as 1862, New York City was under pressure to regulate petroleum storage. Even though the state passed two laws in 1865–66 setting down the conditions under which one could store "Crude petroleum, earth or rock oil, or any of its products," including a fire test of 100 degrees Fahrenheit for "refined petroleum, kerosene or machinery oils," unsafe handling of petroleum in New York City was unabated. The *NYT* reported in August 1869 that the fire commissioners had collected samples of kerosene oil throughout the city, only to find that most of it was very dangerous.[20]

From an economic perspective the source of this danger came from the fact that before the Civil War there was limited demand for naphtha and benzine—the volatile "top cuts" produced by the distillation of raw coal oil or petroleum before kerosene boils off.[21] This tempted some refiners to turn out incompletely refined kerosene or to mix small amounts of these explosive fractions with kerosene, producing a poorly refined, low-quality "burning oil" that was cheap, profitable, and deadly. Drugstores and street vendors sold it to the poor, who often died or were badly burned when it exploded. It was a threat to public health that could not be ignored.[22]

It should be noted, however, that technical improvements in naphtha-gas production soaked up some of the available naphtha-benzine stocks after 1870, taking advantage of the so-called Gale-Rand process that converted naphtha into a permanent gas that did not condense in pipelines and could be burned for light or heat or for enrichment of coal gas. There were still abundant supplies of the light cuts of petroleum to tempt wayward kerosene refiners and suppliers, but the improvements gave the petroleum industry "its first large-volume market" for benzine and naphtha, even though both substances as well as their manufacturing remained dangerous on whatever scale they were processed or consumed as far as fire-insurance underwriters were concerned.[23]

In 1868–70, New York City's Board of Health investigated the sale of dangerous kerosene in the metropolitan area and passed an ordinance covering the retail sale of petroleum, kerosene, or similar substances such as burning fluids (illuminants) unless they passed a flash test of 110 degrees Fahrenheit.[24] A Health Department chemist, Dr. C. F. Chandler of Columbia University, led these investigations and published a lengthy report in 1871 of his findings at the behest of the board in the interest of the "suppression of the traffic in dangerous oils."[25]

Chandler's report was an expansive exploration of the history of petroleum, its chemical composition, distillation, and refining. He explained why most kerosene sold in New York City was dangerous, how it should be tested scientifically, what the Board of Health had tried to do to stop retail sales, and legal obstacles to enforcement.[26] He surveyed the risk-management policies of other cities and states, ending with a portion of the text of an 1871 New York statute that consolidated and augmented the above-cited laws of 1865–66 on petroleum storage under the fire-prevention rubric of "storage and keeping of combustible material."[27]

The risk-management policies of these major cities regarding flammable liquids were part of a legal campaign in many states to control these substances. From 1865 to early 1871 thirteen states, including Ohio, Illinois, Michigan, and Pennsylvania, passed laws that established safety standards, including flash and ignition tests, for coal oil and petroleum products (benzine, naphtha, kerosene), sold or stored within their jurisdictions. There were shortcomings such as voluntary adoption of regulations and no penalties for lax enforcement. Even so, persistent campaigns to tighten restrictions on petroleum hazards built upon this flawed legal platform. Michigan underwriters, for example, petitioned the state in 1893 to amend its laws regarding kerosene testing by adopting "stringent rules and regulations."[28]

The Gulf South also had statutes and ordinances to regulate petroleum hazards that specifically addressed flammable liquids such as distilled turpentine, burning fluids, petroleum, and kerosene. Mobile (1866), Galveston (1871), and New Orleans (by 1881), prompted by coal-oil and kerosene accidents, restricted storage locations and amounts for retail sale, and set down minimum flash or burning temperatures.[29]

FLAMMABLE-LIQUID HAZARDS AND THE INSURANCE INDUSTRY

Fire-insurance companies, boards, and associations also played a major role in the risk culture of the nineteenth century, their attention drawn to any event or condition that posed a fire hazard, including mill and building design; electrical wiring; machinery-manufacturing; distilleries, lumber yards, and gas machines and carburetors; as well as refining, storing, and handling crude petroleum and its products.[30]

In the 1860s the Philadelphia Board of Fire Underwriters (PBFU), out of concern over numerous accidents involving crude oil and mineral-based illuminants (popular substitutes for sperm oil and lard candles), investigated the "Hazardous nature" of petroleum, coal oil, and their products. Retail dealers told the board's secretary, John Williams, that these substances were dangerous only if stored in poorly ventilated buildings. Subsequently, a committee appointed to recalibrate premiums and set restrictions on oil storage recommended maintaining existing rates on benzine, naphtha, or benzole stored in three barrels or less, charging the current rates applied to "specially hazardous risks" for refined coal oil stored in five or less barrels, and assessing progressively higher rates on more than five barrels. No contracts could be issued on crude coal oil, petroleum, rock or earth oil when stored in the "built-up portions" of Philadelphia. If stored in properly vented, isolated buildings or on wharves or in sheds, a minimum 5 percent per annum rate applied.[31]

The committee also urged the underwriters to draft laws prohibiting the storage of dangerous oils in densely occupied areas of the city, confining oil refining and manufacturing to buildings constructed for these purposes. In 1863 the board drew up a memorial to the City Council, asking it to recognize the explosive and flammable nature of petroleum and coal oil and their products, and to pass an ordinance to restrict storage of same within the city limits. In January 1864 the Philadelphia Board increased tariffs on refined oil to 2.5 percent and 7.5 percent on crude petroleum per year.[32]

Hazardous flammable liquids also caught the attention of Zachariah Allen (1795–1882), prominent mill owner, inventor, and insurance leader, who founded the Manufacturers' Mutual and the Rhode Island Mutual Fire Insurance companies in 1835 and 1848, respectively. With the Boston Manufacturers' Mutual Fire Insurance Company (BMMFIC, founded in 1850) the three formed what became the Associated Factory Mutual Fire Insurance Companies (AFMFIC). Allen outfitted Allendale Mill, an early example of so-called "slow-burning" construction, with various innovative fire-prevention features such as fire doors and pumps. In 1861 the Board of Directors of Rhode Island Mutual asked him to investigate the "comparative explosibility of coal oils," and to determine the "consequent probable danger" of using them for illumination in manufacturing plants and homes—a classic expression of risk assessment.[33]

Many fire-insurance companies had, Allen noted, raised rates on coal-oil storage and use in response to accidents. "Practical experiments" distinguished safe from unsafe versions of coal oil on the basis of how much, if any, naphtha or other light products of coal or petroleum was present. Allen found that kerosene distilled from coal or petroleum and mixed with small amounts of naphtha—sometimes called rock oil—produced a bright, nearly smokeless light. However, if proper procedures were not followed, naphtha-laced kerosene was extremely dangerous to "life and property," justifying the classification of such liquids as "specially hazardous."[34]

Pure naphtha was, according to Allen, even more pernicious—an ironic observation considering that naphtha was otherwise nearly worthless—tempting refiners to adulterate kerosene with it. Allen concluded that all light products of coal or petroleum distillation had to be stored properly in warehouses and consumed with "due care and forethought" in all circumstances.[35]

Insurance rules to that effect, reflecting sentiments shared with Allen as well as fears of petroleum hazards in densely populated urban areas, emerged in December 1861. The New York Board of Fire Insurance Companies (NYBFIC) appointed a committee to examine the storage of crude petroleum and its light fractions in public warehouses and buildings in the "compact portions of the city." The committee considered crude oil to be an extremely volatile, flammable liquid, large quantities of which could destroy vast amounts of property. This initiated a campaign to control not just crude petroleum but also its fractional distillates, which were used for lighting and heating when petroleum supplies and prices made oil an economic alternative to coal. And for insurance groups information to prosecute the campaign and formulate rules and regulations rested on a series of committee investigations that relied heavily on accumulated risk knowledge and practical experiments with crude oil and its products.[36]

The NYBFIC committee recommended that the board declare crude petroleum "uninsurable in all buildings in compact portions of the city" as well as in all warehouses designated for "hazardous and extra hazardous merchandise." Agents could grant clients the privilege of placing crude petroleum in detached sheds or warehouses used exclusively for storing such oils or "substances of a similar character" under contracts at a 3 percent premium.[37]

Further, despite being "exceedingly explosive and dangerous," benzine, benzole, and naphtha were stored in the metropolitan area in large quantities. Consequently, they should be kept, the committee advised, in no more than three barrels in any building, and if present in greater quantities, they fell under the same conditions and rates as petroleum. The same strictures applied to refined petroleum oil, kerosene, and all oils made from coal, classified as "extra hazardous" when kept in amounts under ten barrels and "specially hazardous" above that threshold.[38]

The board agreed to these restrictions, setting the pace for dealing with petroleum in step with rising crude production and availability and the search for a broad market for gasoline or naphtha. Insurance companies for the rest of the century specified where and under what conditions crude oil and its distillates could be stored in urban areas. Gulf South cities in 1901–2 used this same formula for mitigating petroleum hazards.

The New York board also gave attention to petroleum hazards through insurance-industry activities. After the Civil War the fire insurance business in the United States faced rising losses, unfavorable state legislation, and ruinous rate-cutting. To address this crisis the NYBFIC, working with representatives of many fire-insurance companies, organized the NBFU in July 1866.[39]

At the NBFU's first annual meeting in February 1867 the Committee on Local Boards, Rates and Commissions reported that it had sent out circulars to local boards urging them to secure municipal action on the "storage of petroleum and its products." A special committee on "gases for lighting made from Petroleum or its products," appointed in December 1866, asked for more time to complete its investigation, and delegates heard a resolution urging the board to ask Congress to make illegal the sale of "Refined Oil, ignitable at less than 100 degrees Fahrenheit."[40]

With regard to policies, one delegate proposed that all NBFU member companies refuse to take or renew any risks on a building where more than five barrels of "Refined Carbon Oil" (coal oil) or more than one barrel of benzine is "stored or kept," or on a block of contiguous stores, or on a single building in such a row where more than ten barrels of coal oil or more than two barrels of benzine are stored or kept, excluding buildings "used *exclusively*" for such purposes. Further, no member company should issue a permit for a portable gas machine that uses vaporized gasoline or benzine as fuel, unless such machines are in "separate brick, stone, or iron magazines" away from any "insured premises." After a discussion, a delegate offered a resolution that required companies to stipulate in their policies that contracts are void unless they contain specific clauses permitting gasoline vaporization.[41]

During the transition from the NYBFIC (a voluntary organization), to the New York Board of Fire Underwriters (a legal corporation) in May 1867, city underwriters called on the state to assist them in dealing with petroleum hazards. First, a NYBFIC committee and then its NBFU counterpart asked the state legislature to establish the office of fire marshal for metropolitan New York with extensive investigative powers; and to authorize the NYBFIC to create a metropolitan bureau of fire prevention that could regulate the "manufacture, storage and sale" of dangerous, combustible substances and flammable liquids such as crude petroleum, benzine, naphtha, and kerosene. Underwriters sought centralized supervision of flammable liquids to match the increasing scale of their hazards.[42]

In December of 1866, five years after its first investigation of petroleum and coal oil hazards, the NYBFIC (soon to be the NYBFU) appointed a second committee to revisit this subject.[43] The NBFU formed a similar committee which endorsed the NYBFIC-NYBFU committee's report. Subsequently, the committees submitted their joint recommendations to the NBFU in July 1867, and its Executive Committee officially adopted them in September.[44]

Following its 1861 predecessor, the joint report denounced "benzine, gasoline, and other kindred, inflammable liquids" as extremely flammable and dangerous, requiring extreme caution "in their use" and very high premiums if used as illuminants, but did not address, as in 1861, storage in public warehouses. Instead the 1867 report focused on the hazardous character of kerosene, benzole, naphtha, or gasoline when used in gas machines. These apparatuses had to be installed at least fifty feet from the insured property to reduce the possibility of fire and equipped with a drainpipe and cut-off valve. The petroleum products to be vaporized also had to be stored "away from the premises to be lighted."[45]

The dangers of loading these flammable liquids on vessels moored at "petroleum yards or depots" came before NBFU's Executive Committee in February 1869, a body composed of the presidents or vice-presidents of important member-companies in cities such as Boston, New York City, Philadelphia, and Chicago. A special subcommittee recommended that all local boards adopt provisional rules based on those used by the New York Board of Marine Underwriters to deal with such hazards. The marine rules focused on ship construction, securing barrels or cases stored below deck, and ventilation. Further, the special committee responded to "several communications" setting forth succinct instructions that addressed problems that demanded immediate attention. They required all member agents to restrict the amount of naphtha and other light products that could be stored in warehouses "not used exclusively" for their storage, something the NYBFIC as early as 1861 and more recently by the NBFU had previously addressed. The report was adopted, but the committee's investigations of oil-storage hazards moved ahead, an urgent task for underwriters in light of poorly enforced municipal ordinances, dangerous gas machines, adulterated kerosene, and numerous explosions and fires that killed or disfigured many women and their servants. Its mature findings appeared two months later at the NBFU's third annual meeting.[46]

At that April 1869 gathering, Vice-President Henry Oakley, chairman of the "Special Committee on the Storage and Sale of Petroleum and Kerosene," presented a long report that superseded the preliminary observations submitted in February. It included a slightly amended version of the marine rules for handling petroleum that appeared two months earlier, as well as extensive excerpts from C. F. Chandler's 1869 investigations on dangerous illuminants in New York City.[47]

Daily reports of accidents and fires as well as Chandler's damning evidence convinced the committee that it faced a national not just a New York City problem and should combat this "great, and . . . unnecessary evil, so inimical" to the common interests of fire underwriters. It prepared a sample ordinance that cities and town could use to regulate the "storage and sale of petroleum and its products." All local boards were directed to push for its passage. The Executive Committee adopted the report but not until members reevaluated contested safety procedures. The NBFU received numerous requests for relief from the strict rules laid out in the joint NYBFU-NBFU report of September 1867, which required gas machines to be outside buildings. If that was not possible, they had to be in a fireproof room and refilled from outside the structure to mitigate the hazards of gasoline, naphtha, and benzine. Further, one carburetor manufacturer wanted to install its apparatuses inside buildings to enrich street gas with the same volatile fuels.[48]

A discussion ensued about the safety of petroleum carbonization of street gas piped into homes for illumination. Henry A. Oakley observed that the "very volatile character" of kerosene or gasoline is the reason why these light-petroleum products enrich petroleum or coal gas, and is the "cause of the danger we contend against." Heald agreed that this was why the NBFU had stipulated that dangerous fluids had to be "introduced into the carburettor on the outside" of buildings, after which members voted to adopt the special committee's report with no changes in gas-machine operations. The board's stance on the careless use of light products of petroleum was consistent with actions taken in 1861, notwithstanding a concession made six months later by its influential ally, the NYBFU, illustrating the unavoidable, ad hoc compromises called for by new or modified technologies.[49]

The New York board, acting in October 1869 to straighten out some misunderstandings about the joint report submitted to the NBFU in the summer of 1867, reiterated its risk perception that the light products of petroleum were extremely dangerous when used to power gas machines. Both fuel and apparatuses should be "far removed from the property insured," and the same strict precautions appropriate when handling these substances also applied when enriching common, illumination gas. However, the NYBFU thought it should be permissible to install a small carburetor or enrichment unit in a cellar near the gas service and meter if fitted with very tight piping and maintained by a commercial firm. Otherwise, carbonizing devices in cellars including appurtenances and piping had to be encased in brick and refueled through an iron pipe from outside the building. The exception was "carbonizing machines," skillfully designed and fabricated for "comparative safety," and carefully maintained by a company or business.

The committee cautioned all board members and their field agents that they could not grant anyone permission to use or sell the light products of petroleum in insured premises unless these flammable substances passed a flash-point test.[50]

Shortly after this action, the NBFU adopted this NYBFU report at the suggestion of Henry Oakley, vice-president of the NBFU, chairman of its special Committee on Gas and Gasoline, and the president of the NYBFU, reconfirming a collaborative commitment to mitigate the extreme hazards presented by dangerous petroleum fractions used for illumination.[51]

During the 1870s, insurance underwriters surveyed municipal firefighting resources such as water supplies, fire brigades, and equipment, and investigated alarms, building-construction materials, and arson.[52] As a result the NBFU withdrew from Chicago, charging that the city's water supply, building laws, fire limits, and Fire Department were inadequate, poorly enforced, or nonexistent as though nothing had been learned from the city's 1871 conflagration. At the same time, the board diligently kept track of hazards through the end of the century that arose from the sale and storage of petroleum products as well as their use in gas machines as part of an effort to identify major sources of fire and devise effective means to eliminate them.[53]

Two factors dictated vigilance. Technical improvements in these devices, particularly the Springfield Gas Machine, forced the NBFU to stay abreast of new safety features that might require adjustments of restrictions. Further, the popularity of such air-gas machines grew rapidly, especially in rural and suburban areas outside municipal gas-light systems. This attracted capital investment that allowed inventors and manufacturers to develop and refine portable light or heat sources that used petroleum fractions for fuel, as shown by the hundreds of patents approved from 1860 to 1880. Gilbert and Barker's Springfield Gas Machine operated automatically on gasoline, luring many consumers from older apparatuses that used coal or rosin oil to make gas. Such machines were widely used in the post–Civil War period.[54]

Technology notwithstanding, fire underwriters viewed any device fueled by the volatile elements of petroleum with suspicion in light of ongoing problems from the 1870s forward with enforcement issues, uncertainty, or deliberate disregard of state laws that established flash or ignition points, and mercantile abuses in the trade and storage of naphtha, benzine, kerosene, and gasoline. They were quite familiar with the trail of accidents attributable to petroleum fractions and had to field many questions from local boards and agents about how to interpret insurance rules and policy clauses covering these substances.[55]

In the case of gas machines that used the so-called cold-process—passing air across the surface of the light products of petroleum to produce vapors fed into burners—fire underwriters recoiled from allowing such dangerous procedures inside buildings served by gas machines. Further, in response to numerous requests from companies for approval of their gas equipment, the NBFU had to scrutinize the technical refinements and improvements in the design and manufacture of gas machines whose makers advertised as completely safe.[56]

Fire insurance literature starting in 1869–78 highlighted the industry's apprehension about the extreme dangers of burning adulterated kerosene as an illuminant and the light products of petroleum used as fuel in gas machines. Insurance writer, editor, and publisher C. C. Hine (1825–1897), set a baseline for permit rules, advising agents to adhere to the strict controls issued by fire-insurance companies and boards such as the NYBFU regarding gas machines and carburetors. Flammable hydrocarbons used for fuel were, he cautioned, highly volatile, "inflammable and explosive . . . dangerous to the last degree" and should never be "stored, handled or emptied, on or near, premises that are insured." Houses or vaults that contain the "generator, receiver, carburetter, or other apparatus containing the liquid" must be made of brick or stone and be 50 feet from insured property or exposed buildings.[57]

Severely hurt by member companies' failures and the financial damage the insurance industry suffered in the wake of the 1871 Chicago and 1872 Boston conflagrations, the NBFU reorganized, campaigned for raising insurance rates to counter the proliferation of questionable fire-risk contracts based on premiums far below loss requirements, and returned to the issue of dangerous petroleum products and gas machines at its 1873 national meeting.[58]

In his presidential address, Henry A. Oakley underlined the importance of reconnecting to state boards and associations; the need for realistic rates, strict building laws, opposition to state taxation of the fire-insurance industry; and the adoption of standard, NBFU policy forms that would protect the rights of customers and insurance companies, rather than inadequate state efforts.[59] The pressing need for building-construction regulations coupled with damning statistics on the Boston fire losses lead logically to the issue of fire causes. Oakley claimed that scientific investigations and the "practical experience of underwriters" had identified many fire sources, showing that "proper regulations" of petroleum and its products were clearly necessary. The widespread use of petroleum in "manufacturing risks" throughout the country and in "patent lighting machines" demanded NBFU scrutiny.[60]

Oakley revisited these topics at the 1874 annual meeting, to which he added related tasks: preventing incendiarism caused by carelessness, and managing dangerous hazards such as exploding kerosene lamps, defective gas fittings and heating apparatuses. During the second day of the meeting the intertwined interests and close relationship between the two boards surfaced, as they grappled with the issues of kerosene illumination, burning fluids, and the light products of petroleum, complicated by gas generators and carburetors that were clearly in play as early as 1867. Edward A. Walton, president of the Citizens' Insurance Company, a member of the NYBFU and the current NBFU committee on gas machines and carburetors, asked the board to adopt an exhaustive NYBFU report on these devices; it was laid over.[61]

The backdrop of Oakley and Walton's remarks took shape in the fall of 1873. In November a large fire in Haverhill, Massachusetts, destroyed several city blocks. A barrel of shoe cement containing alcohol, turpentine, and other flammable ingredients exploded, spreading flames to adjoining property, a hazard the NBFU's Executive Committee had addressed in previous meetings and in a circular regarding the use of cement as well as benzine and naphtha in shoe manufacturing. The committee observed that the fire confirmed that strict controls on the amount of flammable liquids kept in buildings were warranted.[62]

More disasters called attention to the pressing need for careful evaluation of the hazards of gas machines. In early December 1873 gas-machine fuel exploded in a sack factory in Canajoharie, New York, killing several people and inflicting severe burns on others. In a lawsuit two years later Arkell & Smith, plaintiffs and the factory's owners, contended that the Commerce Insurance Company had given oral consent to allow gasoline in the factory, which the court sustained, despite Commerce's objection that the company had clearly forbidden gasoline use.[63]

In similar fashion a gas-machine leak in January 1874 set off an explosion that destroyed much of the H. E. Bradford mill in Bennington, Vermont, killing and injuring many women in this $100,000 loss. It was later believed that an entire tank of vaporized gasoline had escaped through a defective pipe into a ditch underneath the knitting mill and then exploded when a worker investigating the leak dropped a lighted match. These two incidents were classic insurance case studies in hazard identification, carelessness, and misunderstandings over policy limitations, factors that insurance boards and companies repeatedly pointed out to member companies and agents at all levels. Fire-insurance organizations sought to eliminate these troublesome and persistent difficulties through astute and informed analysis of risks and through uniform, and if necessary, properly endorsed contracts that purportedly anticipated all known contingencies surrounding a hazard. Fire-insurance associations and many state governments favored standard policies (a contract), a perfect means to advance the cause of uniform descriptions of policy contents, such as grounds for policy cancellation or coinsurance (indemnification proportional to loss), hazards, and the exclusion of agreed-upon risks thought to be uninsurable or undesirable. Listed in the Massachusetts standard policy (1887) are petroleum products, chemicals, or burning fluids, which if kept or used in insured buildings, "except that what is known as refined petroleum, kerosene, or coal-oil may be used for lighting," voided the policy.[64]

In February 1874, NBFU general agent Thomas Montgomery told the Executive Committee that in light of the explosions caused by leaky gas machines at Canajoharie and Bennington it should reaffirm its sanctions, adopted in September 1867, against installing gas machines inside a lighted building.[65] Montgomery's proposal anticipated the challenges the NBFU faced from erratic markets and the ever-changing claims of gas-machine manufacturers. He solicited the Executive Committee's views on carburetors, directing members to the Woodward patent device that used benzole to enrich gas, which the manufacturer had submitted for the committee's consideration and possible approval for placement inside buildings. The NYBFU endorsed it in January 1870 for city use under specific restrictions on location and installation, although the NBFU's Executive Committee, based on a position statement from December 1872, made it clear that carburetors like gas machines had to be outside of serviced buildings.[66]

The NYBFU and the NBFU had divergent opinions on how to mitigate gas-enrichment hazards, making more discussions on gas generators and carbonization unavoidable. At its 1875 annual meeting the NBFU heard from an investigative committee that had examined several acknowledged sources of fire, including the light products of petroleum and illuminants as well as patented carburetors. Gas machines and carburetors had already become artifacts of technological risk, a condition made clear in the committee's rationale for studying them. They *might* "prevent or arrest" fires, lower liability from accidents, or prove to be inherently dangerous.[67]

The committee advised the NBFU to bar insurance-company agents from granting anyone the privilege of using the light products of petroleum except under the rules and regulations put in force by the Executive Committee. It also offered a resolution calling for the National Board to prohibit the use of any gas machine that used these products, when any part of it, except the service pipes, was inside or adjoined the serviced premises. As for carburetors, the committee laid out detailed stipulations about their construction, which, if they conformed to this standard, could be located in cellars with natural light far from furnaces, a trade-off that would allow carbonization of water or coal gas in large commercial buildings in densely populated urban areas. The committee also recommended passage of an NBFU resolution restraining all local boards and their agents from recommending "any patent or useful device of any kind whatever." Only the national board could recommend a specific device. To be considered it had to have been in use for some time and be demonstrably safe.[68]

The following January, 1876, NBFU officers Henry Oakley and Daniel Heald submitted a short report on gas machines to the board's Executive Committee, replete with language reminiscent of previous NBFU-committee material. They described gasoline as an "extremely hazardous" illuminant, particularly if mixed with kerosene; a substance that should never be allowed inside any building; and as a vapor source for gas machines or carburetors, something to be kept in an underground vault or "small *detached* building of brick or stone" at least forty to fifty feet away from insured property.[69]

Their report included an illustration of a Springfield gas-machine system that the authors gave their "fullest endorsement." This device was fitted with an air pump conveniently installed in the basement of an insured home or building and a generator enclosed in an underground, brick vault some distance away.[70] They contended that this design was durable and safe if closely supervised by manufacturers, a claim that was presumably based on new risk knowledge about air pumps in insured buildings. Such machines featured secure piping and provisions to prevent condensation, double valves to stop gasoline from flowing into the pump while filling the reservoir, and an "*automatic check valve.*" If so constructed, these apparatuses did not violate, Oakley and Heald alleged, existing NBFU rulings, because no parts of the machines that were filled with gasoline were inside the serviced building. Oakley and Heald were important NBFU officers and leaders who knew that up to this point the board had avoided calls to endorse individual manufacturers' apparatuses. They contended, nonetheless, that one or two companies provided conscientious installation and service, permitting the use of their gas machines "without detriment to the interests of insurance companies."[71]

They relied upon their extensive risk knowledge to assess the dangers posed by gas machines. This methodology allowed qualitative estimations of the likelihood of undesirable consequences such as fire or explosion in the absence of long-term statistical data associated with twentieth-century procedures (post–World War II risk management).[72] As Oakley and Heald explained, gas machines had for some time been merely "experimental," forcing underwriters to be tentative until a pattern of "safety or hazard" became clear. Previous committee reports on the subject were sometimes ambiguous or inconsistent, but now the "matter may be fully understood," they insisted, allowing one to set forth definitively how best to control the products of crude petroleum. They pointed out that one company, Springfield, had over one thousand machines, as described in the report in service, some of which had run for eight years without a loss.[73]

Local underwriters could achieve this level of hazard mitigation, the report claimed, by following the NBFU's hazard-mitigation strategy for gas machines, made possible by the board's accumulated experience with technological artifacts as a basis for risk characterization. ("Technological artifacts" is my phrase and is not in the report. Nonetheless, the term expresses my understanding of the report's engagement with the technical features of patented or well-designed gas machines and their capacity to mitigate risk.) Requirements included a detached vault or gas house containing a well-made device and its gasoline or flammable fluids for vaporization, provided that no gasoline or fluids stored in barrels were anywhere on the property. The Executive Committee adopted the report, another contribution to the debate over gas-machine regulations reflecting ongoing technical refinements that emboldened manufactures to plead for adjustments in NBFU regulations.[74]

At the beginning of April 1876, representatives of several gas-machine companies, who wanted to use the gas-generator unit in their machines to enrich municipal gas, met with the NYBFU. They reached an agreement, promising in writing to use only NYBFU-certified workers who had to follow very detailed procedures and act with great care in refilling machines installed in stores or homes. The manufacturers met shortly thereafter to prepare an appeal to the NBFU to approve their apparatuses, claiming that by using naphtha rather than kerosene their machines safely carbonized or enriched "ordinary gas." This was based on the presumption that there was widespread carelessness in handling kerosene. The companies thought their arrangement with the NYBFU might persuade the NBFU, which otherwise regarded gas machines as unsafe, to allow naphtha-charged apparatuses.[75]

Apprehensions about the danger of gas machines, carburetors, and light-petroleum products remained strong from the 1870s to 1901–2 when the NBFU issued technically based rules about installation and use of gas machines, lamps, and systems.[76] The rules divided gas machines into two categories: outside and inside carburetors. Outside carburetors, buried underground at least thirty feet from all buildings, were considered the least hazardous. The rules favored "automatic, self-governing" mixers that produced a "uniform quality of gas under all conditions." The supply tanks for inside carburetors also had to away from all buildings, but if not practical they could be put in a ventilated vault or in a "non-combustible" building. An inside carburetor could contain at most one quart of gasoline, had to be made of brass or copper sheet metal, and outfitted with an "overflow connection" that drained back to the "outside supply tank." By devising separate rules for carburetor placement, the national board resolved the ongoing technical and placement issues that flourished up to 1900 regarding gas machines from many different manufacturers. And to cover any ambiguous cases the rules put all machines that did not have "automatic control" over gas "quantity or quality," or allowed more than one gallon of gasoline to enter a serviced building, or which employed a flame to vaporize this fuel in the especially hazardous class.[77]

Inventors and manufacturers inundated fire underwriters with their proprietary gas machines and carburetors, claiming improved safety, which in one instance involved generating a chemically stable gas (could not be condensed with conventional equipment) rather than an evanescent mixture of air and hydrocarbon vapors which often had uneven quality and was subject to condensation.[78]

The insurance interests firmly insisted in the second half of the nineteenth century that gas machines, their parts and reservoirs remain outside insured buildings and at a safe distance from adjoining properties.[79] The hazards of the light products of petroleum used in commercial processes and services such as dressing leather, shoe-making, producing and mixing paints and varnishes, as well as burning kerosene and gas-generation for illumination and heating, and pressurized gasoline or "vapor lamps" kept these activities and substances under close watch by insurance patrols, inspectors, and agents, adding to our understanding of the role of petroleum in the post–Civil War risk culture.[80]

Up to this point we have examined the efforts of governments, especially states and municipalities, to regulate the storage and use of petroleum and its light products through statutes and ordinances. Insurance companies and associations practiced consensual risk governance through policy exceptions and endorsements, in addition to schedules and tariffs that fixed minimum rates and added extra charges for keeping hazardous petroleum or its products in insured buildings or storage facilities—notably those that were on wharves or in densely built districts or central business areas; inspections and surveys of communities to find fire hazards and recommendations for mitigation; and permits issued to the insured who wanted to keep or acquire fire coverage that would be contingent upon compliance with permit rules judged appropriate for the special or extra-hazardous nature of the risks to be assumed by the company. When kerosene became the favorite fuel for illumination, insurance underwriters were very keen to control its retail sales because street vendors and drugstores often sold cheap kerosene that had been poorly refined and contained dangerous levels of benzine, naphtha, or gasoline to an unsuspecting public.[81]

These actions point to important nineteenth-century interactions among public and private agents, groups, and institutions. They also illustrate the incremental legalization of insurance rules and requirements. Fire-insurance associations and boards lobbied governments for such things as building-construction ordinances or controls on the retail sales of flammable liquids. This politically augmented and accredited their imposition of fire-insurance demands that spread risks among the public. Such private-public engagements facilitated the state use of private associations and companies, and through informal alliances and cooperation the fire-insurance interests in turn made these interactions concrete, social, and political, making possible the private delivery of a public service.[82]

STORING AND BURNING PETROLEUM FOR FUEL

There remained, however, an additional stage in utilizing petroleum as an energy carrier in the late nineteenth century that played a major role in the risk-management problems of our Gulf South cities in 1901–2. The nineteenth-century petroleum industry in Pennsylvania, New York, and Ohio gave birth to new energy markets from New York to Chicago, creating hazards and risks that accompanied experimentation and technical innovations arising from burning petroleum for fuel.

The Gulf South faced the same hazardous constituents that were evident early on in the post-1860 oil industry—production, storage, refining, transporting, and marketing petroleum and its products—and which justified insurance control and municipal regulation. Professional risk experts in insurance and fire protection formulated and refined rules in the nineteenth century based upon a body of experience with and knowledge about oil hazards that had been underway from the 1860s when states, cities, insurance companies, boards, and associations collected information, debated, and responded to the use of petroleum products for illumination. Over several decades this provided a risk-management platform that also suited petroleum as fuel. Municipal authorities in the Gulf South drew upon this risk knowledge to fashion ordinances to regulate the storage and consumption of petroleum in their jurisdictions.[83]

From 1859 to 1900 the Pittsburgh–Cleveland–Chicago area, the Philadelphia–New York City–Boston axis, and the Pacific coast were major areas for the use of petroleum and its products.[84] The search for a hydrocarbon substitute for animal and vegetable oils such as whale oil, tallow, or camphene for illumination generated interest in the rapid growth of petroleum production in Pennsylvania during the 1860s. Surging amounts of crude-oil stocks heralded steady supplies of refined illuminants, during a time when coal oil, paraffin, and gas also provided energy for light, cooking, and heating.[85]

Standard Oil Company (SOC) played a central role in producing and marketing petroleum for illumination, developing markets in Europe and

Asia as well as the United States. Although many other companies refined or distributed petroleum illuminants, SOC had the capital and managerial expertise to organize the gathering, storage, and long-distance pipeline systems required to match oil production with consumption and to create new petroleum products.[86]

During the development of the Pennsylvania oil fields in the 1860s and 1870s, drillers, producers, and refiners experimented with burning petroleum for fuel. As industrialization moved forward in the Pittsburgh-to-Chicago region, as well as St. Louis, manufacturers relied heavily on coal, coke, and to a lesser degree natural and coal-gas. Market fluctuations and labor problems periodically made petroleum an attractively priced alternative to coal, an anticipatable advantage after Ohio oil fields showed promise in the fall of 1885. Annual production subsequently skyrocketed to over 20 million barrels by 1896.[87]

In 1886–87, SOC bought up land around Lima, Ohio, and built a small refinery, taking a great risk buying and storing Lima crude without immediate prospects to dispose of it rapidly enough to stay ahead of production. Even so, the company purchased many Ohio oil companies and independent producers, especially after 1889, so that it dominated Ohio production by 1891. SOC accumulated massive amounts of petroleum on the ground but faced difficulties marketing it as an illuminant because the oil was malodorous, earning it the nickname "polecat oil."[88]

SOC discovered by late 1888 how to remove sulphur from Lima oil, resulting in a clean, good-quality kerosene without an offensive odor. It also laid plans to create new uses for its staggering amounts of Lima oil on the ground. Several decades of experimentation in Russia, Europe, and the United States suggested to manufacturers and engineers that crude petroleum had great potential as fuel in Pittsburgh, Cleveland, and Chicago, as well as in the East, where industrialization had gained traction.[89] Detecting a growing interest in using petroleum for fuel oil in Chicago, SOC completed a 210-mile pipeline in 1888 from Lima to Chicago to satisfy this demand, under the umbrella of the National Transit Company's subsidiary, the Buckeye Pipe Line Company.[90] At the terminus of the line, Standard built giant storage tanks in South Chicago to supply heavy industries using oil-tank cars. Manufacturers used the oil to heat furnaces and ovens and to produce steam. Standard subsequently sought municipal approval for a pipeline extension that would move kerosene and crude petroleum into the city under the aegis of the P. C. Hanford Oil Company.[91]

Adopting fuel oil in Chicago had been underway before SOC committed resources to the market, but company officials had faith in what a well-capitalized, wisely managed company could do on a large scale. In November 1886 Standard opened a fuel-oil department and dispatched representatives to Chicago, Cleveland, St. Louis, and other cities around the country to market SOC's huge crude reserves through below-cost pricing and long-term contracts. Here was an opportunity to promote Ohio crude in industrial areas as an efficient and clean alternative to coal. A Pittsburgh agent claimed the department had clients in "217 cities" in twenty states and territories by April 1888. The fuel-oil department focused on operations that raised steam to power equipment and on energy-intensive activities in metallurgy such as annealing, machining, and heavy fabrication, as well as glass making and iron-steel furnaces. Further, SOC took advantage of Chicago's extensive railroad system to ship huge amounts of fuel oil throughout the Midwest and around the country.[92]

In 1893 the company provided the Columbian Exposition with crude petroleum pumped from the Lima field through a second pipeline to a huge refinery SOC built in 1889–90 in Whiting, Indiana, on the fringes of Chicago along Lake Michigan. SOC built a branch line to the fair, where the power plant and oil house received fuel, using equipment designed by Clayton O. Billow (1860–1945) of the National Supply Company (NSC) of Chicago. Clayton and his brother, Elmer E. Billow, designed, marketed, and installed oil-burning apparatuses known as the Billow System. During the controversies over handling and storing fuel oil in 1901–2 in New Orleans, Clayton Billow testified in support of the oil interests and argued in newspaper interviews and before the City Council for the safety of liquid fuel.[93]

Burning fuel oil for industry was part of a relentless effort to cut production costs and achieve new efficiencies in manufacturing and fabrication. Oil as fuel required local fire underwriters and agents to identify petroleum hazards, assess the potential harm to humans or their property, and devise rules for its use, which were the basic components of the risk management of hazardous petroleum liquids.

Recognizing a growing interest in burning oil for fuel in utilities, manufacturing, and heavy industries—iron and steel works in particular—and

SOC's activities, the Chicago Fire Underwriters' Association investigated the issue in the spring of 1888. The underwriters moved in July to "regulate the use of the fuel as to reduce the fire hazard." A board-appointed committee drafted "rules and regulations for fuel for generating steam" that set out technical specifications and arrangements that became fundamental to managing the hazards of handling, storing, and consuming fuel oil. The association adopted a final version as permit requirements to burn crude petroleum for fuel on November 14, 1889.[94]

The rules for outside oil-plants made an enduring distinction between large storage tanks and small "daily supply tanks." Storage had to be in large, vented-iron tanks that were one hundred or more feet from all buildings, surrounded by an embankment to contain oil from tank leaks, overflows, or ruptures, and a water-drainage ditch. Similarly constructed supply tanks that fed the boilers could contain no more than five barrels of oil and had to rest outside of all buildings upon iron bracing or masonry. A pump moved oil from the storage to supply tanks through an underground pipe and returned unused oil the same way.[95]

Inside plants could only use "reduced" or refined fuel oil, probably residuum, that passed a burning test of 205 degrees Fahrenheit. Storage tanks constructed as described above had to be located outside all buildings and buried at least two feet below grade. Supply tanks inside the building could not exceed 250 gallons capacity per boiler and had to be equipped with a two-inch (minimum diameter) return pipe, rigged with shut-off valves at both ends, that the plant owners warranted would send unused oil to the storage tank. If there was no storage tank, the supply tank, not to exceed 1,000 gallons capacity, had to be enclosed in "two feet of solid masonry, cemented and under ground," outside of all buildings and at least twenty feet from the "mouth of the furnace" using the oil. Plant operators also had to warrant that petroleum fires in the boilers would be started using wood or coal before oil was introduced. Extra rates applied in all cases to buildings exposed to storage or supply tanks.[96]

Although critics complained that the extra rates would discourage oil use, they privately thought that the underwriters would, having retreated from stringent regulation of using naphtha and gasoline, eventually do the same for burning oil for fuel. This was reasonable. Insurance rules evolved in tandem with practical experience, "improved methods," and technological advancements. Embryonic forms of fire-protection engineering in the nineteenth century produced risk knowledge about handling petroleum and its products that called for modified techniques in managing such hazards.[97]

But there was no quick relief from the proposed rules. Chicago's mayor, commissioner of public works, and chief of the Fire Department accepted the fire underwriters' position that fuel oil was dangerous and promised not to issue permits indiscriminately for its use. The underwriters in turn promised not to write fire contracts on property where fuel oil was used, unless the owner had a permit signed by the commissioner of public works that was accompanied by written proof that the Board of Underwriters' superintendent of survey had inspected the premises. This illustrated the coercive, semigovernmental power the fire-insurance companies exercised. The city allowed storing, selling, or consuming oil in a particular location, but if one wanted indemnification for loss, the fire underwriters had to approve of the conditions of use.[98]

In the 1880–1900 period, expanding production along with reliable transportation, distribution, and storage of petroleum lured many manufacturers to substitute oil for coal. An increasing number of companies in Chicago, as well as in California,[99] the East, and the Midwest, adopted crude petroleum or residuum, attracted by price advantages over coal and labor-maintenance savings. In 1891 the Chicago Fire Underwriters' Association determined that, in light of the widespread adoption of petroleum fuel, the association should expedite its use while mitigating petroleum hazards and reducing risks to property under the new circumstances.[100]

A committee tackled this project, working for six months in the belief that safe handling of crude oil was an important issue for fire underwriters in all large cities. The association promulgated the recommended rules in January 1892 and printed or explained them in several publications, making it possible for underwriters in densely populated areas throughout the country to use the general stipulations and conditions thought appropriate for Chicago in their communities. The rules also proved valuable in 1892–93 when the Columbian Exposition leaders decided to burn fuel in the fair's power plant rather than wood or coal.[101]

The new rules displayed clearly the hands-on knowledge of technically astute consumers, technological practitioners, and experienced engineers such as Clayton and Elmer Billow. Oil tanks had to be put in water-tight, brick enclosures belowground with sloping floors to collect seepage, covered by a top made of iron-supported masonry overlaid with at least twelve inches of dirt and fitted with manholes and ventilation pipes.[102] The committee spent a great deal of time on the details of tank construction, and after consultations with engineers and oil experts they settled on a basic, cold-riveted, boiler-iron tank similar to the tank required in the 1888 rules. No more than two tanks could be put in a vault, each no more than eight feet in diameter and twenty-five feet in length (1,257 cubic feet per tank). Specifics for tank accessories and related equipment covered flanges, filling pipes, manholes, pump-supply and steam connections, overflow pipes, pipe and connection dimensions, tank mountings, gauges, pumps, and standpipes, details not presented in the older rules. The vault specifications and tank capacity suggest that the committee had in mind standards appropriate to large-scale operations such as the Columbian Exposition.[103]

Activities by many companies in the Atlantic-coast markets such as Philadelphia, New York, and Boston also contributed to the evolution of insurance rules about using liquid fuels. After the development of the Lima field, SOC's fuel-oil department targeted New England's mills, factories, and utilities as potential bulk purchasers of the company's huge oil reserves. New England manufacturers sought cheap energy, expressing a regional concern to compete with the West—Pittsburgh, Cleveland, and Chicago—to which some industries such as glass making had migrated to be near the coalfields. Edward Atkinson noted in retrospect that New England had turned to technology and manufacturing to compensate for sparse inland resources. Trials of oil-burning apparatuses in Lowell, Massachusetts, and other New England locations confirm that industries and towns that had used water, gas, or coal for energy and power were interested in the possibilities offered by petroleum.[104]

Burning petroleum to raise steam was not a novel idea. In the 1860s manufacturers, inventors, and boiler engineers conducted tests to determine if liquid fuel could be used efficiently and economically in boilers. In 1865 veteran engineer Julius W. Adams, acting in behalf of the Petroleum Light Company, attached an oil-burning apparatus to a boiler supplied by the Morgan Iron Works of New York City and demonstrated that petroleum could be successfully used to raise steam.[105]

The U.S. Navy experimented in 1866–67 with petroleum under marine boilers but subsequently rejected the idea. In London, C. J. Richardson conducted more extensive tests for the same purpose at the Woolwich Dockyard and came to the same conclusions as had Adams but declined to endorse fuel oil for ships, believing "commercial considerations" would determine that issue.[106]

Practical experiments with petroleum continued in the 1870s in varied circumstances, including a brass foundry in San Francisco where a local inventor demonstrated an oil-fired furnace for metallurgy and steamship propulsion, and the Illinois Central Railroad's machine shop in Chicago, in which the railroad conducted tests to see if it could substitute petroleum for coal in locomotives.[107]

In the 1880–1900 period a number of New England inventors, plant engineers, and companies set up appliances to burn petroleum to power industrial processes, clear examples of experimentalism and invention intertwined with engineering problem-solving that contributed to the nineteenth-century, petroleum-risk culture.[108] Working against a backdrop of uncertainties over coal supplies and skepticism about the comparative efficiency of petroleum versus coal, they explored and expanded mechanical practices, hoping to provide mills and utilities with an energy carrier that would sustain economic growth. One newspaper thought fuel oil held out the possibility of a "complete revolution in the method of furnishing motive power for our great manufacturing plants." It would mean "freedom from uncertain supplies of coal fuel, absence of storage at the mills, clean boiler rooms, and more economic management."[109]

Agents of oil companies and distributors, such as SOC's fuel-oil department, blanketed the Northeast offering petroleum—reputedly safe to fire boilers—to mill operators and superintendents. Fuel-oil hazards raised red flags among fire-insurance agents and associations, which in this case was the AFMFIC, actively assisted in technical matters by the BMMFIC led by Edward Atkinson, who also participated in the New England Cotton Manufacturers' Association (NECMA). The NECMA investigated and discussed textile-machinery lubrication including mineral-based substances from 1868 into the early twentieth century.[110]

Atkinson was admirably suited to addressing fuel-oil issues. He made important contributions to nineteenth-century risk culture through his activist role in the insurance industry and participation in the NECMA, serving as a director, 1865–78, and president, 1878–1905, of the BMMFIC. After analyzing William B. Whiting's records of Mutual fires and their causes, supplemented by questionnaires sent to members, Atkinson identified pressing risk issues about which he issued numerous circulars on mill hazards. He also hired as an inspector MIT graduate C. J. H. Woodbury, who had five years of practical experience in mills. Woodbury and Atkinson demonstrated the value of "technical education in industry" and "technical research procedures in insurance work for industry." Woodbury, acting under Atkinson's patronage, took the first steps to create a "new branch of applied science—that of loss prevention engineering."[111]

In the early 1880s under Atkinson's leadership the BMMFIC's Boston office became the "engineering headquarters" of the AFMFIC companies, sending out "recommendations and rules for the use of fire protective equipment, and the removal of fire hazards." Member companies also turned over to Atkinson the costs of inspections, "engineering, appraisal and adjustment," for which individual companies were responsible "in proportion to the amount at risk." In 1886 Atkinson hired John R. Freeman, also an MIT graduate, to work in the BMMFIC's Inspection Department. Freeman increased the details recorded on inspection-engineering reports, more complete than ever before, and conducted independent research and testing on automatic sprinklers, hydrants, and valves. He became the assistant chief in 1894 and chief of the department in 1894, on his way to a brilliant career as an engineer.[112]

One example of Atkinson's campaign to introduce technical research as the basis of risk management of insured mills and factories is his search for reliable and safe lubricants for textile machinery. Atkinson oversaw experiments starting in 1878 on friction and lubricating oils—animal, vegetable, and petroleum or mixtures thereof—to determine if failed lubricants caused fire loss.[113] He contracted for a major analytical report on this subject with Prof. John M. Ordway of MIT, made innumerable comments at NECMA meetings, and at the end of his life cited lubrication as one of the important subjects he had investigated in his role as insurance official and textile-industry savant.[114]

In addition, Atkinson examined mill illumination and the safe use of kerosene or gas and took on the issue of burning petroleum to raise steam for manufacturers in a region with few natural resources other than the ocean. He encouraged eastern manufacturers to stay abreast of concepts and methods to identify and calculate the true costs of operating mills and plants, looking at water, coal, electricity, and by the late 1880s, petroleum for fuel, an effort the NECMA supported by entertaining papers on topics such as the transmission and distribution of power to and within plants.[115]

In 1887–88 as Ohio crude oil came to market in the East, first for metallurgy in foundries and forges, as well as for power in mills, chemical and wood-pulp plants, superintendents and owners insured by AFMFIC sought guidance on safe storage and consumption of fuel oil. Atkinson sent out a circular or bulletin in February 1888 to Factory Mutual members informing them that burning fuel oil was not objectionable as long as it was stored in iron tanks safely away from nearby buildings and the piping properly configured to prevent siphoning and detect oil discharges. These conditions, the circular noted, coincided with views held by the highly regarded Hartford Steam Boiler Inspection and Insurance Company.[116]

Five months later, Atkinson released a special report that outlined technical features and methods for hazard mitigation of petroleum risks. This included underground iron tanks for storage that did not pose threats to contiguous property, surrounded if aboveground by dikes or embankments, and recharged by gravity if filled from tank cars. He instructed Factory Mutual members to install small supply tanks to feed burners and to add cutoff valves and other provisions to return unused oil to the tanks. It was also made clear that these "requirements" could be modified in circumstances such as industrial metal work.[117]

The New England Insurance Exchange (NEIEx), formed in 1883 by special agents to rate paper mills and other hazards outside Boston, and its counterpart, the Boston Board of Fire Underwriters, had fuel-oil investigations underway by the summer of 1888. The exchange's findings, printed a year later under the title "Petroleum Fuel System," identified the important components of oil-burning plants that any rules or regulations had to cover: tank construction and materials to prevent leaks or overflow; tank location, grade, and distance from insured buildings; quality of oil and capacity of storage and supply tanks; appurtenances for delivering oil to these tanks and burners such as piping, pumps, valves, and shutoff cocks; and efficient burners.[118]

Resorting to what I call the "Ohio" narrative—anchored in SOC, its huge Ohio-oil production unsuitable for illumination, full-bore sales campaigns to sell it in New England and the Northeast for fuel, and the energy needs of New England—the report claimed that there was a "great diversity of opinion" among insurance men in the West (Pittsburgh to Chicago, south to Cincinnati) regarding details such as whether to use gravity or pumps to convey oil to tanks or burners. There was no "generally accepted code of rules" among Western fire-insurance companies, who preferred to accept, the report argued, specific fuel-oil risks based on the judgment of their special agents, making the NEIEx hesitant to adopt rigid rules.[119]

The professional aims of the NEIEx, made up of field men and special agents representing many fire-insurance companies, shaped the report. Working through committees, exchange members classified and oversaw major categories of risks and worked with local boards to set standard rates, thereby establishing close contact with insured-mill owners, operators, and superintendents through inspections and hazard assessments. Informed by this first-hand knowledge, the petroleum committee recommended that the exchange should, in light of the different circumstances of "widely scattered" oil-fuel plants in New England, employ a "competent inspector" to visit them and to prepare surveys from which agents could devise flexible, site-specific rules to mitigate the hazards associated with burning fuel oil.[120]

In the wake of the AFMFIC and NEIEx inquiries, Edward Atkinson learned from several mutual companies that agents and oil suppliers—some of whom carried SOC products—had offered to supply them with residuum, even allowing interested companies to test oil-burning apparatuses. Residuum is a heavy residue of petroleum, extensively refined to remove dangerous and volatile cuts such as benzine, naphtha, or gasoline, that requires relatively high temperatures to ignite. Subsequently, customers found that suppliers had shipped them crude petroleum that emitted dangerous vapors, causing several accidents and discouraging adoption of fuel oil. Atkinson warned member companies that they should not use this crude oil to raise steam unless they understood the risks it entailed and the possible diminution or loss of fire-insurance coverage if they proceeded without Factory Mutual consent.[121]

In response to a newspaper claim that he had quarreled with SOC over dangerous petroleum—something he denied—Atkinson insisted that the AFMFIC had cordial relations with SOC, having successfully persuaded the company to change its distillation methods in the early 1880s to produce paraffin-oil lubricants that met the quality standards for safe machine- lubrication in mills. Atkinson spearheaded investigations to mitigate oil hazards in mills and plants that produced the lubricant benchmarks. Further, Atkinson pointed out that fire underwriters were investigating the safe use of liquid fuel, a project SOC wholeheartedly supported. (SOC subsequently fought local, oil-storage regulations and inspections in court throughout the United States into the twentieth century.) The *Boston Herald* tried to clear the air, claiming the "only unpleasantness" came when a SOC salesman tried to convince Atkinson that fuel oil was "not inflammable"![122]

Atkinson, acting in behalf of the BMMFIC and all members of the AFMFIC, and the NEIEx, as well as underwriters in Hartford, Philadelphia, and surrounding states, advanced the systematization of the experiences, risk knowledge, and technology that fire-insurance companies, associations, local boards, and inspection-rating bureaus developed or promoted through hazard-mitigation methods and procedures.[123] Drawing upon their clients' experiences with petroleum, underwriters responded with appropriate techniques to prevent fires and reduce losses. Risk agents—those who gathered and applied basic principles of risk knowledge through this network of industry and fire-insurance advocates—distilled their understanding of what it took to handle and consume fuel-oil safely into rules and regulations.[124]

This penultimate stage in the evolution of a nineteenth-century, petroleum-risk culture grew out of decades of state, municipal, and insurance efforts to manage the risks of petroleum and its products, a process which by 1900 benefited from the growing influence of fire-prevention engineering, embraced and encouraged by fire underwriters and technical groups such as the NYBFU, NBFU, and NFPA.

The NBFU had a long track record of addressing oil hazards, expressing concern in 1867 about the "storage of petroleum and its products," and again in 1869 through a committee that focused on the sale and storage of petroleum and kerosene. An important report on gas machines adopted in 1876 observed that crude petroleum contained a lot of "volatile oil," most notably kerosene, benzine, and gasoline.[125]

However, there was a shift in emphasis in 1889 from volatile petroleum products to burning oil under boilers that arose from a surging industrial-manufacturing interest in the cost-benefits of crude petroleum or residuum. The NBFU entertained a report from the Committee on Lighting, Heating and Patents regarding electric lighting, natural gas, and "petroleum as fuel." It recommended adopting the simple rules for petroleum use originally devised by the NYBFU in 1887 (and possibly by the Western Union as well). They specified high-fire-test petroleum, outside tanks set underground below the grade of the oil-supply line and made out of boiler iron—the earliest underwriter rules to surface in this study.[126]

This rationalization of petroleum hazards took place in a broad context.

In engineering, industry, business, and insurance, rationalization identifies techniques, conditions, or processes and organizes them logically into abstract categories or groups, based on performance or characteristics, for some purpose or end such as uniformity or standardization. This reconstitution of information makes tacit (informal) knowledge explicit to achieve maximum utility or economic gain, which in the fire-insurance business is loss prevention. Rationalization as a social-economic activity includes critical decision-making, efficiency, regulation, management, control, and the identification of specific objectives to the exclusion of all others. Rationalization of manufacturing means actions such as organizing plant-floor layouts for efficiency; instituting cost controls; defining tasks that unskilled workers can execute; and establishing quality objectives and control over expenses, particularly labor costs; all apply to the petroleum industry as a socio-technical system, including its regulation and the spreading of risks inherent in petroleum and its products by fire-insurance companies.[127]

With regard to the rationalization of fire insurance, petroleum hazards and risks, it is clear that the NBFU and other insurance associations had a long-standing concern over uniformity and standards in the classification of hazards, rates, commissions, tariff schedules, fire and ignition testing of oils, and procedures to be used by local underwriters' boards, as well as surveys and inspections by local, state, and special agents and inspectors.[128] As early as 1870 the NBFU noted its "efforts to systematize and organize" the fire insurance business, but achieving uniformity in the above matters was often elusive, fraught with protracted political struggles, internal controversies, and endless debate, especially on the sensitive issues of rates and whether to have uniform state laws or national supervision to regulate practices.[129]

Surveys, inspections, and rating schedules were tools that advanced the accumulation and rationalization of risk knowledge about many hazards, including petroleum and its light products. In November 1871, in the middle of protracted battles over flammable petroleum products, the NYBFU learned from an individual that an enormous stockpile of empty gasoline and oil drums threatened warehouses stuffed with valuable materials and insured goods, a situation calling for immediate legal action. Sound information was the solution the NYBFU offered for this problem. In 1877 the board charged its Committee on Surveys to survey all buildings "that the interests of members require," to make its findings available to members, and to examine regularly the "public warehouses" in the city and Brooklyn, reporting all violations of the board's "standing rules . . . governing such warehouses." This data could be used in annual revisions of the board's established classes of hazards that included petroleum storage.[130]

An NBFU committee report outlined in 1875 possible strategies to curtail the dangerous use of petroleum and its lighter products, noting that it would require state and municipal inspections and possibly a "tariff of charges" for the retail sale, consumption, and storage of these special hazards. Nearly two decades later the NEIEx invoked its system of detailed classification of risks, also known as a "schedule," that covered hundreds of hazards in various categories such as mills and factories or "manufacturing risks." Careful inspection by local boards, exchange committees, and individuals identified and assessed every risk included in the schedule, allowing hazard investigations and risk inspections to be conducted on a "uniform basis." It also allowed the exchange to study construction, hazards, and protection, which are basic elements and methods of risk assessment.[131]

This was familiar insurance procedure. Schedules that date to the Civil War era promulgated basic rates on risks classified by group or degree of hazard and specific function or activity, such as mills, building features and construction, and physical characteristics (fuel, drugs, and oils; provisions in barrels; iron-clad warehouses, and so forth). In some instances these itemized lists showed additional amounts to be charged, extra costs that reflected inspections, experiences with and professional judgments about various hazardous, making it possible for agencies or insurance companies to raise or lower rates for individual clients relative to a known price point. Insurers could plausibly justify advancing or lowering rates on the basis of specific circumstances or conditions, such as the quality of local fire departments, reliable water sources, unregulated building construction, or extra hazards arising from exposure to adjacent property, that changed the degree of risk.[132]

Using information about hazards provided by boards or associations, making judgments on the severity and dangers posed by hazardous substances or conditions, and devising fixed schedules for a great range of risks in major areas of concern to the insurance interests created an array of people—field men, special agents, surveyors—and organizations that distilled from this matrix of experiences, perceptions, and procedures credible and rational risk knowledge that included the products, technology, and infrastructure of petroleum.[133]

The logic of uniform rates demanded consistent identification and classification of hazards. Agents who surveyed or inspected manufacturing plants, machinery, building construction, and electrical installations needed objective criteria for making risk assessments, establishing tariffs, and setting rates. NBFU-approved policy forms, hazard classifications, and tariff schedules, it was hoped, would provide a set of revisable benchmarks for these crucial fire-insurance procedures.[134]

To that end the NBFU and the NFPA participated in a campaign to mitigate fire hazards related to electric lighting, generation, and wiring. The quest for uniform electrical rules, started as early as 1881 by the NYBFU and the BMMFIC,[135] attracted the support of the National Electric Lighting Association (NELA), the NEIEx, and other agencies and associations, acting together to rationalize fire-insurance surveys and municipal inspections through consistent, countrywide identification of hazards and application of mitigation techniques to eliminate electricity as a fire source.[136]

Many NELA members operated or managed central-power-generation stations which many fire-insurance companies considered uninsurable or insurable only at exorbitant rates.[137] To persuade fire-insurance companies that these facilities were generally safe, the NELA formed a committee to harmonize their hazard mitigation efforts with the demands of the companies.[138] Additionally, NELA members heard several papers at annual meetings on how to use fuel oil properly to raise steam in central stations, accompanied by explanations of the applicable insurance rules.[139] One solution to the tensions between electricity producers and the insurance companies was to recruit, train, and certify workers and inspectors who would examine all electrical installations. This would reassure insurance skeptics that the electric companies had devised standard operating rules that addressed all hazardous conditions in their central stations, including knowledge of and adherence to insurance rules on fuel oil. Other paths included proper construction of central stations and well-trained workers.[140]

In 1896 the NBFU issued up-to-date electrical standards recommended by the Underwriters' National Electric Association, which contributed to the realization of uniform, national standards and rules in cooperation with numerous professional organizations, including the AFMFIC and the NELA, culminating in the promulgation in March 1897 of the National Electric Code.[141]

It therefore followed that when the NBFU, working with the NFPA, took up the issue of fuel oil, it addressed this hazard in a framework of uniformity and standardization, seeking scientific insight and technical knowledge from engineers and experiential knowledge from so-called practical men such as special agents, inspectors, and surveyors who had an interest in establishing acceptable ways of dealing with hazards, including those arising from the storage and handling of petroleum fuel.[142]

The drive for uniformity gained ground through the efforts of insurance leaders who sought technical and scientific legitimation of contemporary fire-hazard mitigation measures through centralization, standardization, and cooperation. This was quite evident in the formation of the NFPA. It grew out of meetings in 1895 and 1896 to establish an organization in the interests of stock fire-insurance companies for the "standardization of materials, devices and methods," initially formulating "uniform rules and conditions . . . governing automatic sprinkler installation" but subsequently embracing blower systems, water hydrants, fire alarms, and fire doors as well as other contrivances.[143]

In the same vein, NBFU President E. C. Irvin sent out Circular no. 542 in March 1899 to all board members, in response to concerns expressed by the board's Executive Committee, officers, and committee chairmen about the irregularities in local treatment of fire hazards. The circular proposed the creation of a scientifically informed "central testing bureau" in New York City that would offer the NBFU's executive and standing committees the legal and scientific findings required for accurate, complete, and authoritative rulings. This department would enable and complement the centralization and standardization of inspection procedures and "tests of devices, appliances, and materials" such as lighting apparatuses, fire alarms, and sprinklers connected to fire-hazard mitigation. The board would assume jurisdiction over forms, legislative issues, and fire-department and water-supply inspections, with the power to promulgate "all rulings on these important matters."[144]

Following an appeal by Irvin at the board's annual meeting in May, members authorized the Executive Committee and several standing committees to set the centralization plan in motion, as well as creating a special committee to consider the details of Irvin's recommendations and to report.[145] In December 1899, after discussions with various fire-underwriters associations, the Executive Committee approved creation of a "Board of Consulting Engineers" (BCE), members of which would represent every "general rate-making association" as well as local boards in big cities, and secure "uniform rules" for "devices and materials" linked to fire hazards that were at the heart of Irvin's recommendations. NBFU members ratified this decision in May 1900.[146]

The December consultations also suggested to attendees that it would be advantageous for the NBFU to cooperate with the NFPA and "adopt its standards and promulgate them," resulting in a conference in February 1900 that recommended this action. In addition the NBFU would print NFPA standards as NBFU rules and requirements, as well as NFPA reports and annual proceedings. Board members agreed to this at the annual meeting in May.[147] In this relationship the NFPA's recommendations complement and reinforce the BCE's experimentalism and empiricism in the attainment of uniform and unassailable standards. NFPA leaders were convinced that through "mechanical and scientific treatment" their organization could objectively define fire hazards and thereby attenuate possible losses, a way of approaching risk management that one prominent insurance figure called "the science of preventing loss by fire."[148]

Cooperation produced a suite of hazard-mitigation initiatives. In 1901 the NBFU issued several "Rules and Requirements" based on NFPA standards: wiring for fire and sprinkler alarms; stationary and hand-held fire extinguishers; cotton, rubber-lined fire hose for mill use; construction and installation of fire doors and shutters; automatic sprinkler equipment; watchmen's time-recording apparatuses; use of wired and prism glass as a fire retardant; and acetylene gas generation for lighting.[149]

The rules on alarm wiring rested on an NFPA standard from 1899 created before the NBFU agreed to promulgate NFPA material as its own. The NBFU compiled guidelines on the organization and equipment for mill fire-brigades as "Suggestions" from circulars that several mutual fire-insurance companies sent out. Rules for gasoline-vapor lighting machines and systems came from the NFBU's "Committee of Consulting Engineers" (that is, BCE), the bailiwick of which was special hazards (by agreement with the NFPA) related to lighting or heating excluding electricity.[150]

The NBFU's Committee on Lighting developed acetylene rules in April 1898 which were widely distributed, even though a meeting of insurance associations in June failed to produce a consensus on the long-standing controversy over installing gas generators inside buildings. By December a special subcommittee decided to allow them—short of the uniformity and "concert of action" so valued by contemporary insurance leaders—inside buildings in "special circumstances." In 1900 the BCE drew up "Rules and Requirements" for constructing and installing acetylene gas machines, garnering acceptance as a "uniform standard throughout the country."[151]

In pursuit of uniformity and consensus, the BCE also developed rules for storing and using fuel oil as well as the construction and installation of oil-fired equipment or systems. These requirements evolved incrementally, a regulatory form of self-correction in step with technical advances and accumulated risk knowledge. The rules also reflected more than a decade of field observations and inquiries from people or companies who were already burning petroleum or wanted to try it. Fire-insurance agents and engineers contributed observations about and acquired valuable insights into the design and function of oil-burning apparatuses or systems, sometimes derived from information published in technical periodicals. This exchange of technical knowledge often took place in industrial or commercial concerns where owners or supervisors had installed oil-burning equipment to conduct trial runs to calculate the economics of replacing coal with petroleum in specific applications such as heating metallurgical furnaces. Trials exposed design deficiencies or construction glitches, requiring ad hoc mechanical modifications and experimentation to overcome these problems.[152]

Over time and in parallel with equipment refinements, common sense and reasonable precaution suggested segregating large storage tanks from buildings containing the point of ignition, setting up small supply tanks below the oil burners, and adopting piping configurations that drained unused oil away from the burners when they were not in use or had appropriate cut-off valves to prevent liquid fuel from spilling into fire boxes and starting a fire. Insurance agents and inspectors investigated the "use of crude petroleum" in various situations, observed successful and safe practices, and framed their rules accordingly.[153]

This network of risk knowledge expanded, drawing the attention of fire-insurance groups and companies such as the NYBFU, AFMFIC, and the Chicago Fire Underwriters' Association. Between 1887 and 1892 they formulated rules based on fieldwork, inspections of fuel-oil plants, engineering expertise, and the technically informed loss-prevention methods of mutual companies. They codified observations and hazard-management techniques, established a shared language of risk categorization and management, and diffused practical know-how or tacit knowledge—reconfigured as engineering routines and processes susceptible to scientific corroboration—through the fire-insurance, government, and industrial communities. This body of risk knowledge distilled into mandatory fire-insurance rules devised in the late nineteenth century constituted the pedigree of the BCE rules.[154]

Before the BCE came into existence, the NBFU acknowledged the growing use of petroleum but did not view it as an immediate threat if used prudently. At its twenty-fourth annual meeting in New York City in 1890 the NBFU's Committee on Lighting, Heating, and Patents declared that using "petroleum fuel was still largely experimental," but not "unduly dangerous" if consumed under three simple rules: burn oil with a high flash point; put tanks a "safe distance from the building in which it is used"; and use pumps, not gravity, to feed burners through piping kitted out with backflow and automatic shut-off valves to prevent liquid fuel from entering the furnace.[155]

The obvious hazards of raising steam with petroleum challenged the NBFU by 1900 to standardize hazard-mitigation measures. The NBFU-NFPA alliance and the subsequent creation of the BCE granted these confederates in fire-protection engineering joint access to the resources of William H. Merrill's Underwriters' Laboratories in Chicago and the NFPA's investigations of fire-protection appliances, flammable substances, and hazard mitigation. Their cooperation made credible the NBFU's determination to standardize their rules and requirements and ground them in reliable scientific and technical advice, making it necessary to employ adequately trained men to test devices and materials, investigate systems, and report their findings.[156]

One such person involved in this quest for technical certitude was Allen M. Schoen (1869–1928), graduate of the Virginian Military Institute, BCE member, and the electrical inspector for the SETA.[157] In early 1902 the consulting engineers sent him to the Gulf South to gather information in the oil fields and major cities so that it could fine-tune its work on technical rules for storing, handling, and burning fuel oil. This was the penultimate stage of an investigation that was underway at least as early as November 1901, when the Chemical Hazards Committee of the NEIEx announced, in response to a "recently made request," that it was deferring to the NBFU on the matter of "formulating rules for fuel oil systems." Similarly, the BCE claimed that it had compiled rules on fuel oil after examining a number of "oil-burning equipments" and in response to "continued requests from the oil fields."[158]

Schoen finished his investigations in early July, agreed in late October to BCE recommendations submitted to the NBFU's Executive Committee, and witnessed the promulgation of official rules and requirements in December regarding fuel oil.[159]

Extracts from his report show that Schoen traveled to Mobile, Beaumont, the SpindleTop oil field, Port Arthur, as well as New Orleans, Galveston, Houston, and Dallas. He took careful note of various petroleum-fire hazards that involved familiar risk-management issues such as mandatory distances between ignitable liquids and sources of ignition, illustrated by the placement of red-hot boilers downwind from drilling rigs and producing wells in the densely packed Beaumont oil field that had already seen several destructive fires in 1901.[160]

He also found that local engineers with excellent reputations disagreed among themselves about hazard mitigation, specifically putting out large oil-tank fires, making it hard to fine-tune proposed regulations. Jesse C. McDowell, vice-president of J. M. Guffey Petroleum Company, favored moats around tanks and a concomitant reduction of the tentative BCE-tank-separation rule from five hundred to three hundred feet. This stipulation confirms that the BCE had drafted rules before Schoen's tour for him to use as a baseline for his investigation.[161]

Schoen met in New Orleans with engineers from several equipment suppliers, including Worthington (pumps), NSC (oil-burning systems) and Rockwell Engineering (boilers). They favored small standpipes to maintain oil pressure but considered steam coils in storage tanks to be superfluous.[162]

He toured Guffey's big refinery at Port Arthur and visited commercial oil-burning plants in Houston and Galveston. This raised the issue of burners or jets, which were available in so many different designs that he refused to make "recommendations or comments," an ironic suspension of the central purpose of his investigations. He also found that some engineers favored pop-off valves to regulate oil pressure, while others turned to standpipes for steady oil pressure at the burners. Multifarious practices make exacting specifications difficult to establish, illustrating the complexities of mapping technical artifacts and real-world procedures to conform to risk-management protocols.[163]

In Dallas, Schoen met with the Fuel Oil Committee of the Texas Fire Prevention Association and found some disagreements over hazard-mitigation techniques. The committee thought wood or cement tanks were safe, but Schoen did not agree. Committee members also wanted underwriters to allow small supply tanks under sidewalks and "overground tanks" where there was adequate "ground space" as well as appropriate conditions. Even so, Schoen was satisfied that in most details the Fuel Oil Committee agreed with the BCE on installation methods. In retrospect he had reasons to be pleased. Schoen's recommendations, "elaborated and embodied" in the NBFU's *Rules and Requirements,* became the "basis of the forms to be issued by the National Board," which circulated throughout the United States, ending, as one insurance journalist claimed, the national confusion brought on by a "multiplicity of forms, as are now in use." Thus, Schoen contributed risk knowledge to the hazard management of petroleum, advancing the campaign for uniformity and standardization in the fire-insurance industry.[164]

The 1902 *Rules and Requirements* was a milestone in the nineteenth-century evolution of the risk management of petroleum hazards and an exemplification of fire-protection engineering that embodied standardization and codification of hazard mitigation. Carrying forward and expanding upon the technical features found in the insurance rules of the post–Civil War period, this document set down a consistent set of rules that covered the storage, delivery, and consumption of fuel oil in commercial and residential installations through a variety of technical specifications. Large oil-storage tanks could not exceed 37,500-barrel capacity or thirty-foot height; they had to be made of "steel or iron plates" riveted together and braced with an inflammable, ventilated top; located at least five hundred feet from "burnable property" or from a water front, and three hundred feet from other oil tanks; surrounded by an embankment creating a reservoir that could hold 150 percent of a tank's capacity; furnished with a steam pipe feeding into the tank above the surface of the stored oil; having "supply and discharge" in separate, underground piping leading to and from pumps outside the tank/reservoir and fully functional even in case of a tank fire; and rigged out with gate and control valves.[165]

From the perspective of the BCE, oil-burning equipment raised the level of risk beyond that associated with large-scale oil storage. The *Rules* for such apparatuses were also technical and prescriptive and applied only to oil with a flash test of at least 160 degrees Fahrenheit that a small supply tank, capacity no more than 12,600 gallons, fed to the burners. If the supply tank was filled directly from a tank car, as would be the case in densely built urban zones, consumers had to have two receiving-supply tanks—each equal to the capacity of a tank car—so that the car could fill one while the other fed the burners. The tank(s) should be installed at least two feet underground, thirty feet from all buildings, and below the grade level of the lowest pipe in the oil-burning apparatus. If underwriters having jurisdiction over the apparatus agreed, the tanks could be placed in a brick or concrete vault.[166]

If the same fire underwriters sanctioned an aboveground tank or tanks, they had to be two hundred feet from "burnable property" at the lowest grade available and no more than 1,500 barrels (63,000 gallons) capacity. Tanks underground had to be made of at least three-eighths-inch iron or steel and a quarter-inch for vaults or aboveground tanks, painted with tar or some other anti-corrosive material, ventilated, and equipped with an oil-level indicator. And there were more specifications dealing with a manhole to access the tank when required, filling pipes and hose, a "duplicate system of oil pumps" designed to limit oil pressure to a maximum of sixty pounds and configured so that, if one pump went out of service, the other would keep the apparatus operative, as well as provisions for oil to drain back into the supply tank when the pumps stopped and shut-off valves in case of an accident that might release oil outside the system.[167]

These details are important because they point to recognized hazards that had for decades been the source of explosions, oil spills, and fires—fire boxes inundated by liquid fuel due to erratic oil pressure; gas trapped in faulty piping arrangements; pump failures that allowed hot oil to drain into supply tanks—noted and acted upon by fire underwriters, surveyors, and insurance boards who incrementally refined their fuel-oil rules. The specifications gave the NBFU and its member boards and associations a formal, semi-official control over the "effort and resources" to be expended by the insured if they wanted to raise steam with petroleum, while indirectly supervising the "events and conditions" of storing and burning oil through an engineered system that prevented loss and established what level of risk the fire underwriters would tolerate in an "operational system." In many instances, local enforcement was inconsistent or incomplete, but the rules put fire-insurance customers on notice that they could not ignore hazard mitigation if they wanted insurance.[168]

In the nineteenth century a risk culture was established in association with the mitigation of petroleum hazards. Federal, state, and municipal governments set down minimum flash and ignition temperatures for storage, transportation, retail sale, and industrial uses of petroleum and its products. The insurance industry paid close attention to storing and handling petroleum and was concerned about devices—gas machines, lamps—that used petroleum's lighter products such as kerosene, naphtha, and gasoline for illumination and heating. In the late 1880s the petroleum-risk focus shifted from illumination to burning oil to raise steam. Fire underwriters and their associations accumulated and rationalized risk knowledge by 1900, receiving help from the emergence of fire-protection engineering.

This led to NBFU regulations, based on NFPA research as well as the activities of the board's consulting engineers, culminating in rules for storing, handling, and consuming fuel oil. The four Gulf South cities of this study translated these voluntary, consensual insurance rules into legally enforceable ordinances, in the risk-management tradition laid down by fire underwriters and municipal officials in Chicago, San Francisco, and the Northeast, where industries and businesses experimented with and adopted liquid fuel.[169]

The marriage of risk knowledge and municipal governance in the Gulf South emerged from nineteenth-century efforts to control local fire losses and a concomitant maturation of hazard recognition. By recording, reporting, and analyzing events, processes, or equipment, those who worked with petroleum or its products gradually acquired some measure of clarity about what could go wrong and how to avoid dangerous errors or mistakes while handling or using oil. From a socio-technical perspective, Gulf South cities participated in the development of a nascent network or technical system marked by human interaction with energy resources and participation in energy transitions, increased efficiency through risk management, and growing complexity arising from efforts to solve social and economic problems. While there are numerous historical antecedents, this comprehensive and complex system is a distinctly modern accomplishment in which the Gulf South participated.[170]

These techniques were eventually distilled into informal rules that fire-insurance companies and agents used to prosecute their business and then translated into general notions—ultimately codified and standardized—that businessmen or plant supervisors could understand and act upon. Our four Gulf South cities drew upon this legacy to manage the petroleum risks faced by oil distributors, pipeline operators, manufacturers, and businesses. It is appropriate, then, to examine the events in Mobile, Alabama, where the City Council in 1901–2 restricted large-scale petroleum storage through the politics of risk management to areas away from the city's riverfront and central business district.

2

FIRE HAZARDS AND PETROLEUM RISKS IN MOBILE, ALABAMA, 1894–1910

The work of risk assessment, research, dissemination of codes, and promotion of technical standards from 1880 to 1920 by risk-aware individuals, groups such as the NBFU and the NFPA, and fire underwriters had a grand design: to prevent fires and to reduce fire damage in the interests of economic efficiency, good stewardship of property, and protecting human life. This struggle against fire dated to the eighteenth century in densely populated areas such as New York, Boston, and Philadelphia. The Chicago fire in 1871 was a massive catastrophe that underscored what a lack of building standards or their enforcement, inadequate fire services, and municipal corruption could do to a community. Keeping statistics about fires, their causes, monetary damages, and especially hazardous sources of ignition became an important task in the opinion of many fire-insurance executives, leaders, and safety advocates by 1900.[1]

While developing regulations to deal with large-scale oil storage, the city of Mobile benefited from the accumulated risk knowledge of major risk experts and organizations in the Northeast through engineer Allen Schoen, who visited Mobile in February 1902 as a fire-protection advisor in the hire of the local fire-insurance interests.[2] He contributed to the NBFU's *Rules and Requirements*—NFPA codes 30 and 31—on storing flammable liquids and oil-burning apparatuses that in greatly expanded form are still used by federal, state, and local governments to mitigate fire hazards, guide municipal risk-management policies, and protect human life. Mobile officials pursued these major goals in the early twentieth century through the politics of risk management.

Fire hazards have been a major source of risk in the United States, requiring extensive risk-management efforts to mitigate loss of life and property. Southern ports in the nineteenth century suffered destructive fires as a result of wooden building construction, crowded conditions, inadequate water supplies, poorly equipped voluntary fire departments, and lax enforcement of city ordinances that stipulated fire limits or construction codes.[3]

Newspapers reported that fires incinerated millions of dollars of cotton bales and other commodities and goods stored on or near the municipal wharves or in the cargo holds of moored vessels. Approximately forty-one major fires took place in New Orleans from 1830 to 1856. Six of these destroyed a combined 31,100 cotton bales. Savannah had at least fourteen major fires between 1820 and 1854, of which five destroyed approximately 10,000 cotton bales. One of the five reduced a Savannah cotton warehouse in April 4, 1852, and its 5,000 stored bales to cinders and burned an English ship, the *Jane Hammond,* just loaded with cotton, "to the water's edge." Similarly, New Orleans; Baltimore; Charleston; Jacksonville, Florida; and Wilmington, North Carolina, suffered extensive fire damage that ruined entire city blocks, clusters of businesses and residences, as well as railroad rolling stock and moored vessels vital to local economies.[4]

Fires in Mobile before 1860 followed this pattern. A fire on April 12, 1820, left the city a "scene of smoking ruins." Even worse, on October 21, 1827, a wind-driven fire destroyed the "heart of the town" including three hotels, "brick warehouses," and several cotton presses, the "most business part of the town," which together totaled two hundred structures at a loss of $1,000,000. Fires in November 1837 and June 1838 caused losses of $30,000 and $60,000 each. A disastrous fire on October 7, 1839, again driven by wind, destroyed "fourteen squares" of the city "containing six hundred buildings" at a loss of $2,000,000. A theater fire in October 1839, along with several dwellings, burned $40,000 of property, and six months later in March 1840 fire consumed five "large stores and their contents." In June 1848 fire destroyed four hundred bales of cotton, an iron foundry, and a machine shop—another $40,000 loss. Mobile fires in November 1852 and September 1854 reduced $160,000 worth of property to ashes.[5]

Reducing the threat of fire, particularly massively destructive, unmanageable conflagrations, was an important challenge to the professional and commercial oligarchy that ruled post–Civil War Mobile. To address this issue, especially between 1880 and 1910, city officials, often prodded by insurers, took steps to mitigate fire hazards and risks. They established a fire department, sought reasonably reliable water supplies and fire alarms, passed street-specific, code-based construction standards also known as "fire limits," restricted the storage of explosives, and banned kindling fires around cotton wharfs.[6]

Flammable petroleum products presented particular policy challenges to Mobile officials. Cheap, poorly refined kerosene, because of its low flash point around spark or flame, caused innumerable accidents in homes and businesses in the nineteenth century. By 1871 thirteen states, beginning with Maine and Ohio in 1867, set minimum flash points. Similarly, fuel oil, which could be anything from unrefined crude to a residuum that might still contain a significant level of volatile distillates, often released extremely flammable vapors in storage. Fear of petroleum-related tragedies, coupled with public awareness of how difficult it could be to contain large fires, made many people apprehensive about the storage of great quantities of flammable liquids.[7]

Mobile citizens had good reason to worry. Their concern that burning oil might set local stockpiles of combustible cotton, lumber, and naval stores afire and destroy the entire city was well founded in light of the town's record of destructive fires. In February 1878 the oil warehouses of Chess, Carley & Company as well as Voss, Taylor & Company burned, destroying 6,000 gallons of bulk oil, 1,500 cases of kerosene, 500 barrels of turpentine, as well as animals, equipment, and "five large oil tanks." Sparks ignited fourteen cotton bales, completely destroying two. Two months later another fire took down two furniture stores, a liquor business, and a clothing shop. In June 1879 another major fire in Mobile consumed several important shops and stores. Strong winter winds swept a fire through downtown Mobile in January 1886. T. G. Bush & Company, a huge wholesale grocery operation, caught fire, spreading destruction to several buildings housing commission merchants, other wholesale grocers, and a grain dealer.[8]

These were only a harbinger of a major catastrophe. On Sunday, October 26, 1890, fire broke out at the Stewart & Butt shingle plant on One-Mile Creek on the city's northern edge, sending aloft "white hot coals and flaming cinders," which set ablaze Goodman's Warehouse at the corner of Commerce and Lipscomb streets near the river. Fanned by high winds, the fire destroyed most of the northeastern section of the city's business district along with steamers *Ruth, Mary Elizabeth,* and *Jewel*; vast amounts of baled cotton; four cotton presses; several warehouses; a cotton-seed oil mill; and other properties and buildings. Miraculously, the Fire Department prevented the fire from spreading to the "immense oil tanks of the Standard Oil Company," which were near the shingle mill. Had the wind blown from true north to south, contemporaries speculated that most of the town would have burned, but instead the flames arched eastward out over the wharves into the river away from the heart of the city. This was not enough, however, to spare the Gulf City Oil Mill (cottonseed) where a large, burning oil tank provided spectators with "the grandest sight of the fire." A twirling, intertwined, "immense rope . . . of lurid flame" and smoke rose one hundred feet in the air. Approximately 6,916 bales of cotton went up in flames as well, which underscored how dangerous fires could be in a town overflowing with cotton and oil combustibles.[9]

PETROLEUM AND PROPERTY

Mobile had experience with stored petroleum products in the immediate post–Civil War era. Chess, Carley & Company, a major oil, turpentine, and naval-stores supplier from Louisville, Kentucky, and an independently managed participant in the Standard Oil Trust, had by September 1876 an office and sales facility in Mobile. North of the business district, "on the outskirts of the city," the company constructed "huge tanks . . . used as the place of storage of oils." When Standard Oil of Kentucky gained control of Chess, Carley & Company in 1886, the only apparent change in Mobile was the name on the distribution and storage equipment. In 1901 the Standard Oil tank yard appeared in this same outlying region on a survey map prepared by city engineers. The tank yard was well north of the city's "central core" near the Mobile & Ohio Railroad complex and Bayou Marmotte, considered a convenient catch basin to drain burning oil.[10]

However, the first important oil-as-hazard incident involved one of Standard Oil's small, regional competitors and fed the concern that stored oil was a fire hazard and a menace to property. In October 1894, W. T. Jones, manager of People's Oil Company, an Atlanta-based independent marketer of kerosene (illuminating oil), petitioned the Mobile General Council to permit the construction of a "Storage Tank for the safe Storage of our oil" large enough to hold one railroad car of product (approximately 83 to 160 forty-two-gallon barrels or about 3,500 to 6,720 gallons) at the intersection of Bloodgood and St. Joseph streets, the location of the company's oil-storage warehouse. (See map 1.) Jones attached a drawing of an excavated brick compound for the tank, similar to oil warehouses required in New

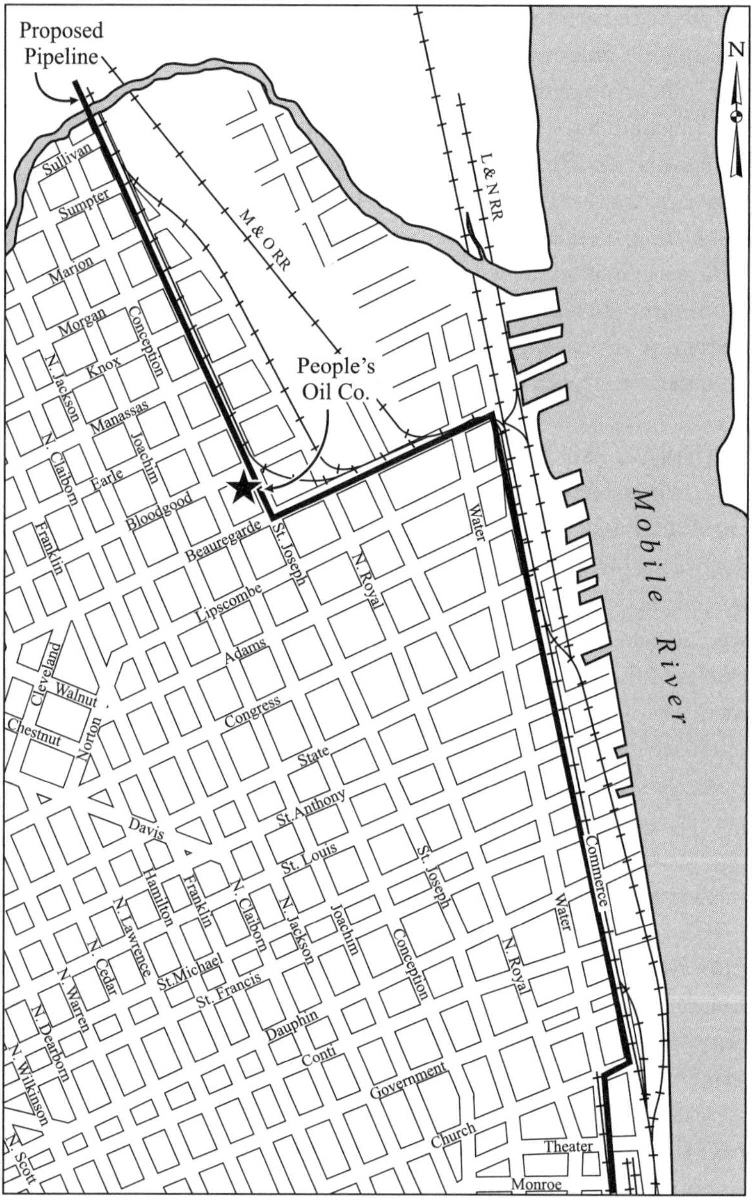

Map 1. Mobile: path of a pipeline franchise granted to J. M. Guffey Petroleum Co., May 1902; also the site of People's Oil Co., whose manager, W. T. Jones, successfully petitioned the City Council in 1894–95 for permission to install iron tanks to store kerosene.

York City and Chicago by 1881. Jones asked the council to note the cheap insurance rate tendered for the property next door to the proposed tank, arguing that it demonstrated the acceptability of the project to the insurance interests. Furthermore, the proposed location was outside the fire limits—that portion of the city where new wooden structures were banned—and thought to be especially vulnerable to fire. On October 5 the council referred Jones's petition to their Committee on the Fire Department and Fire Protection for additional comments and review.[11]

Local opposition to this request arose from people troubled by the perceived threat to property and safety. First Ward residents led by H. H. Porter had petitioned Mayor C. Lawrence Lavretta in September to close the People's Oil Company warehouse on the southeast corner of St. Joseph and Bloodgood streets—the same warehouse referenced in W. T. Jones's petition. When Porter returned to Mobile from a trip, he learned of Jones's petition to build a storage tank and appealed to First Ward Councilman Robert L. Maupin to "use all your influence against permission for the storage of this oil upon this location." Porter explained that he and his neighbors considered the oil "hazardous and dangerous to our lives and property." They opposed locating the tank in such a "thickly settled a neighborhood as ours."[12]

City leaders took these misgivings seriously but not as insurmountable objections. The Fire Protection Committee, chaired by Alderman Daniel E. Huger, a socially prominent cotton trader, reported favorably on the project in early November, stipulating that certain conditions must be met. First, Jones had to "build a solid parapet wall" to shield "all oil that may be kept on hand at any time." Further, People's Oil Company had to stop storing oil in the "small warehouse" which First Ward residents had asked to be removed in September. Last, the company had to recognize the city's right to "revoke this permit at any time." Councilman Maupin urged the General Council to "proceed slowly." Despite his warning that an accident would burn down everything in the four-block span east from St. Joseph Street to the Mobile River, his colleagues sanctioned the project.[13]

Jones made a second request six months later in May 1895. He informed the mayor and general council that market opportunities—"the Jobbing & Export trade"—required an increase in storage capacity at his Bloodgood–St. Joseph streets facility and asked for approval to place an "Iron tank of about one [railroad tank] car capacity" alongside the tank approved in 1894. He proposed to enclose both tanks in an earthen pit lined with cement and brick walls and covered outside with iron, promising to follow any Fire Department instructions about "proper safe construction." In effect Jones had offered to build a vault for these tanks in step with a developing consensus among fire underwriters about oil-tank safety. Elements of this stance were in the permit rules which the Chicago Fire Underwriters' Association advanced in 1889 about the placement of underground tanks at a distance from ignition points.[14]

These conditions attached to fire-insurance policies applied to burning fuel oil in heavy industry, while Jones's business was primarily in illuminating oil; but a gradual transition was underway from a focus on the safety of kerosene to concern over fuel oil, leading eventually to the need for specific ordinances that incorporated underwriters' rules about tank construction and placement for petroleum storage and consumption. Permission for Jones's big tank came about a month after his request with no recorded council discussions; nor was there controversy or uncertainty about how to proceed. Jones had preempted serious objections by carefully drafting his request with safeguards in mind.

Specificity in municipal regulations often elicited legal battles over the meaning and scope of terms, and their application in unanticipated or new circumstances. The Jones petition on behalf of People's Oil Company dovetailed with a controversy over municipal fire regulations designed to prevent conflagrations. A bitter quarrel broke out in the fall of 1894 between Francisco Gomez, local tinware manufacturer, and George B. Thomas, real estate investor and warehouse owner, over construction of an iron-clad, wooden building inside the fire limits that Thomas believed would lower the market value of his adjacent property. City Engineer D. M. N. Ross with the support of Mayor Lavretta and the General Council approved Gomez's project. Ross and Lavretta not only allowed Gomez to put up the building but also refused to let Thomas put up a similar building.[15]

Thomas's attorneys complained that the mayor had not fulfilled his executive duty to enforce city regulations against putting up wooden buildings inside the fire limits and that the city engineer had unlawfully discriminated against their client. They also protested that the General Council had no authority to delegate mayoral powers to the council committee that had approved the Gomez permit, thereby violating the plain language of the city's charter and ordinances. To confuse the issue even further, reports circulated that city authorities had deliberately ignored the law. Their excuse for granting permits to build frame buildings in the fire limits was that no one objected, and wood construction was cheap, encouraging builders to fill in vacant lots.[16]

CHARTER CLARIFICATION

Correspondence exchanged during the Gomez-Thomas controversy shows that City Attorney Peter J. Hamilton, joined by the editor of the *Mobile Daily Register* and the General Council, questioned (on the basis of a state supreme-court case) whether Mobile's charter explicitly granted the city government the authority to control fire hazards through building codes and fire limits. The newspaper contended the ordinance in question was more of a penalty than an absolute prohibition, since you could erect and use a wood building as long as you were willing to pay a fifteen-dollar fine per day. And if one failed in such a case to obtain a permit from the city engineer, the only consequence was a fifty-dollar fine.[17]

Uncertainty notwithstanding, George Thomas's attorneys contended that Section 21 of the city charter gave the General Council the "full power and authority" to regulate "fire wards," the "field for operation" for which certainly included designating "districts of the city . . . to regulate and control the character of structures permitted in such districts." This was the legal foundation, they claimed, of the city ordinance that banned wooden construction in the fire limits.[18]

To remove all doubts about the city's authority to regulate fire hazards, the General Council created a Special Committee on Legislation, composed of councilmen John Craft, Charles E. McLean, and Robert Savage. They proposed charter amendments in November 1894 that would allegedly foster efficient exercise of city powers and increase tax revenues. The influential *Mobile Daily Register* advised readers that the amendments would shore up any perceived gaps in the city's legal powers in matters of fire safety and storage of hazardous materials, providing "additions to the provisions of the charter now in force, perfecting that instrument where . . . it has been found deficient."[19]

Under the projected modifications, Section 21 would give the city the authority to regulate the "establishment and regulation of fire wards, fire companies . . . fire department[s]," and fire limits. Further, the state would recognize the legitimacy of existing laws on fire limits until new, explicit ordinances could be passed, a nod to those who had questioned the city's absolute legal power to ban wood construction inside the fire limits. Mobile state representative Edward M. Robinson introduced the proposed amendments to the legislature in December 1894, and that body approved the changes to seven sections of the city charter in February 1895.[20]

Allowing metal-clad, wooden buildings inside the fire limits also surfaced during the Thomas-Gomez dispute. In a revised fire-limits ordinance adopted in 1897 the General Council determined that frame buildings "roofed and covered within and without with metal" could be constructed in the fire limits "only by express sanction" of the council. In addition to a fine for violating this law, the mayor could order a noncompliant building to be "removed at the expense of the owner," constraints that formally eliminated a construction technique that Francisco Gomez used to his advantage and satisfied doubts expressed in the *Press Register.*[21]

The 1894 charter amendments also touched upon oil storage, giving Mobile the power to regulate the "storing and handling of gun powder, dynamite, oil and all other inflammable or hazardous substances." City ordinances published in 1866 gave fire wardens, and by 1897 the chief of the Fire Department, the power (with mayoral approval) to designate "at such place or places" where "rosin, tar, pitch, spirits of turpentine, petroleum oil, and kerosene oil" could be stored. The only proviso was that merchants could keep five barrels of each substance on retail sites. Further, city ordinances published from 1859 to 1897 also restricted possession of "coal, wood, pales, planks . . . lumber or any fuel of any description, combustibles, or materials that may occasion hazard, or danger of fire" in Mobile unless stored or secured safely in the judgment of the mayor or general councilman.[22]

It was quite clear that the city could control retail petroleum storage. However, legal hermeneutics that produced doubts about regulating fire limits in 1894 might also trigger suspicions about regulating large-scale petroleum storage and consumption. For example, city ordinances in place by January 1901 did not mention crude petroleum or burning petroleum for fuel; the dimensions, fabrication, or location of oil tanks; or piping and levees—all of which had to be addressed specifically in regulatory ordinances to establish the city's authority over such matters. Even so, Mobile's 1895 charter amendments clearly empowered the city to control hazardous petroleum goods and provided a legal platform from which the city could manage the risks posed by storing and handling the petroleum that became available following the January 1901 Spindletop strike at Beaumont, Texas. What follows is an examination of efforts to bring fuel oil to Mobile and the city's actions to regulate its storage and consumption, drawing freely from risk-management rules imposed by fire underwriters as early as 1892 on using petroleum.[23]

REGULATING OIL AFTER SPINDLETOP

In the wake of the Spindletop discoveries, investors incorporated over ninety-eight companies by mid-1901 to acquire land and to drill for oil near Beaumont, Sour Lake, and Jennings, Louisiana, while others planned to transport the oil from source to market. Aside from Standard Oil and Shell Transport & Trading Company, several relative newcomers or revitalized companies grew quickly. Sun Oil, Texas Fuel Company (March 1901), later the Texas Company (April 1902), and the J. M. Guffey Petroleum Company (May 1901), backed by the Mellon family of Pittsburgh and later known as Gulf Oil Company, collected oil from producers, built refineries and pipeline systems, and operated means of transport: tankers, barges and tugs.

Texas crude was relatively abundant. Production went from 65,975 barrels in 1897 to 18,083,658 barrels in 1902, 11,215,007 barrels of which were available for local consumption or shipment out of state. Further, Beaumont oil was high in sulphur and considered more suited for fuel rather than illumination, although one analyst believed excellent lubricants could be recovered from it. In terms of calorific value, the federal government estimated that 3.0 to 4.14 barrels of Beaumont oil (approximately 1,000 to 1,333 pounds) equaled one ton of coal. At prevailing prices, fuel oil could be 33 to 50 percent cheaper than coal.[24]

With such ample supplies on hand, the fuel-oil market in the Gulf South in 1900–1901 was poised to expand rapidly. The thirteen steamer lines that called at the port of Mobile used coal in their ships, providing candidates for burning fuel oil. Numerous sawmills and other small manufacturers that operated on the fringes of the city and inland needed energy sources. Alabama Iron Works, the Electric Lighting Company, Mobile Light & Railway Company, and Mobile Street Railway Company had facilities capable of using fuel oil or were considering converting to it. Enterprises that consumed large amounts of coal in a city where receipts for 1899 totaled approximately 189,300 tons were clearly potential oil customers. Existing energy requirements convinced city visionaries and entrepreneurs to develop transportation and supply systems to bring oil into Mobile. The Fuel Pipe Line & Mineral Oil Company was an early effort, although its defective petition for an oil pipeline scuttled the sought-after franchise.[25]

Large storage tanks, usually constructed on or near the river both to offload fuel oil from barges or tankers and to be accessible to railcars, required municipal approval. The Fuel Oil Transit Company was organized in early May 1901 for the explicit purpose of transporting "crude oil from the Texas oil fields to the markets of this and other countries, and to introduce the crude oil into the Mobile market."[26] William Chase Spotswood,[27] government contractor and supplier,[28] tugboat owner-operator, manager of the Mobile Towboat Association, and president-general manager of Fuel Oil Transit Company,[29] acted quickly and petitioned the mayor and General Council for permission to "erect reservoirs or tanks . . . for the storage of fuel oil" at the riverine foot of Palmetto Street. (See map 2.) Fuel Oil Transit Company leased or owned tugs, including the huge seagoing *E. L. Russell,* and the barges *Coleen, Goodwin,* and *Helen Buck.*[30] Spotswood believed fuel oil would displace coal as the energy source for his boats, and tested the notion on the working tug *Adelaide,* hoping the Mobile Towboat Association fleet would take notice and follow suit.[31]

Spotswood's petition informed the city that he and his associates also planned to apply for permission to "lay, maintain and operate a pipe line" to supply customers with oil. Mobile code required publication of the "full text of the [proposed] ordinance or resolution twenty days prior to filing for a franchise." Accordingly, Spotswood posted a proposed ordinance, May 24, in the *Daily Register,* granting his company a twenty-five-year franchise or right to lay a "fuel oil pipe line" along and under Water Street, to be connected to "tanks or reservoirs" at the foot of Palmetto Street. Construction of the pipeline had to satisfy the chief of the Fire Department and the city engineer, and the company could use the line to deliver fuel oil to customers.[32]

Although the General Council accepted Fuel Oil Transit's franchise application "to lay fuel oil pipes on Water Street" in June, Spotswood subsequently withdrew his franchise bid, having decided that it was incomplete and did "not exactly cover the lines contemplated." The proffered motion did not completely or correctly describe what he had decided to do, so he backed out.[33]

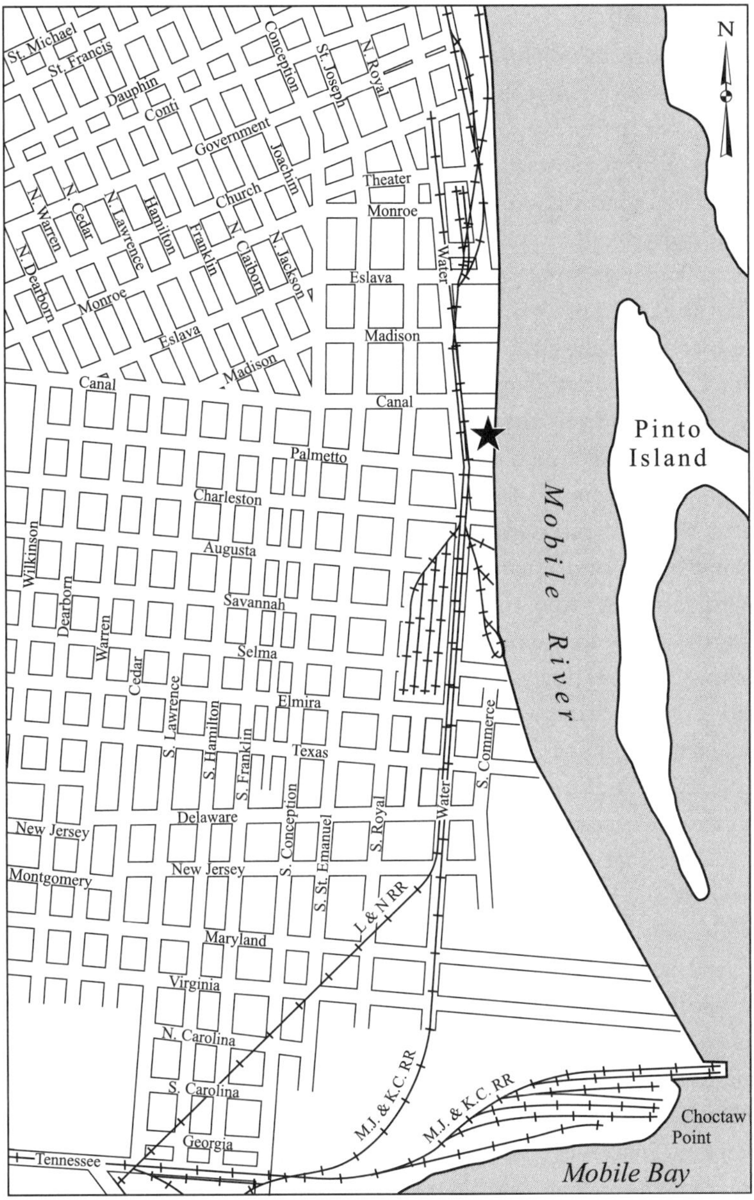

Map 2. Mobile, 1901: site of proposed J. M. Guffey oil-storage tank, defined by Canal, Palmetto, Water streets, and river; showing also the area south to Tennessee Street and Choctaw Point, wherein the city allowed petroleum-storage tanks.

RISING TENSIONS

Fuel Oil Transit Company's attempt to enter the Mobile market signaled the beginning of tensions between aggressive efforts to satisfy anticipated demand for fuel oil and protection of property. Awareness of the rising volume of oil coming from Port Arthur and Sabine, Texas, to other Gulf ports raised concerns among insurance brokers and agents about the safety of oil transportation and storage. On August 19, 1901, the Mobile Board of Fire Underwriters appointed a committee to "look into the question of the storage of petroleum in the vicinity of Mobile" and "to obtain such legislation from the city authorities as will be necessary to protect property from a conflagration from this source." The committee wrote Mayor T. S. Fry asking him to inform the city's General Council of its mission, so the council could "take such action as may be necessary." The insurance committee offered to meet with any General Council committee to which the matter might be referred.[34]

First Ward Councilman Cary W. Butt, partner in Stewart & Butt, a shingle manufacturer and dealer, moved in a council meeting of August 21, 1901, to receive the communication and have it "taken up under . . . new Business." Interestingly, Butt's mill and storage yard were a block from the head of Water Street, precisely where Spotswood's aborted oil pipeline would have extended. In response, the General Council authorized standing Committee No. 2 to meet with the underwriters' committee on the issue of the "storage of petroleum."[35]

An examination of the underwriters' committee members reveals the reason for their interest in storing flammable liquids. Their members represented many of the largest New England and London fire underwriters who carried heavy risks in marine liability and cotton underwriting and stood to lose heavily should stored fuel oil spread fire over the Mobile harbor front.[36]

James K. Glennon, since 1886 a real estate and insurance agent, was president of the Mobile Board of Fire Underwriters; John Gaillard, president of Gaillard-Johnston Coal Company, was the secretary of Stonewall Insurance Company; Charles R. Garnett owned an insurance agency. Gaillard and Garnett were on the Executive Committee of the Board of Fire Underwriters. The fourth member, Gustave B. Thames, owned a fire-insurance agency. His partner, Lloyd D. Batre, was also on the Executive Committee.[37]

The underwriters' concerns had a firm basis in both sentiment and fact. Fires in Mobile during the years 1878–79, 1886, and 1890 damaged or destroyed offices, shops, manufacturing plants, oil storage, docks, and ships. Smoldering cotton bales contributed substantially to the losses. These blazes cost insurance companies roughly $773,000. Such a dubious record of incineration justified anxieties about property, insurance claims, and financial ruin. Apprehensions over conflagration not unexpectedly surfaced again in 1901. In late March, fifteen railroad tank cars wrecked on a hillside above the village of Glen Gardner, New Jersey, releasing a "flood of blazing oil" that swept through the community, burning eleven buildings. On May 3 a major fire destroyed much of Jacksonville, Florida, initiating a large increase in insurance rates for "mercantile risks" based on a new schedule for "first class cities" that was immediately applied to Jacksonville.[38]

In June, Mobile experienced another near-disaster on the riverfront just north of the city. The "waste and oil house" in the Mobile & Ohio Railroad yard burned, drawing a "large crowd" of "hot, fatigued and disappointed" onlookers who thought the Standard Oil Works was on fire. Only two days before Mobile's important council meeting, August 21, when the insurance underwriters' committee on fuel oil defined its mission, a catastrophic blaze consumed part of the massive Atlantic Refining Company facility in the Point Breeze section of Philadelphia. Firemen built earthworks to prevent the burning oil from running into the Schuylkill River, thereby spreading the disaster. This was the precise scenario subsequently advanced by insurance men who opposed building big storage tanks in downtown Mobile.[39]

The envisioned risk did not remain merely hypothetical. On August 9, roughly two weeks before the August 21 council meeting, the Fuel Oil Transit Company brought the first load of Beaumont crude to the city in the barge *Helen Buck*. Only a "lack of tankage" prevented unloading her right away. The first ripple of a possible flood tide of hazardous fuel oil had arrived on Mobile's doorstep with no clear risk-management policies in hand.[40]

SAFETY CONCERNS OVER LARGE OIL TANKS

Potential dangers to property left other parties unfazed. Organized in Texas in May 1901, the Mississippi-Texas Oil Company owned land on Spindletop and in San Jacinto County, Texas, as well as other parcels of oil-bearing property. The company promoters reportedly also planned to lease land to drill for oil in the eastern coastal counties of Texas and to construct facilities for storing and transporting crude oil.[41] The company petitioned the Mobile General Council on October, 15, 1901, for permission to erect a steel storage tank of 37,500 barrels capacity. Committee No. 4 prepared a report on the petition, to be presented to the General Council in a special meeting on December 30.[42]

The size of the proposed tank had intensified worries over property. On December 13, two days before the Mississippi-Texas Company submitted its petition, the underwriters' delegation notified Committee No. 4 that they had carefully considered the oil-storage situation and had invited an "expert" to examine and to report on the Mobile situation. In light of "the many fires occurring on account of petroleum," the underwriters were convinced it would be a "gross error to permit the storage of oil in large quantities on the river front within the limits of the city of Mobile." However, if the city was to permit storage, they recommended an enclosure, south of the main business district and away from the river, defined by Water, Texas, Conception, and Virginia streets. (See map 2.) Any tanks erected inside this circuit should be underground and connected to the riverfront through pipes that conveyed oil by means of gravity and set in place with cutoff valves. Finally, the underwriter's special committee asked that oil-bearing vessels that lay in the port of Mobile be given only twenty-four hours to off-load their cargo.[43]

The recommended site also reflected social and economic circumstances in Mobile. It was within a larger area, south of Canal and east of Broad streets over to the river, that has been traditionally known as "down the Bay" among Mobilians. It contained wharves, warehouses, neighborhood stores, small shops, as well as modest residences and working-class boarding houses. Blacks, Irish, Italians, creoles, and mulatto people inhabited this space. It was well below the opulent standards of Dauphin or Government streets which were inside Mobile's proscribed fire limits and lined with the residences of the city's ruling orders.[44]

But a decision on the site came only after an unforeseen change of plans. A frustrating lack of quorum put the December 30 council meeting off by one day.[45] Led by its chairman, Walter F. Walsh, Committee No. 4 reported that the Mississippi-Texas Oil Company had abandoned its plans to construct a storage tank in Mobile, giving way to a petition from the J. M. Guffey Petroleum Company to commence tank construction.[46] J. M. Guffey Petroleum, incorporated May 1901 as successor to the J. M. Guffey Company (a partnership of A.F. Lucas, John Galey, and J.M. Guffey that drilled the first Spindletop gusher), was led by prominent Pittsburgh politician and oil man James Guffey. He used Mellon money to buy out partners John Galey and Anthony F. Lucas and consolidate their holdings under control of himself and the Mellons. He and Andrew W. Mellon also formed the Gulf Refining Company in November 1901.[47]

J. M. GUFFEY PETROLEUM COMPANY

With ambitious and competitive leadership, the J. M. Guffey Petroleum Company leased the Palmetto Street land from Fuel Oil Transit Company principals W. C. Spotswood and Harry T. Hartwell.[48] This was part of a comprehensive marketing strategy. Guffey had contracted in May and June 1901 to supply fuel oil to Shell Transport & Trading Company at Port Arthur, Texas, for export to overseas markets. The Guffey company planned to establish a domestic network of oil transportation and storage facilities on the Mississippi River, the Gulf Coast, and the Atlantic Seaboard. It also made arrangements late in May 1901 with the Riter-Conley Manufacturing Company of Pittsburgh, Pennsylvania, for ten steel tanks, 115,000 barrels each, for oil storage.[49] By June 1902, with its Port Arthur refinery nearing completion, Guffey had assembled a fleet of large tankers, tugs, and barges for this system.[50]

In its petition to Committee No. 4, the Guffey company attached a set of "Plans and specifications" describing the construction of the tank and measures to avoid fire. The tank was to be a "steel or iron fuel oil storage tank," not exceeding 37,500 barrels capacity, and would be used for the "receipt, storage and distribution" of fuel oil on a lot marked off by Canal, Water, and Palmetto streets and the Mobile River. (See map 2.) This circumscribed the site leased from Spotswood and Hartwell, on the southern, or downstream, edge of the city's riverfront-business area and fire limits. The city reserved the right to lay down further precautions to insure the safe storage and delivery of oil, all of which would take effect upon council approval of Committee No. 4's attached resolution.[51]

In an adjourned meeting, December 31, 1901, the City Council heard concerns about threats to property values posed by the proposed oil tank. Councilman Daniel P. Brown from the Fifth Ward, which contained the down-the-bay area, argued that the petition was "detrimental to the public's interest," in particular property owners south of Canal Street, where the city projected future commercial and residential growth. Those affected should be allowed, Brown contended, to come before the General Council or Committee No. 4 to express their opinions. The tank construction, urged Brown, would depreciate property values in the area, just as happened in the neighborhood around the Standard Oil tank complex, where the land was currently "practically worthless." Brown thought that all the oil companies should be required to operate "across the river" and "lay a pipe under the river to the city." He asked the council to postpone action until the next monthly meeting. This construction should be investigated "from all sides," Brown pleaded, and the council should "go slow in this matter" before granting such privileges as had been requested. His motion failed for lack of a second, and the council approved Guffey's petition.[52]

Having allowed Guffey to proceed with its tank, the General Council then approved of Committee No. 4's draft ordinance entitled "An Ordinance To Regulate the Storage and Use of Fuel Oils in the City of Mobile," to govern future conflicts between the economic advantages of bulk oil storage and requisite property safety. It confined "general storage tanks," parked railroad tank cars, and vessels off-loading oil, to another down-the-bay area sketched out by the intersection of Canal and Water streets, along Water south to Elmira, west to Conception, south to Tennessee Street, and then east to the riverfront. (See map 2.) No one was authorized to construct large tanks for storing fuel or illuminating oil in this precinct without first submitting "plans and specifications" to the General Council. Further, anyone who planned to build "individual fuel oil supply tanks" had to apply to the city engineer for a permit, and provide all requisite plans conforming to the ordinance's specifications. These small tanks had to be outside all buildings, belowground but not beneath any sidewalks, enclosed in a brick vault unless more than fifty feet from a building, fitted with a vent pipe with a minimum diameter of 1.5 inches, and set up with automatic pumps—not gravity—that would feed oil to burners. Finally, railroad tank cars had to be "discharged" or unloaded reasonably close to large storage tanks, and vessels to be unloaded had to be moored along the fifteen-block stretch between Canal and Tennessee streets. Even so, the northern tip of the approved area abutted the southern edge of the fire limits and the city's business district, an adjacency which alarmed local underwriters.[53]

Ending its 1901 work, the General Council passed two controversial risk-management acts, one that gave J. M. Guffey Petroleum permission to construct a large oil-storage tank close to the business district, and the other that regulated the storage and use of fuel oil. The insurance interests early in 1902 contested some of the details of the oil ordinance and campaigned for recension of Guffey's privileges, aided by Allen Schoen who was an important source of risk knowledge, insurance concerns, and fire-protection engineering.

BULK OIL: MENACE OR PROGRESS?

The oil-storage ordinance was a setback for the insurance interests which had urged the Mobile General Council to restrict oil storage to areas away from the business center of town. The insurance groups in 1902 had two objectives: first, to rally their forces to block completion of the Guffey tank or to prevent its use; second, to persuade the General Council to move the area for large storage tanks well south of Canal Street. Mobile insurance agents benefited in part from similar controversies in New Orleans and Galveston.[54]

Many large American and English insurance companies controlled the Executive Committee of the Atlanta-based South-Eastern Tariff Association (SETA)—one of the original members of the NFPA in 1896—which they engaged to oppose unregulated storage of fuel oil in New Orleans and other southern cities. New Orleans insurance agencies cited telegrams from these out-of-town companies confirming that underwriters were frightened and would dramatically increase insurance rates if the city failed to segregate large-scale oil storage from valuable business properties. The insurance-inspired oil ordinance reflected many hazard-mitigation measures that insurance companies and groups in the Northeast favored, rules and standards the SETA passed on to local agents and fire underwriters.[55]

The underwriters' liaison committee brought Allen Schoen to Mobile to help them. He coordinated the SETA's 1901–2 campaign to control fuel-oil storage in densely populated areas, including Savannah, Pensacola, and Tampa among others. Schoen visited Mobile during the weekend of February 15–16, 1902, one stop on a trip through the Gulf South—New Orleans, Houston, Galveston—on behalf of the consulting engineers to gather information to be used in devising NBFU rules on handling fuel oil, which the NBFU issued later in the year as guidelines—"rules and requirements"—that covered large storage tanks, industrial oil-burning equipment, marine vessels, and domestic heating and cooking. This cemented a risk-management bond between Gulf South communities such as Mobile and the more mature risk industry led by the NBFU, the NFPA, associations and boards of local agents and underwriters in large northern cities, and the Factory

Mutuals who supplied organization, engineering expertise, dissemination of risk knowledge, and advocacy on the state and municipal levels on behalf of uniform fire-prevention and building codes.[56]

Having assessed the risks presented by J. M. Guffey's Palmetto Street oil tank, Schoen immediately notified a number of insurance companies, who, in turn, contacted their Mobile representatives and urged them to lobby the General Council to revise its policies and to move oil-storage tanks well south of the riverside business district. Schoen had to be careful, however, because some state legislatures, including those of Mississippi and Alabama, viewed tariff associations as a restraint of trade, an issue which, if it became entangled in the oil controversy, might derail the insurance drive to restrict storage.[57]

An editorial, February 20, in the *Daily Register* rehashed the Guffey petition, took note of telegrams from insurance companies expressing opposition to erecting a "fuel oil tank . . . at any point near the river front," and local insurance efforts to lobby the General Council to embrace "safe storage." The newspaper called the Guffey plan a "great menace to the business interests of Mobile" that "merchants and owners of wharf properties" should stop immediately.[58]

Attorney Harry Pillans replied to the editorial in a letter echoing a long-standing Anglo-American belief that efficient harnessing of energy was a "prerequisite" to creating an advanced, "complex industrial" society marked by "individual and social well-being" even to the point of enjoying perpetual leisure and abundance.[59] He urged that "we cannot . . . lag behind in this race of progress." He complained that the editorial planted "unreasonable apprehension in the minds of the people of Mobile." Mobile's "self-constituted guardians, the fire underwriters," Pillans charged, had fed the *Daily Register* "erroneous" information about the "storage and handling of fuel oil." He warned readers that Mobile's economic future was at stake, because affordable fuel oil brought "factories and industries." Pillans was certain that the General Council had adopted reasonable restraints to insure a safe supply of fuel oil by following the New Orleans example of keeping storage tanks away from the center of town. The council's fuel-oil precautions far exceeded existing practices in handling highly flammable kerosene, turpentine, or gasoline, which Standard Oil and other dealers stored within Mobile in "considerable quantities."[60]

Pillans concluded that the underwriters had been given ample opportunity to raise objections and had remained silent at the council meeting when the Guffey petition was approved. Their current grumbling was, he charged, an "eleventh hour opposition" inspired by New Orleans fire underwriters but not in Mobile's best interests. Specifically, he pointed to, but did not name, a New Orleans agent of a "foreign insurance company" who, Pillans contended, had failed to stop the introduction of fuel oil in that city. There are two possible candidates for Pillans's censure: first, J. G. Pepper, chairman of the New Orleans Board of Fire Underwriters' committee appointed to deal with the oil storage problem, who represented Liverpool, London & Globe Insurance Company, one of the world's biggest underwriters; and second, M. J. Sanders, agent for Reliance Marine Insurance Company (Liverpool), who distributed telegrams from various big insurance companies to the insurance opposition in Mobile, expressing concern over hazardous oil storage. Both were prominent, outspoken enemies of storing oil on the riverfront in New Orleans.[61]

The *Daily Register* was not impressed. It rejected Pillans's gibe that underwriter threats to withhold coverage of tanks located on riverfront property was a venerable insurance weapon that could be safely ignored. But if the threat was carried out, the newspaper rejoined, it would be disastrous for the city. While one might resent the power of the insurance interests, they could withdraw from the Mobile market if they found it unprofitable to carry risks on property exposed to stored fuel oil. It made no difference if Pillans was reassured by the care exercised in the storage of oil, the editor insisted. What counted was the insurance companies' perspective. It would do no good to worry about Mobile progress or keeping up with sister cities if property values were not protected by insurance. The sooner Mobile stored the fuel oil according to the "ideas of the insurance men," the better for everyone. To reinforce his case, the editor printed a letter from the Mobile real estate, insurance, and money-lending partnership McDermott, Duggan & Taylor and from the Factors & Traders Mutual Insurance Company of Mobile. McDermott enclosed copies of telegrams from Norwich Union Fire Insurance and Scottish Union & National Insurance companies advising that building an oil-storage tank on the riverfront "within the city limits" would constitute a hazard demanding rate hikes. The Factors and Traders Mutual Insurance Company of Mobile produced a similar admonition from its insurance allies in Hartford, Connecticut, illustrating the close ties between the Mobile insurance men and the risk-knowledge networks of fire-insurance companies, insurance associations, engineers, and fire-safety advocates.[62]

Pillans's skeptical reply appeared the next day. He pointed to the wording "as informed" in the above-noted telegram to the Factors and Traders from Hartford, as proof that the "sudden outburst" of telegrams sent to Mobile, dated February 20, came about because someone had "misrepresented the situation," falsely claiming that Guffey's oil tank was a hazard to property. But Pillans insisted that there was no fire threat, since the tanks were to be placed inside levees as ordered by the General Council and would be "far removed from the business district." Pillans openly wondered, "who is this enemy of Mobile and its progress" that has been sending out misleading, damaging reports to the insurance companies?[63]

Insurance agents Gustave B. Thames and Lloyd D. Batre (Thames and Batre) rebutted Pillans's initial letter, noting that he was attorney for W. C. Spotswood and Harry T. Hartwell, owners of the land that Guffey Petroleum had leased. Thames and Batre denied that outsiders had patched together the current opposition to the oil tank, reminding readers that in 1901 a delegation from the local Board of Fire Underwriters met with Standing Committee No. 4 of the General Council, and left believing that the council would restrict oil tanks to a place south of Delaware Street (see map 2), where in the event of a fire "oil would flow off into Choctaw Swamp." Thus, the insurance interests were surprised when the city approved a different location for the Guffey tank. Furthermore, they asserted, only two questions were at issue. First, how dangerous is a 37,500-barrel tank of crude petroleum to Mobile's commercial riverfront? Common sense says it is a "terrific menace." Lightning-induced petroleum fires that spread up and down a river destroyed both Titusville and Oil City, Pennsylvania.[64] Second, how should the insurance companies protect themselves from such a risk? Forty telegrams sent to Mobile agents indicate the companies were quite aware of the dangers posed by storing fuel oil in downtown Mobile.[65]

Even so, it was not crude oil as such to which the underwriters objected. In New Orleans similar storage tanks, they contended, were 4,600 feet from the Mississippi River and overflowed into a swamp. The insurance establishment would be satisfied if Mobile were to adopt similar arrangements. If not, Thames and Batre averred, no company they represent would write insurance for Water, Commerce, or Front streets, and they doubted if other companies would be any more liberal. An oil tank on Palmetto Street would put businesses and warehouses on these principal north-south streets that run through downtown Mobile at risk. Thames and Batre cautioned readers that crude petroleum burned on moving water, and it was, they claimed, the "only fire, of ordinary origin, which cannot be extinguished." If stored where currently proposed, it would only be a matter of time before the "roar of conflagration bellows" over Mobile, "carrying terror and destruction in its path."[66]

The issue came to a head at the Mobile General Council's meeting March 14, 1902. Council members anticipated considering a license ordinance, but faced by a delegation from the Mobile Board of Fire Underwriters, they suspended the rules and gave the insurance faction time to present its concerns. James K. Glennon, president of the underwriters' board; A. M. Schoen, the SETA's designated technical consultant on fuel-oil storage; and E. W. Gould, a steamship company official, argued that fuel oil was an especially dangerous fire hazard when stored close to a river. After asking many questions, the General Council called a special meeting for March 21, so all parties in this dispute could be heard.[67]

Others were afraid of the location of the Guffey tank, too. Someone styling himself "Protection" described for the *Daily Register* a recent railroad accident in Virginia. Oil tank cars plunged into the Roanoke River, exploded, and set the bridge ablaze. Oil burned on the surface of the river for almost two miles. "Protection" asked readers to imagine "how great a fire" there will be "when the big tank now building on the bank of the Mobile river explodes and burns." Roughly one hundred Mobile merchants, commercial men, and professionals presented a petition to the General Council on March 19, 1902, to "rescind the permission given for the erection of the oil tank" and move it to a remote location. Otherwise, the Palmetto Street tank jeopardized the entire downtown business district and would make insurance very expensive.[68]

RESCISSION: A VICTORY FOR PROPERTY?

On Friday evening, March 21, the General Council endured almost four hours of testimony and speeches. James K. Glennon reiterated the local fire underwriters' position. Storing fuel oil on Palmetto Street was dangerous and presaged high insurance rates or cancellation. Attorney Gregory L. Smith assured the council that it had the legal right to rescind its December decision regarding the Guffey tank. E. L. Russell, chief counsel for the

Mobile & Ohio Railroad, pointed out the "danger to the property interests he represented." Mobile lawyers Zebulon M. P. Inge and J. W. Black as well as Lloyd D. Batre voiced concerns about the menace the oil allegedly posed to "property and shipping."[69]

On the other side, Guffey attorney Harry Pillans urged the General Council not to rescind its decision and cited the bright economic prospects that cheap fuel oil held out. Pillans contended that fuel oil was the least dangerous of "all mineral oils." He described at length the precautions taken to "prevent the oil from getting into the river or . . . outside of the revetment" around the tank. After some discussion the council voted eleven to three to suspend the rules and to reconsider the fuel-oil ordinance "as originally adopted." Councilman Butt then moved to repeal the oil ordinance, but council member John Craft, an opponent of reconsideration, offered an amendment, asking that Guffey be "allowed to complete their tank," after which the city would call in the insurance people "to investigate it." The amendment lost, and Butt's original motion to repeal carried. At the suggestion of City Attorney B. B. Boone, the council specifically rescinded "action on the petition and all privileges" formerly granted to the J. M. Guffey Petroleum Company.[70]

Reaction to this change of heart was mixed. The *Daily Register* triumphantly called the rescission a "victory for the property interests of Mobile." Lest consumers feel ignored, the newspaper advised, the action should not be considered an "injury" to a promising business. The General Council will, the editorial said reassuringly, cheerfully "grant . . . permission to build on another site." Directors of the J. M. Guffey Petroleum Company were not deterred. On the following Monday, March 24, 1902, in Beaumont, they passed a resolution to transact business in Alabama under state laws governing foreign corporations. They designated Mobile as their "known place of business" and William C. Spotswood as their "authorized agent."[71]

Rescission demonstrated that obtaining a legal dispensation to store fuel oil in Mobile was fraught with obstacles, equivocal judgments, and revocable decisions. A central hurdle was the municipal franchise. Franchises affected oil distribution because giant storage tanks required pipeline networks running under streets and sidewalks to supply the street-car company, ice plants, factories, mills, and small distributors. Pipelines routinely crossed public and private property boundaries, which legally required an easement from the city. This could be a litigious process. Mobile Light &

Railway Company, Mobile Gas Light & Coke Company, and Southern Bell Telephone & Telegraph asked for franchises from the city to run rolling stock, track, pipe, and wiring across and under Mobile streets. Faultfinders raised doubts about their validity or conditions of operation, making securing or retaining franchises contentious.[72]

GUFFEY PUSHES AHEAD

Guffey addressed this potential difficulty in early April 1902 by advertising two additional ordinances. The first would grant the company a thirty-year franchise to lay and operate an oil pipeline system from the south end of Water Street, north to Church Street, across to Commerce, north to Beauregard, west to St. Joseph, and then north on St. Joseph to the city limits. (See map 1.) The suggested ordinance would also permit Guffey to install customer-service connections and branch lines not exceeding two thousand feet in length along and under streets other than those already named. In return the city would receive an annual fee. On April 15, 1902, the Guffey company filed its advertised ordinance with the General Council, which directed the proposal to Standing Committee No. 3, the designated council recipient of all franchise requests, to consider the petition.[73]

The second advertised ordinance, also a thirty-year, oil-pipeline franchise, projected construction along Maryland Street between Conception Street and the Mobile River as well as lateral feed lines to customers. This put the pipeline roughly two blocks north of the Mobile, Jackson, and Kansas City Railroad Company's marshaling yard at Choctaw Point. (See map 2.) It extended to the riverfront, where it could be used to off-load oil from barges or tankers or connect to large storage tanks. On May 15, Guffey asked the General Council to mark out the area within the city where oil tanks could be erected and connected by pipelines to the riverfront. The company then requested the right to construct such a system. On motion by Alderman P. J. Lyons, one of the three council members to vote in favor of allowing J. M. Guffey to retain its now-rescinded right to put up an oil tank, this petition went to Committee No. 3, which would report to the General Council at a special meeting.[74]

Guffey's perseverance brought mixed results. At the conference, May 20, the council passed a general ordinance "to regulate the storage and use of oils in the city of Mobile," a revision of the ordinance it had passed on December 31, 1901. Section 1 of the new ordinance reduced the boundaries within which suppliers might mount oil storage tanks, from an area south of Canal Street and bound by the Mobile River, Water, Elmira, Conception, and Tennessee streets (see map 2), to a smaller down-the-bay rectangle delineated by Water, Delaware, Conception, and Virginia streets (see map 2), where burning oil could drain into Choctaw swamp, precisely as the Mobile Board of Fire Underwriters delegation had urged in an early interview with a committee of the General Council in the fall of 1901.[75]

Newly approved Section 2 stipulated that "within those limits" any tank larger than 2,500 barrels had to be enclosed in a levee or embankment capable of holding the tank's capacity plus a 10 percent margin of error. Revised Section 4 confined the area in which oil barges or tankers could unload on the riverfront to a domain just north of the Mobile, Jackson, & Kansas City Railroad yard at Choctaw Point, between Texas and Virginia streets. (See map 2.) Previously, such vessels could off-load anywhere between Canal and Tennessee, a zone that had extended an additional four blocks south and seven blocks north. More important, however, was the fact that the larger area had included the foot of Palmetto Street, the hotly contested riverfront site of Guffey's 37,500-barrel tank deemed dangerously close to the business-district perimeter. Insurance interests scored a major victory with the revised ordinance, convincing city officials that the Palmetto Street tank posed such a major fire hazard that large-scale storage must be moved further south to prevent burning oil from setting fire to adjacent business operations or floating on the river and destroying city wharves loaded with cotton and lumber. On the plus side for the Guffey company, the General Council approved both of its pipeline-franchise requests along Commerce and Maryland streets, from the "trunk line" starting on the south end of Water, down Commerce, and ending up on the northern city limits via St. Joseph, on the condition that fees accrued to the city based on pipeline volume. On the downside, the city required Guffey to remove the big tank already built at the foot of Palmetto Street within ninety days.[76]

The impact of this new ordinance on Guffey's business fortunes was not immediately clear. The Guffey company had, as part of its March 1902 expansion, purchased twenty-five acres of "deep-water front" property in Bayonne, New Jersey. Here, it built a refinery to serve the upper Atlantic seacoast market. During the late summer of 1902, Beaumont production fell precipitously, reducing the oil supply from producers with whom Guffey had contracted. Along with these uncertainties, Guffey was unhappy over the disposition of the Palmetto Street tank. Despite the city's decision to move storage to a different area, J. M. Guffey had finished the tank but had put no oil in it as of August 1, 1902. Joseph Cullinan, who managed the Texas Company out of Beaumont, visited Mobile the last week of August 1902. He reported to Arnold Schlaet, a company official in New York City, that J. M. Guffey Company had put a tank on the Mobile riverfront, "in the heart of the shipping district," but had been barred from using it. He surmised that Guffey planned to move it to a more satisfactory location and then make Mobile a distribution site.[77]

Guffey sought an alternative location near Maryland and Conception streets. Initially, the Louisville & Nashville Railroad refused to build a sidetrack to the lot and would not allow Guffey to build loading racks along the right-of-way of the siding. Guffey considered dropping the entire project at this point. However, after investigating the matter in more detail, the railroad changed its position and agreed to both the siding and the loading racks.[78]

William Spotswood, Guffey's Mobile agent, complained bitterly about the situation. He claimed that the effort to move the Palmetto Street tank was going to cost Guffey $6,000 to $7,000, in addition to the $10,000 or $12,000 originally spent to construct it. The extra cost exasperated the company, which considered the Palmetto site as the best in the city. In northern cities, Spotswood contended, insurance companies had underwritten property adjoining tanks in locations similar to the Palmetto Street plot. Guffey was willing, Spotswood claimed, to put up a dirt levee around the existing Palmetto tank to confine completely any oil that might escape and catch fire. Spotswood appealed to the "insurance people" to change their minds; otherwise, he argued, they would force Mobile to ban what other cities had allowed. He emphasized that fuel oil was a booming industry. If Mobile had no way to provide oil to visiting ships, the carrying trade to and from Latin America and the Caribbean would shift to New Orleans or other ports. Spotswood also called upon Mobile businessmen to convince the insurance companies of the importance of this issue. If something was not done, he warned, Guffey might abandon Mobile rather than shoulder heavy and pointless expenses.[79]

It was not an idle threat. Near the end of October 1902, the company dismantled the big tank on Palmetto Street. Frustrated by the situation and facing difficulty satisfying its contracts owing to production shortfalls in the Beaumont field, Guffey completely pulled out of the Mobile market.

The company shipped the Palmetto Street tank to the New York City area, likely its Bayonne facility, for reassembly.[80] Also, Guffey reported in late November that demand for oil was growing in the "interior of the Mississippi Delta," particularly among sugar planters. This probably explains why, by November 1, 1902, it had bought out the Fuel Oil Transit Company (FOTC), including the tug *Captain Sam*, and directed FOTC to focus its efforts on the Mississippi River. The FOTC had been moving oil for Guffey for some time, so its vessels would continue to ply the Gulf Coast, but now under Guffey direction. The company would also keep operating under the Fuel Oil Transit name, and presumably, William Spotswood would continue as manager.[81]

After Guffey's departure, inconsistent oil supplies starting in 1902 and coal from northern Alabama kept Mobile a coal-bunkering port for many more years.[82] Globe Oil Company, originally a Minneapolis dealer in lubricants and kerosene, and since 1887 a "secret" subsidiary of Standard Oil, worked the Mobile market, though whether Globe sold fuel oil in quantity is uncertain.[83] Globe lost a Mobile oil tank to fire on January 3, 1904.

According to newspaper reports, a fire destroyed a building on Franklin Street that was "said to be used as an oil depot."[84] Only three days later Globe asked the General Council to change slightly the storage limits established two years before (May 1902) in the Guffey case. Globe wanted to "erect a new tank in place of the one recently burned down" yet conform to the Guffey regulations. The General Council conceded the change, moving the southern border two blocks from Virginia to South Carolina Street, and shifted the western boundary one block west from Conception to Franklin Street (see map 2), the exact site of the previously damaged oil depot. Led by Aldermen E. B. Kirkbride and Cary W. Butt, the council approved the modification without rancor or controversy. By July 1910 the General Council moved the southern boundary still two more blocks south to Tennessee Street (see map 2) to accommodate the tank farm of the Texas Company.[85]

The Guffey organization returned to Mobile several years later, having been reorganized by the Mellon banking interests in Pittsburgh, Pennsylvania. Between July and October 1902 the Mellon banking group forced James M. Guffey to transfer day-to-day direction of J. M. Guffey Petroleum Company to Mellon-appointed managers, who reorganized field drilling and concentrated on distribution to the Atlantic coast, Caribbean, and Europe. In June 1905 the Gulf Refining Company registered with the Alabama secretary of state to do business in Mobile as an out-of-state corporation. It named Benjamin A. Rowley, a veteran Mobile oil salesman, as resident agent, who by 1906 operated out of an office on Mobile's South Conception Street, within the tract prescribed by the General Council for oil storage.[86]

To take advantage of oil deposits in Caddo Lake in northwest Louisiana, William L. Mellon and several associates incorporated the Gulf Refining Company of Louisiana in New Orleans, December 22, 1905. This Louisiana incarnation of Gulf Refining appeared in the Mobile city directory of active businesses in 1908. To avoid conflicts with the city, the company built a storage station at Magazine Point, on the west side of the Mobile River at Chickasaw Bayou, well north of town and beyond the jurisdiction of the city's fuel-oil ordinance.[87]

The Guffey-Gulf Refining Company's odyssey through a maze of insurance opposition and local fears of massive fire destruction precipitated Mobile's turn-of-the-century struggle to control the use of fuel oil within the city. The resolution of pre-1900, legal uncertainties over the storage of oil in Mobile established the city's authority to delineate fire limits, to require fire-proof building construction, and to regulate the storage of flammable liquids. Thereafter, even though enforcement of the fire limits was haphazard and inconsistent, Mobile's task was to superintend risk.[88]

The question was not whether the city had the authority to set aside locations for oil storage, but which sites were sufficiently distant from expensive business inventories and the private residences of the civic elite to reduce or to eliminate fire risks. In a broader sense it was a question of how to safeguard the city from fire hazards through appropriate risk management, while simultaneously allowing Mobile to enjoy the economic boom that oil promised. In May 1905 inspectors from the NBFU visited Mobile and decried inadequate water supply and fire equipment, in addition to inconsistent enforcement of fire regulations, including storage of flammable liquids, but this had no impact on the development of the oil-storage industry in Mobile.[89]

The restrictions imposed on Guffey in 1902 underwent minor modifications, but provoked neither quarreling within the General Council nor expressions of concern by insurance agents or the Fire Department. Once the city established its regulatory power, applying it through the spatial management of risk became uncontested policy. Without question, insurance forces played a central role in achieving this outcome. They challenged a major municipal decision to allow Guffey to proceed with a Palmetto Street oil tank, crafted a detailed policy alternative, and convinced Mobile officials to adopt it.

The fire underwriters received crucial backing as well as risk information on hazard mitigation from many fire insurance companies with risks on their books from Mobile and the port; they benefited from A. M. Schoen's visit under the sponsorship of the NBFU. He appeared before the General Council as an "expert" advocate for strict oil regulation, confirming a link between the city and the SETA. The association noted in 1903 that it had agreed to the request of the NBFU to let Schoen, its "Chief Electrician and Engineer," conduct an investigation, resulting in a report that was sent out to many fire-insurance companies and regarded as a "most valuable and interesting document." It was Schoen who notified many fire-insurance companies of the hazards posed by riverside oil storage, passing this concern on to insurance men in New Orleans; the Mobile fire-insurance interests also formed a committee to represent their case before the city government, and they inserted rules in their proposed ordinances for oil-burning equipment already in play in many cities under the sponsorship of local fire underwriters and large companies such as L&L&G.[90]

In guarding their interests, the insurance men set down a foundation for the city's twentieth-century participation in fire codes and fire-protection engineering, demonstrating that risk management of petroleum hazards to protect property, without hindering capital accumulation or thwarting business growth, was a critical component in the shaping of Mobile's regulation of petroleum. Protection of property triumphed over consumer convenience, efficient distribution, and whatever private advantage might have been gained from unregulated storage.

In the long term, the public struggle in Mobile from 1894 to 1910 to define and to apply equitably its legal power to control fire hazards related to the storage and use of fuel oil, along with other regulatory actions, put the city roughly in step with the late-nineteenth-century trend for local government to develop a nascent, bureaucratic concern for continuity, rationality, regularity, and management. It illuminated the hesitant, incremental, and often difficult transition from informal community rule to "managerial government" to which Mobile would contribute as it pursued risk management of flammable liquids and combustibles throughout the twentieth century. We will return to how that took place in comparison to Houston, New Orleans, and Galveston in the epilogue.[91]

Charles J. H. Woodbury (1851–1916), MIT class of 1873, first technically trained engineer hired by BMMFIC, 1878, as an inspector and by default fire-protection engineer for the AFMFIC. Yorke, *Able Men of Boston* (1950), 155. Courtesy of FM Global.

Edward Atkinson (1827–1905), president, BMMFIC, 1878, considered by some to be the father of fire-protection engineering; played a major role in mill-hazard mitigation for the New England Cotton Manufacturers' Association; advocate for technically based risk inspections; an opinionated polymath who published extensively on many subjects, including abolition, economics, tariffs, and finance. Yorke, *Able Men of Boston* (1950), 201. Courtesy of FM Global.

Jesse C. McDowell (1853–1926), vice-president, J. M. Guffey Petroleum Co., and general manager, Gulf Refining Co., 1901; testified before New Orleans City Council for the oil interests; sharp critic of fire underwriters and their demands for strict regulation of petroleum. *Romance of American Petroleum and Gas,* ed. A.R. Crum, vol. 1 (New York: Romance of American Petroleum and Gas Co., 1911), plate following p. 298.

Allen M. Schoen (1869–1928), member, NBFU Board of Consulting Engineers; electrician, South-Eastern Tariff Association; toured Gulf South cities, 1902, conducting interviews that contributed to NBFU's "Rules and Regulations" for storage and use of fuel oil (NFPA 30). Reproduced with permission from the archives of the National Fire Protection Association, Quincy, Mass.

L&L&G officials in New Orleans, including Henry V. Ogden, Clarence F. Low, and John G. Pepper (1864–1934) who supported strict regulation of storage and handling of fuel oil, 1901–2. *1848–1898 Fifty Years of Work in the United States of America* (1898), plate facing p. 68.

Binz Building, Houston; owner Jacob Binz, 1902, contracted with Elmer E. Billow, National Supply Company, to install a fuel-oil plant in the basement, leading to problems with city rules on handling fuel oil and a protracted lawsuit by Billow to recover costs for materials and work on the apparatus. Courtesy of the Houston Public Library, HMRC, MSS 0114-0883.

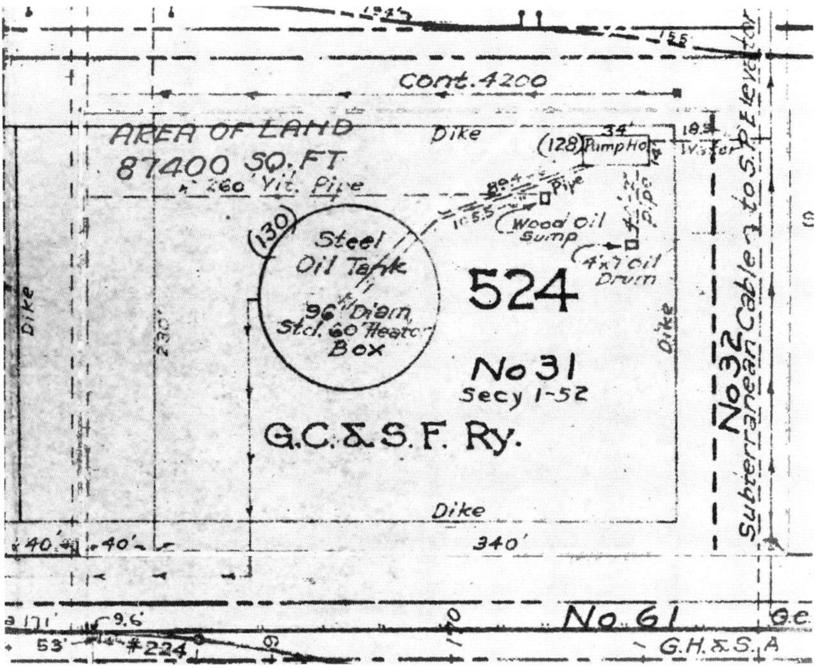

Block 524, blueprint, GCSFR yard, Galveston, site of massive oil tank, part of the 1902 fuel-oil controversies; following storm damage, 1929, railroad officials wanted to fix the tank if they could remain in compliance with municipal rules on storing petroleum. Courtesy of the Houston Public Library, Santa Fe Railroad Collection, HMRC, RG D 004.

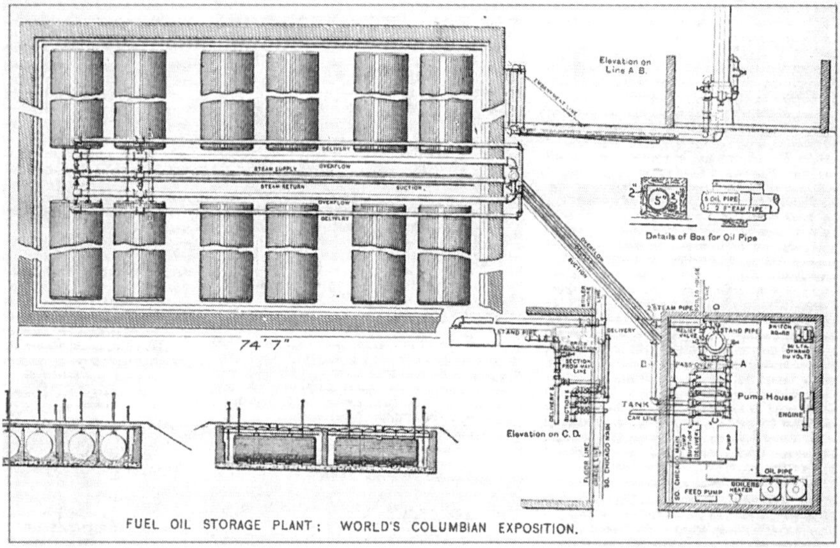

Fuel-oil storage plant, designed and installed by engineer Clayton Billow (1860–1945); it reflects Chicago insurance rules that required the twelve large oil tanks to be in an underground vault, with piping and pump appurtenances, to supply fuel for the Columbian Exposition's power plant. *Engineering News,* April 13, 1893, 343.

3

RISK MANAGEMENT IN THE BAYOU CITY

Fuel Oil in Houston, 1901–1914

Gulf Coast crude oil had exceptional potential as a marketable product for fuel and as an illuminant, which had been the dominant use of petroleum distillates in the late nineteenth century. Business leaders, entrepreneurs, and ambitious regional boosters recognized that oil was an endless source of energy to build up the industrial base of the Gulf South, to stimulate commerce and trade through Gulf ports, and to power streetcar systems, electric light plants, ice plants, and waterworks.[1] Following the 1901 discovery of oil near Beaumont, Texas, investors formed companies, solicited capital, and transported fuel oil to local and foreign markets, while canvassing local manufacturers, sugar-mill operators, sawmill owners, and rice farmers to determine their fuel needs.[2] By late 1908 speculators, drillers, prospectors, and fieldworkers, in addition to many out-of-state participants, had found, captured, and brought to market a vast amount of crude petroleum in successive strikes at Sour Lake, Saratoga, Batson, Humble, and Goose Creek.[3]

Fuel oil, starting in the 1860s, grew to become an important and relatively cheap energy source in the United States, even though it did not seriously challenge coal as a general power source nationally until after World War I. The Beaumont production area and its successors made unprecedented amounts of cheap petroleum available to Gulf South consumers—where boiler plants traditionally had consumed expensive coal, lignite, and wood—thereby transforming the regional energy market. Early production and utilization of oil by commercial groups in Houston made the city a pioneer in the development of this new energy market, a role that deserves examination, since fuel oil was feared to be explosive, making its use problematic. The city's experience illustrated how a southern municipality devised policies to regulate a potentially dangerous new fuel supply and its consumption. In the midst of heated public debates, Houston officials identified the legal basis and economic justification for protecting their community's built environment. They engaged in the politics of risk management.[4]

As an increasingly important financial and administrative center for the Texas petroleum industry, Houston constructed a rational interface between energy production and consumption which clarified an important feature of post-1900 southern urban history. Discussions and debates between fire underwriters and oil men exposed city officials to risk knowledge and fire-prevention information that conduit organizations such as the Texas Fire Prevention Association (TFPA) circulated throughout the state. The TFPA, founded in 1901, conducted fire-hazard inspections through the state, catching the attention of the NFPA and enjoying full membership in 1901–2. The NFPA was an organization that crafted scientific, engineering, and experience-based standards. When codified and published by its progenitor, the NBFU, these voluntary standards could be incorporated into city ordinances or referenced in a federal law, state statute, or municipal ordinance as mandatory requirements in the mitigation of a broad spectrum of fire hazards such as petroleum storage, consumption, and refining.[5]

Houston's rational interface integrated or brought together regulators or municipal officials with oil producers, transporters, suppliers, and consumers to promote the public interest. The interface embodied accepted requirements such as NFPA, UL, or ANSI rules, standards, and techniques for hazard mitigation, recognized by government agencies, as well as professional, scientific, and industrial associations or laboratories. The legal and political implements of this interface were state laws or municipal ordinances. They incorporated or made reference to fire-prevention or building codes that rested on normative standards certified by scientists, engineers, architects, and fire-service communities. Finally, the interface took into consideration the possible consequences of petroleum hazards if they initiated an explosion or ignited a conflagration that caused catastrophic damage. This was part of the chain of municipal risk management: identifying hazards, determining what risk they posed, and devising politically viable strategies to eliminate the hazards or mitigate their consequences.[6]

Galveston, New Orleans, and Mobile, all energy-consuming communities, otherwise quite different in size, population, and regional status, also faced problems posed by petroleum storage and consumption. Understanding what happened in Houston invites comparative analysis of these southern cities to establish who supported or opposed large-scale oil storage and what regulatory tactics and strategies these towns developed to deal with energy-market issues at a time when cities across the nation were expanding their reliance upon expert knowledge and bureaucratic supervision. Appreciation of Houston's fuel-oil experience expands our understanding of a significant episode in energy adoption in the southern United States.[7]

By 1901, Houston had survived a decades-long competition with Galveston for preeminence in maritime trade and, as a railroad center, had become by design an entrepôt for East Texas and its vast timber, rice, and cotton production. Buffalo Bayou was a lifeline to the coast, and with federal aid this watercourse would later be widened and deepened to allow seagoing ships to come close enough to Houston to establish it as a major port. Cheap fuel oil quickly became a foundational component within this constellation of city promotion, commercial dominance, and well-designed transportation facilities.[8]

With the opportunity for commercial hegemony within sight, Houston had to deal with a basic issue: how to make petroleum available to meet anticipated energy demands. The city had to establish conditions of safe storage, distribution, and consumption, which involved selective adaptation of energy-delivery technologies to Houston's particular circumstances and the city's community-defined safety standards. Both contributed to the creation of a new energy market that brought together buyers and sellers of oil. Informed by necessary rules, or "systems of regulation," that specified how to marry energy supply with new energy demands, consumers would be free to substitute oil for traditional energy sources.[9]

Deciding how to proceed raised questions for Houston leaders, dilemmas that required thoughtful answers. Should Houston subordinate the safety of its infrastructure—roads, buildings, wharves, manufacturing districts, railroad networks, and channel access to the Gulf of Mexico—to accommodate oil investors, producers, and distributors? If city leaders shaped public policy to suit oil promoters, what would the effect be on Houston citizens and their natural and built environments? What standards should prevail in making such decisions—safety, convenience, commercial goals, or aesthetics? Finally, did city officials and interested citizens have adequate knowledge of the oil business and sufficient information about petroleum safety hazards to balance "the interests of risk creators and risk bearers"?[10]

To determine the community's questions and analyze the answers that resulted, this chapter examines Houston's roles as both observer of and participant in the Texas Gulf Coast oil fields, primarily in 1901 and 1902. These areas produced enormous amounts of crude oil, which stoked the city's raging desire to use this apparently inexhaustible and relatively inexpensive energy source to achieve commercial and manufacturing dominance in the Southwest. (Galveston, its longtime rival, was too devastated by the 1900 hurricane to compete for leadership in the new age of oil or to build a substantial manufacturing base.) Houston's struggle to match energy supplies to consumption illustrated how a turn-of-the-century municipality embraced fuel oil under terms developed through serious debate and negotiation. Houston leaders, reflecting public perceptions and ambitions, fashioned a regulatory compromise agreed to by various interested parties whose property, commercial aspirations, and fears were at stake in the creation of a new energy market.[11]

The regulatory authority which the Houston administration brought to bear on these matters rested upon the city's police powers to regulate and control fuel-oil storage. One major goal was to avoid a disaster such as a massive oil fire or explosion. In the framework of the regulatory and environmental history of the South, Houston's mediation between energy supply and consumption was an exercise in environmental adaptation devised through the politics of risk analysis and management. The issue was how to take economic advantage of a promising new energy source without putting human life and urban property at risk.[12]

Houston—in particular its commercial and civic elite and its political leaders—aided in the creation of an energy market by mediating the relationship between a readily available energy supply new in terms of quantity and price to the Southwest and the interests of industrial consumers and the public at large. Mediation meant delineating through police powers the terms or conditions under which the city would allow consumption of fuel oil.

Petroleum was a natural energy carrier but when consumed by humans became a man-made hazard, subject to human control. This happened quickly in the aftermath of the 1901 Spindletop strike, when Houston's opportunistic commercial set recognized the potential of fuel oil to power industrial plants.[13] In pursuit of this energy prize, they took part in the development of the Gulf Coast oil fields, contributing to the city's nascent financial and administrative leadership in the oil business.

Of the thirty-three oil companies with headquarters in Houston listed in the 1902–3 city directory, many had Houston investors, directors, and organizers who raised money for drilling, formed companies, obtained charters, and directed company affairs in the oil fields. Their practical experience and knowledge enabled them to offer credible opinions on the matter of safe storage of fuel oil within the city.[14]

Oil delivery and storage quickly became pressing issues because specific enterprises, such as breweries, the Houston Electric Street Railway Company, the Cleveland Compress and Warehouse Company, C. R. Cummings Lumber Company, and the Hutchins House Hotel, installed oil burners or began considering adopting fuel oil as their primary energy source. Even the city took part when the City Council voted on June 3, 1901, to outfit the city's new sewer pumping station "for the burning of oil for fuel." When limited test operations began in December 1901, the station burned oil drawn from a large underground tank. Citizens' Electric Light plant changed over from coal to fuel oil in July, and by August 1, 1901, it consumed roughly ninety barrels of oil every twenty-four hours drawn from three twelve-thousand-gallon underground oil tanks. Houston's oil age had begun.[15]

When Houston firms considered refitting their plants and machinery to burn fuel oil, acting out of expectations of competitive advantage and market growth, owners and managers wanted assurances that there was a reliable and adequate supply to meet potential market demands. Once convinced that there was sufficient oil, the next step was to purchase shop-built tanks or to contract for on-site construction. To make Houston's fuel-oil consumption a safe activity, civic leaders needed credible and reliable information about storing and handling oil when they considered safety and nuisance issues posed by requests for permits to install small on-site tanks or huge distribution containers. Experiences gleaned from different types of oil storage in the production fields carried weight with City Council members, their advisers, and friends.[16]

Fires and spills in the oil fields confirmed to the insurance interests that oil was a hazard requiring careful handling. Periodically, oil-field vats caught fire, often from lightning or slipshod field practices. Even though these fires led an association of well owners and operators to institute strict fire-prevention protocols, a destructive fire occurred at Spindletop in September 1902. It would be hard to persuade cautious insurance underwriters and owners of large stocks of flammable commodities such as cotton, cottonseed oil, or timber that they should stand aside and allow unsupervised oil storage within the Houston city limits. The idea of putting oil-filled tanks, which had proved to be vulnerable to fire despite safety measures, inside the city must have sounded to some Houston observers like a failure of common sense and a completely unacceptable risk.[17]

It is equally implausible that many of the city's residents or merchants would approve of safety and environmental production risks such as dumping waste oil or oil-field brine, drilling mud, or well sand into pits, shallow lakes, creeks, and marshes, or simply on the bare ground inside the city of Houston (although such actions were quite common elsewhere in the absence of a major health crisis or threat to commercial viability). These practices, if carried out in Houston, would add another level of fire risk in addition to sanitary nuisances.[18]

Oil companies built storage facilities as fast as available capital, skilled labor, and tank fabricators could respond to the need, but putting up tanks in Houston was a complicated affair and undertaken in uncertain financial and commercial circumstances. Houston's inability in the 1880s and 1890s to secure tax revenues sufficient to provide basic city services, plus endless wrangling over streetcar, water, and gas franchises, and political tensions between reform-minded "Metropolitans" and city-focused ward politicians tempered optimism about Houston's growth potential.[19]

Yet, in the broad context of progressive ideals such as efficient management, businesslike accounting methods, and a concern for community welfare, many turn-of-the-century visionaries in the Business League and on the City Council regarded cheap fuel oil as an indispensable key to future commercial and manufacturing growth. Readily available fuel would sustain all campaigns for social amelioration, progressive government, and community stability. "Any step" taken "to cheapen the cost of fuel," one commentator observed, would benefit Texas consumers.[20] A newspaper article on Houston industry asserted in September 1901 that the discovery of oil so near the city had contributed to "industrial development," which was a "branch of progress." Further, the article claimed, oil production had led to at least six plants using it for fuel and many others making arrangements to do so.[21]

In late January and early February 1901, soon after the Spindletop strike, the City Council granted pipeline franchises to several groups of investors led by local businessmen to furnish oil or gas "or the products thereof" to "any factory, residence and business places."[22] With these franchises formally in place, the investors needed a pipeline from the oil fields to Houston so that distributors could pump crude directly from wellhead or settling tank to storage farms and refineries in the city.[23] The Houston Business League, led by George P. Brown, set in motion a campaign to attract pipeline investors. Texas law limited construction-condemnation proceedings to pipelines eight inches in diameter or less. In response to lobbying by Houston leaders, the legislature reportedly authorized a pipeline to the city but failed to specify the size of the line. This inaction left city promoters no closer to getting a pipeline started than before because knowledgeable investors considered the allowable pipe size to be too small to recover the enormous capital investment needed to finish a pipeline from Beaumont.[24]

However, some citizens, such as nervous insurance underwriters and property owners, considered petroleum a risky substance to bring into a relatively densely populated area filled with homes, businesses, warehouses, and vulnerable caches of flammable agricultural commodities. At several crucial points in the oil-storage battle, ordinary Houston residents made their opposition known through petitions to block tank emplacements. They were not deviant cranks or paranoid faultfinders; their apprehensions about the safety of fuel oil were companions to public disquiet about the likelihood of fire damage, unease made unmistakably clear by pleas these city dwellers made to the City Council for water service, supply extensions, outlets, and fire hydrants.[25]

Uncertainty about fire protection reached a high point on June 24, 1901, when flames destroyed the Market House in which City Hall was located, regarded as the principal building in Houston. Alderman James A. Thompson accused the Houston Water Company of neglect and incompetence for failing to maintain sufficient water pressure to suppress the conflagration. This implied that citizens should not count on adequate water pressure to extinguish a major fire, even if they lived or worked near a water main.[26]

Houston, like other cities, traditionally dealt with threats to community well-being, such as fire or explosion dangers, through an extensive array of state-granted police powers that related to public health, safety, and order. These powers restrained individuals in the exercise of their private rights when those rights infringed upon others or threatened the "general comfort, health and prosperity" of society.[27] States defined and enumerated such powers in municipal charters, as was evident in the Houston municipal charters granted by the Texas legislature from 1862 to 1899. Houston's 1862 charter conferred upon the mayor and City Council "full power and authority" to provide safe streets, city guards, and adequate street lighting, as well as "to determine in what part of the city wooden buildings" could be erected. The city could also prevent storage of gunpowder "in the city or suburbs in such quantity as to endanger the public safety," inspect "all combustibles sold publicly," and "define and suppress nuisances."[28]

The 1866 incorporation act added control over railroad tracks within the city. The 1874 charter then expanded the gunpowder clause to include petroleum distillates and other flammable liquids, allowing officials to "prevent gunpowder or other explosive material, kerosene oil or other inflammable oils being stored within the city limits in such quantities as to endanger the safety of adjacent property."[29] These powers survived in subsequent charter consolidation or amendment acts passed in 1879, 1881, 1897, and 1899. The phrase covering gunpowder, kerosene, and flammable liquids (oils) remained the same in each iteration.[30] The city of Houston cited these same points in a 1906 lawsuit over the installation of an oil-supply tank in a building basement (1902). The city invoked its police powers to protect the safety, health, and public order of Houston and argued that the building was inside the fire limits in which the city could "direct, require and prescribe . . . fire-proof materials," and that it could "prohibit the keeping or maintaining fluids or materials calculated to burn up the said City" within areas determined by same.[31]

By the summer of 1901, Houston's successive charters had not specifically addressed the issue of using fuel oil, although the city's power to control flammable substances such as gunpowder and kerosene remained unchanged. City authorities had three possible legal avenues to regulate fuel oil: the authority to control flammable substances, nuisance abatement (which might allow the city to label fuel oil a threat to public health and well-being), and an 1897 charter stipulation that building modifications in the city limits required a permit from the city engineer.[32]

While nuisance law could be used to oppose construction of oil-storage tanks on the grounds of bad odors or health risks, the building-permit requirement offered a procedural template that could be applied to the risk management of flammable liquids. The city of Houston could require institutions, workshops, commercial firms, and manufacturers that operated boiler-fired power plants to obtain permits if these consumers wanted to install oil tanks or to convert equipment to burn oil. In the absence of an ordinance that delineated how these users should proceed, business inquiries about installation permits put pressure on the Houston City Council to pass ordinances to regulate fuel-oil storage and consumption.

The path from charter authorization to regulate flammable liquids, to a municipal ordinance that exercised this power in a variety of situations (with assistance from state and local fire-insurance interests), started on July 15, 1901, with Lone Star and Crescent Oil Company (LSCOC), a major enterprise backed by New Orleans capital. The company approached the Houston City Council for permission to erect a steel oil tank covered with galvanized iron on property extending from the corner of Wood and Vine streets about one-half of a city block to White Oak Bayou. The company envisioned oil transportation between its Spindletop wells and New Orleans, either through a pipeline or by barge, and wanted to service Houston also. Alderman James A. Thompson moved that the council refer this request to the fire committee and the fire chief "with power to act." Thompson also asked that the council ensure that the city would be held blameless in the event of any damage claims arising from placement of the tank.[33]

At the next council meeting a group of citizens filed a counter-petition to protest the Lone Star project. The petitioners, "citizens and tax payers owning property and residing near the proposed site" in the Fifth Ward, alleged that oil was dangerous and emitted an unpleasant odor. The appellants asked the council to clear Vine Street of all obstructions and to reassert municipal claim to the passageway to "protect our homes, lives and property." The endorsements and dated notations on the original petition indicate that on July 29, 1901, the document went to key groups and figures in the regulatory process—the council fire committee, the city engineer, the city attorney, and Fifth Ward aldermen. Two weeks later they reported to the council that they had sided with the citizens' protest. The council adopted the report, effectively blocking construction of the tank.[34]

Lone Star's agent in Houston was R. W. Gamble and Company, a major supplier of oil-field equipment and steel storage tanks and later an oil-land and stock brokerage. Company communications with Mayor J. D. Woolford and the City Council shed a great deal of light on the business arrangements and legal snares associated with gaining municipal permission to store fuel oil. A letter from the company to the mayor, dated July 30, 1901, recounted how Lone Star, through Gamble, sought to place an oil tank in

Houston. Gamble claimed that on July 5, 1901, it had leased land along Wood Street near White Oak Bayou, between blocks 58 and 59 on the city map, and in turn subleased the site to LSCOC.[35]

Yet, the letter to Mayor Woolford noted that, on July 29, City Engineer I. Austin Miller had notified Gamble that the city "would not permit the occupancy of said land" since Houston had possession of or would make claim to possess it. The original lessor insisted that he never conveyed the land to the city; that there was no grant to prove any such conveyance; and that the city never "opened, improved, layed off and occupied said land as a City Street," which meant that any title the city may have had at one time was void and forfeit. In addition, Gamble contended that not being allowed for two weeks to occupy the land caused the company "material monetary loss." The company asked that the city legal department examine the situation and "designate . . . that portion, if any, which the City possess title to."[36]

Gamble's letter explains why the formal statement delivered August 12, 1901, by City Engineer Miller and the fire committee sided with local opponents of the proposed Lone Star tank. The relevant county deed records showed that Vine Street was "open between blocks 58 & 59." Therefore, Gamble had subleased to Lone Star a passageway claimed by the city for public use and within the city limits. The law firm of Stewart, Stewart and Lockett, acting as the city attorney, noted that Section 34 of the city's charter granted the City Council the power to "prevent or regulate the storing of inflammable oils within the City limits" (the lawyers omitted the remainder of the charter phrase: "within the city limits in such quantity as to endanger the safety of adjacent property"). A tank placed in or on Vine Street would abut other property, thus putting the question of siting it under municipal jurisdiction. If the proposed tank posed a threat to nearby property, there would be grounds to disqualify the site. By adopting the recommendation, the City Council affirmed the view that, under existing police powers, even without a specific statute dealing with fuel oil, the municipality claimed to have valid legal authority to control oil use inside Houston.[37]

In a petition delivered to Mayor Woolford and the City Council a week later on August 19, Gamble renewed its request to "erect an oil tank on Wood street." The company complained that it had conferred in advance with the mayor, the city attorney, and the fire committee about "possible objections on the part of the city to our erecting a tank" on the Vine Street property roughly seventy-five feet from Wood Street on a slope leading to

White Oak Bayou. The mayor had assured Gamble that there was "no possible objection." Gamble also alleged that, prior to the City Council meeting of August 12, when the adverse report came before the aldermen, the company had questioned the city attorney about the Vine Street property and that neither he nor members of the fire committee had claimed this land on behalf of the city. Therefore, Gamble contended, the site on Vine and Wood streets was a perfect location for an oil tank. It fulfilled "all the precautionary requirements and conditions" prescribed by the "insurance companies in regard to the erection of said tank." A crescent-shaped brick-and-earth retaining wall to be constructed between the heavy steel tank and the bayou would make it "absolutely impossible for any of this oil to get into the Bayou" in case of fire. This oil catchment, Gamble assured the city, could handle, together with "excavations which are to be made," four to six times the tank's capacity. In addition, Gamble pointed out, a pumping station would connect the tank to railroad loading racks during delivery so that oil would never come into contact with the atmosphere or be exposed to "light, fire or flame anywhere near it." The tank would also have attached steam pipes for fire suppression in case lightning ignited the oil.[38]

As additional evidence in favor of the Vine Street site, Gamble also introduced approval from "the special committee of Insurance Agents," with whom company officials had visited on August 12, 1901, the same day that the Houston City Council had blocked the tank. While Houston sources do not identify the members, the insurance committee, judging from similar events in Mobile and New Orleans, was probably composed of prominent insurance brokers, agents, or company representatives specifically selected to lobby the city for an oil-storage ordinance in harmony with the industry's safety concerns. Gamble alleged that this special committee had called the proposed tank location an "ideal spot." Closing the arguments, Gamble pointed out that the Lone Star oil tank, placed in a "hollow space" and surrounded by "natural and artificial fire precaution," would essentially be a "buried tank," a configuration that matched insurance demands for underground tanks. "We feel that we have a right to complain at the treatment we have received," a Gamble official announced. The City Council had not responsibly considered the company's request; Gamble was "legally, justly, & equitably" entitled to a permit.[39]

Risk management was important to the city's fire-insurance agents, since the companies they represented were contractually obligated to indemnify many property owners against fire loss. The agents insisted that the city, when creating an interface between energy supply and demand, should take into consideration their concerns about fuel oil. Their envoy J. B. Cochran asked the City Council to prescribe special permits for those who wanted to set up fuel-oil apparatuses. This bureaucratization of expert risk knowledge would placate the insurance interests as well as make the permit process a routine governed by commonly understood standards. The council, aware that it would take a new ordinance to establish specific regulations on using fuel oil in Houston, referred the insurers' request to the city attorney and the fire and ordinance committees.[40]

The underwriters' petition was a classic statement of compliance-based regulation, in that it offered "detailed prescriptive safety requirements" for controlling fuel-oil usage with which engineers, equipment erectors, suppliers, and consumers had to comply. The criteria could be observed, measured, and evaluated in terms of physical dimensions, types and arrangements of equipment, and tank placement. For example, supply tanks, designed to feed a boiler or hold several hundred gallons of oil for daily industrial use, had to be put underground and made of galvanized iron, steel, or boiler iron that was at least an eighth of an inch thick. For the storage of fuel oil in an existing tank that was above or partially above the ground or in a new tank that could not be put underground, the insurance regulations demanded that the tank be at least one hundred feet "from any building, platform or shed" and be oriented so that oil flowing out of it would not run toward any nearby structures. Underground tanks would have to be at least ten feet from the foundation of any buildings, and tank tops would have to be at least four feet below the surface. The highest point of the tank would need to be at least two feet below any device in which oil burned. There were also provisions for how to convey fuel oil to the burner apparatuses and a requirement for iron standpipes.[41]

The insurers also stipulated safety standards for storing larger amounts of fuel oil. Massive aboveground oil-storage tanks erected in Houston, holding two to ten thousand barrels (with forty-two gallons per barrel), would have to be at least 250 feet from Buffalo Bayou and 150 feet, according to the newspaper article, from "any combustible property." The underwriters argued that such restrictions would not impose hardships on anyone but would, "at the same time," preserve the "safety and security" of city property. They devised the bayou provision to prevent property owners along the channel from suffering a loss if lightning set oil tanks on fire and burning oil flowed down the bayou.[42]

The suggested standards were similar to insurance rules on using "crude petroleum for fuel in steam plants," outlined in a letter dated February 2, 1901, from the Houston insurance firm Childress and Taylor to the *Houston Daily Post*. The rules, in turn, reflect technical criteria drawn up by Chicago underwriters in 1892. Faced with unusual, unexpected, or adverse risks, agents sometimes wrote special conditions or exceptions into a contract, generally at the end, as an attachment or endorsement. Insurance companies standardized these conditions and exceptions (risk knowledge that applied to various situations), codified, printed, and distributed them.

This allowed agents, who were normally not trained to detect unique risks, to attach the appropriate exception to a standard policy, as defined by special forms that contained wording applicable to a particular hazard or risk with blank spaces to record the details of time, place, and so forth. Since Houston ordinances in 1901 did not specifically address fuel-oil use, insurance policy riders and special forms were useful as quasi-governmental restraints to regulate the looming tide of available oil in the Gulf Coast region. In the absence of a storage ordinance, potential oil users sought approval from their insurance agents in addition to a building improvement or reconstruction permit from the city engineer before putting in an oil-storage tank.[43]

Through the petition of August 19, Houston insurance underwriters hoped to shape Houston's risk-management policies through oil-storage and oil-consumption standards, incorporating the risk knowledge, gained from engineering and the experiences of fire-insurance companies with petroleum hazards, distilled in special policy provisions applicable to oil-burning apparatuses. Encapsulated in an ordinance, this information would shift the burden of approval upon the city, standardize the process of obtaining permits to use oil as fuel, and lay the groundwork for a risk-mitigation debate in Houston over fuel-oil dangers and constraints.[44]

In early September 1901 more firms submitted requests for the city's permission to erect oil-storage tanks. The Texas and New Orleans Railroad Company (TNORC) petitioned the Houston City Council for approval for a fuel-oil storage tank on a vacant block in the Fifth Ward. The Gulf, Colorado and Santa Fe Railway Company (GCSFR) also requested approval to put up a steel oil tank in the middle of a Houston city block. The council referred both petitions to the city engineer and the two committees on fire and streets and on bridges.[45]

Facing pressure from both citizens and businesses, the city aldermen took action on issues at the council meeting on Monday, September 16, 1901. Houston residents who had assessed the oil-storage risks sent the City Council an unfavorable petition opposing installation of the GCSFR tank. In addition, the fire and ordinance committees had recommended that the council refer the August 19 petition from insurance underwriters to the city attorney with instructions to draft an ordinance controlling oil-tank placement in the city. The full council addressed this recommendation as new business. The city attorney had anticipated the committee report and introduced an "Ordinance governing the construction and building of tanks for fuel oil and at what distances they may be located from surrounding property." Alderman Malcolm D. Bennett asked that the ordinance committee consider the proposal immediately during a fifteen-minute recess and then report to the council. Alderman Louis E. Miller tried to sidetrack this expeditious review, asking that the ordinance "take its regular course." The council rejected this appeal but subsequently agreed to a motion from Miller that the ordinance committee delay its report until the next week.[46]

This roundabout maneuvering produced a detailed ordinance that was a template for the city's future mediation between energy supply and demand. Ordinance particulars provide concrete evidence of how technical skill and information shaped Houston's material culture. In most instances in which the fuel-oil ordinance underwent change arising out of debate by the City Council, the city either sought the well-informed opinions of the city engineer, the chief of the Fire Department, or the city boiler inspector, drawing as well on the advice of technicians, surveyors, and inspectors from the insurance ranks. This illustrated a developing municipal reliance on professional expertise in the face of increasingly complex problems of governance. Engineers, knowledgeable specialists, and men of practical experience represented "systems of technical accomplishment," used increasingly in the late nineteenth and early twentieth centuries to organize sizable areas of the built environment.[47]

At a meeting on September 30, 1901, Alderman Hiram L. Robinson brought before the council an "Ordinance amending the Oil Tank Ordinance," the title of which indicates that it was a redaction of the city attorney's original proposal two weeks earlier. The council consigned Robinson's tendered measure to committee scrutiny.[48] Even though there is no known text of Robinson's ordinance, two closely related extant variants of the City Council's final enactment provide some of the fine points of its content and evolution. The first rendition, "An Ordinance to Regulate the Use and Storage of Crude Petroleum and Fuel Oil by Consumers in the City of Houston," contained in thirteen sections a staggering amount of detail on construction, materials, placement of appendages and attached instruments or devices, the diameters of supply, suction, and overflow pipes, and the size of the reservoir in which a tank must rest.[49]

Sections 1 through 8 dealt with relatively small supply tanks, perhaps ten to eight hundred barrels, which typically would be placed just outside a building whose boilers burned fuel oil. Section 1 ordered that tanks for storing oil in Houston had to be made of boiler or tank iron that was three-sixteenths of an inch thick. Tanks had to be "placed underground in all cases." A tank fifty feet or less from a building had to have a ventilation pipe that ran underground to and then up an exterior wall to a point six feet above the roof.[50] Further, engineers relied upon ventilation pipes to dissipate sulfurous oil fumes.[51] Tanks inside a vault had to have a steam pipe at least one inch in diameter running from the boilers to the top of the vault to extinguish burning oil.[52]

To make allowances for consumers who wanted to use tanks already in place, the ordinance specified that such a tank "wholly or inpart [*sic*] above ground" had to be at least seventy-five feet (changed from one hundred) from nearby buildings and oriented in such a way that the contents of a ruptured tank flowed away from any buildings or "combustible property." The tank also had to be inside a "brick vault or earth embankment" forming a reservoir whose volume was twice the tank capacity. Obviously, concerns about vapor explosions, vessel fractures, and fire damage to property animated these provisions.[53]

The ordinance stipulated that underground tanks with a capacity larger than twenty-five hundred gallons (59.5 barrels) had to be at least ten feet from the foundation of any building in which boilers burned crude oil and at least two feet below grade. Furthermore, the tank had to be situated so that no oil could flow by gravity through pumps, standpipes, or piping to the combustion device. "[A]rtificial pressure or suction . . . whether by pump, vacuum or other means" had to move oil from such tanks to the ignition point, accomplished by placing an iron standpipe between the supply tank and the burners, located outside the building and bolted to the building foundation. Standpipes, whether open or closed, supported fluid pressure in a conveyance system, without which there would be no flow.[54]

In the case of moving fuel oil to boiler burners, engineers connected a much smaller pipe to the fuel transfer line, open at the highest end to insure through atmospheric pressure a steady network flow and to prevent unexpected or unwanted flow reversals that could occur due to siphoning. Depending upon space limitations or design considerations, one might substitute a small tank or "iron accumulator" for a standpipe, fitted with a relief valve and overflow pipe.[55]

Also extant is a second version of this ordinance that is very similar in wording to the above-noted rendition. On October 7, 1901, the ordinance committee recommended to the City Council the "passage of the within ordinance as a substitute for the original with the same caption," but the council debated this draft through December.[56]

In the second version, sections 9 through 13 were similar to the original and covered "large storage tanks." Section 9 limited the capacity of a single bulk tank inside the city of Houston to ten thousand barrels (amended from two thousand), or 420,000 gallons. Although there was no limit on the number that could be erected, no two tanks could be closer than 2,000 feet. Further, each tank had to be at least 250 feet from any "combustible property" and inside a dugout or entrenchment walled in by a brick-and-earth embankment that could hold twice the tank's capacity. These constrictions ruled out tank farms with hundreds of tanks, except perhaps on empty property on the fringes of the city. No large tank, under any circumstances, could be placed at a distance of 250 feet or less from Buffalo or White Oak bayous.[57] Consonant with other applications of municipal police powers such as building regulations, any person or corporation that wanted to burn crude petroleum or fuel oil in Houston had to obtain a permit from the city engineer before installing a storage tank. Section 11 made the chief of the Fire Department responsible for certifying that the installed oil plant complied "with all of the above regulations before being used."[58]

Although council observers assumed aldermen would vote October 15 on final passage of the amended fuel-oil ordinance, there was a delay which bred frustration. At the council meeting of October 21, Alderman H. A. Halverton offered several new amendments to the oil-tank ordinance, pleading for the council to amend and pass the ordinance. This was a pressing matter, Halverton urged, "as the storage of oil for fuel purposes was becoming a matter of great importance to the manufacturers' interest of the city." Storage was also of interest to him since he was the president of the Houston Drilling and Supply Company, organized the previous spring to drill oil wells and to supply "oil well machinery" to other companies. The council instead decided to refer the fuel-oil measure to the ordinance committee again to consider his proposal.[59]

At the last meeting in October, the council carried a motion by alderman and metal manufacturer Malcolm D. Bennett ordering the secretary of the council to request copies of fuel-oil storage ordinances from cities such as Pittsburgh, Philadelphia, Cleveland, and San Francisco. Calling for risk knowledge to inform the debate allowed Houston to compare what it had under consideration with existing regulations in major cities. Precedents from other municipalities would blunt any claims that the city's leaders were strangling fuel-oil usage and sidetracking Houston's march to prosperity out of ignorance of accepted practices. In early December the ordinance committee submitted an amended "Oil Tank Ordinance," but Alderman Louis E. Miller, an electrical engineer and coal dealer, declared the draft ordinance was defective, leading to referring it to W. E. Gale, the city boiler inspector, for a final assessment based on his technical understanding of the hazards of steam production and fuel.[60]

At this point the *Houston Chronicle* joined the battle and raised the issue of subjective versus quantitative risk assessment, asking whether impressionistic belief or experienced-informed opinion should determine the likelihood of hazardous events such as fire or explosion. Or does one need the number of events in a specific time period to calculate the probability of a future incident? In an editorial published December 7, the paper pointed out that the expanding use of fuel oil was a "new form of fire hazard" that had to be assessed by "use and experiment." Insurance underwriters did not have broad experience with fuel oil or statistics to demonstrate a correlation between this energy source and specific hazardous incidents, making it difficult for insurers to "adjust their rates" to match the quantifiable risks of fuel oil. To exploit this calculative shortfall, the writer recommended that Houston officials quickly pass ordinances that minimized fire risks from fuel oil, thereby making it hard for insurance companies to raise rates unless they could show that fuel oil constituted a special hazard.[61]

The essay argued that crafting oil-storage regulations should rely upon a methodology employed by life-insurance companies who had access to multigenerational (actuarial) mortality data. This insight anticipated the extensive use of quantitative risk assessment (QRA) in late-twentieth-century reliability engineering, safety analysis, fire underwriting and fire-prevention applications. The editorial illustrated that basic concepts of hazard mitigation and risk management framed the 1901–2 debates over petroleum storage and consumption, and that it would be advantageous to have more information on risks if a municipality wanted to regulate them.

Ten days before Christmas, Inspector Gale advised the City Council that oil-supply pumps should be fitted with a pressure regulator to prevent excessive flow-rates leading to accidents in the ignition chamber. Pump-fed, unpressurized standpipes should have float regulators to restrain excessive flow on the supply-tank-to-standpipe side of the conveyance systems. But one proposed addition would prove controversial and would ultimately obstruct Houston's search for acceptable regulation. "In no case," read an addendum, "will Oil tanks be allowed under side walks or in streets." Not anticipating a great deal of trouble passing the ordinance—given the time spent and the changes incorporated into it—the council once again referred the amended fuel ordinance to its ordinance committee but without comment.[62]

The committee reported positively on the oil ordinance in early January 1902. In addition to the emendations that the city boiler inspector suggested several weeks earlier, the committee made other changes such as reinstating in section 3 a requirement that aboveground supply tanks had to be 100 feet "from any building or other exposure." Section 8 required that the accumulator tank, installed in place of a vertical standpipe, had to be put outside the building where the burners it served were located. The reported ordinance retained the stipulation that large bulk-storage tanks had to be at least 250 feet from "any combustible property."[63] The *Houston Daily Post* reported that the oil-tank ordinance (in its second version) had a five-thousand-barrel limit, but council members changed this threshold to ten thousand barrels just before final passage, finally deciding how much oil was too much to be stored in Houston wards and neighborhoods.[64]

After months of discussion and maneuvers, Houston had a fuel-oil ordinance, and companies, already straining at the bit to use oil, acted quickly. Three days after the passage of the ordinance, the Santa Fe Railroad announced that it was constructing a 10,000-barrel tank on St. Emanuel Street. The company had asked permission in September 1901 to build it, but in the absence of a specific ordinance on storing fuel oil, nothing concrete came out of the request.[65] Southern Pacific Railroad (SPR), a major line that served Houston, deliberated at length about converting its locomotives to fuel oil. Soon after passage of Houston's oil-tank ordinance, railroad observers reported that SPR planned to build a string of oil tanks from Houston west to El Paso. By August 1902, the retooling of its locomotives to burn oil was underway. Officials estimated that at completion the motive fleet would consume 10,000 barrels of oil per day.[66] One local oil supplier, Southwestern Oil Company, had a crew of men at its Houston Heights refinery set down the "foundations for new storage tanks" as part of a plant addition. The company anticipated that by April 1, 1902, its refinery would be turning out 1,000 barrels of refined oil per day. Penman Steel Tank Company of Chicago closed a deal with SPR for twenty-five to thirty very large tanks, ranging in size from 37,500 to 50,000 barrels, at a total cost of $250,000, at least one of which would serve Houston.

The issue of aboveground and underground supply-tank configurations was an unchanging factor in the campaign to pass a fuel-oil ordinance, the parameters of which included tank volume, distances from buildings or foundations, ground depth, and placement inside or behind vaults or brick walls. The most controversial element proved to be putting supply tanks under sidewalks, which raised questions about municipal liability for accidents.

Liability was a legal issue central to the nature of municipalities. Their police powers were used to protect the community but also entailed various obligations. If passersby sustained injury or loss because of a fire or explosion originating in an under-the-sidewalk oil reservoir, they might charge the city with failure to impose reasonable restrictions on oil storage. Municipal power over the streets was complete, unless the legislature by charter or special statute directed otherwise. This meant city governments had control over the sidewalks, since in common law they were considered part of the street. Power over streets imposed a duty to keep the roadway, curbs, and sidewalks in repair. Even if a city required lot owners to pay all or part of the costs of constructing sidewalks, the city remained liable if they fell into disrepair. According to its 1897 charter, Houston enjoyed almost absolute power over, and responsibility for, its streets and sidewalks. This charter invested in the city the responsibility "to regulate the use, and abate and remove encroachments and obstructions thereon." Storage of fuel oil under city sidewalks was an important legal matter, since injuries arising from it implied city negligence in the repair and maintenance of streets.[67]

This problem surfaced as early as March 1901, about nine months before the creation of a fuel-oil ordinance. Candy manufacturer Charles Heim asked the City Council for permission to "erect a storage reservoir for fuel oil" under the sidewalk in front of his factory at 405–7 Sixth Street. The finance, streets and bridges, ordinance, and fire committees, as well as the city attorney and city engineer, took the matter under review. Sending the petition around for review by municipal officers suggested there was some uncertainty, and for good reasons, about storing oil under sidewalks.[68] At the next council meeting, April 1, the fire committee reported favorably on the petition, provided that no one could hold the city liable for any damages "that may occur from whatever cause." The council's finance committee and the city attorney concurred with the report, although it is not certain that the council ratified the reports and legal opinions. It is clear, however, that the city, even before the oil-ordinance debates in the council, was cautious, in light of potential liability claims, about storing oil under sidewalks.[69]

Sidewalk tank storage came up again January 20, 1902, when Jacob Binz, owner of the Binz Building in downtown Houston, petitioned the City Council, under the name Binz & Settegast, for permission to "construct [an] oil reservoir" in the building's basement along Texas Avenue. The council, following guidelines in the recently passed oil-tank ordinance, referred the petition to its fire and ordinance committees, the city engineer, and the chief of the Fire Department. How they would react to this petition was not clear since the new oil ordinance (effective January 10, 1902) prohibited storing oil under sidewalks. The Fire and Ordinance committees and the chief of the Fire Department recommended, February 3, that the council approve installation of Binz's oil-supply tank if its placement conformed to submitted blueprints. Although the council adopted the report, it did not pass an ordinance to this effect.[70]

What was the reasoning behind this favorable report, considering that what Jacob Binz wanted on the face of it was unlawful? One answer is that there was an ambiguity in the city's new regulations. Section five of the oil-tank ordinance allowed underground tanks to be as close as ten feet from a building foundation. A building owner could claim that a reservoir set down in the center of a basement was more than ten feet from any foundation or, alternatively, that a tank placed under a broad sidewalk was still ten or more feet from the foundation. This made sense since any extension of a building basement under a sidewalk was by law part of the basement. These hermeneutical qualifications may have persuaded officials that Binz's petition rested upon a valid interpretation of the fuel-oil ordinance, but the alternative was to interpret strictly Houston's oil-tank ordinance as a categorical prohibition of oil storage under sidewalks, which would have prevented Houston's business-district property owners from using oil.

If some clarification of the ordinance was in order, the question was whether the best course was to abolish, leave alone, moderate, or toughen tank-placement requisites. To make the oil-storage regulations palatable to all sides, Alderman Halverton introduced at the council meeting on February 10, 1902, an amendment to section 5 of the oil-tank ordinance. The change proposed to allow tanks under sidewalks, still at least ten feet away from the building foundation, if the tanks were encased in a twelve-inch brick or concrete wall. The City Council approved this ordinance a week later, allowing consumers in densely built areas to use fuel oil.[71]

Despite this concession, not everyone was happy with the city's risk-management track record on fuel oil. George P. Brown, influential secretary of the Houston Business League, complained to the *Houston Chronicle* that the state and the city of Houston had imposed too many restrictions on the use of fuel oil. Beaumont oil, Brown contended, was not the same as other flammable liquids—poorly refined kerosene or kerosene mixed illicitly with benzine, naphtha, or gasoline—whose "careless use" had caused so many accidents around the country. He argued that "proper precautions and ordinary care" made fuel oil safe to store anywhere in Houston. Existing oil-storage ordinances delivered a "grievous injury to the oil interests," impeding large-capital expenditures earmarked for pipelines and oil tanks. It was, Brown charged, as though some "ulterior influence" was "at work to hinder and hamper the use of fuel oil" in Houston—likely a reference to the insurance interests that had played a major role in devising the restrictions on the storage and use of oil. Brown proposed that the City Council have an open hearing on this subject and make some concessions on controls on "fuel oil, storage and conveyance pipes."[72]

Brown disliked the city's application of the precautionary principle in its management of fire hazards. In such cases a concern for the "magnitude or severity of the consequences of actions *not* taken" shapes municipal policy on fire hazards. Having a responsibility for the health, safety, and welfare of citizens, Houston authorities believed they had to act with "prudence and anticipation." Not having objective, statistical information on the risks of public use of fuel oil, the city relied on a vast body of anecdotal, experiential, and narrative material about oil as a fire hazard. It was sufficiently convincing for some leaders, insurance underwriters, and residents to act upon, rather than arguing over the credibility of evidence of hazards or weighing various regulatory outcomes against the possibility of a major fuel-oil accident. The latter would have amounted to an inchoate "risk, cost and benefit analysis." In August 1901 local insurance underwriters had referred to an anticipated oil-storage ordinance as a safeguard, not as calculative "models of events" that predicted quantitatively the "probability of an adverse outcome" from using oil. They wanted an enactment to be a "legally mandated precaution" that positively, absolutely reduced fire dangers, notwithstanding the lack of statistically probative knowledge about the exact perils of fuel-oil usage. Note, however, that Brown's rhetoric of reasonable care, proper precautions, concession, and compromise suggested that he wanted city officials to incorporate commonsense causation and likelihood into their deliberations. They could mitigate acceptable risk by balancing "potential gains or losses," moderating restrictions on using fuel oil.[73]

Brown's recommendations brought council action once again. Less than a week after Brown's first newspaper interview, Alderman W. W. Thomas proposed that the major parties in this matter—the Business League, the Cotton Exchange, the Houston Manufacturers' Association, the "insurance agents," the chief of the Fire Department, the City Council ordinance committee, and the mayor—produce an ordinance that would meet the "demands of the users of fuel oil without causing an advance in the rates of insurance . . . or creating a menace generally to the people of Houston." Although the motion carried, nothing happened at the outset, allegedly due to lobbying from the Dallas-based Texas Fire Prevention Association (TFPA). This organization had visited Houston in September 1901 and questioned the Fire Department's readiness and proficiency. The *Houston Chronicle* published a letter from the TFPA to local underwriters, claiming that agents of the "oil and railroad interests" were leading the effort to water down Houston's existing oil-tank ordinance. The association asked Houston agents to oppose "strenuously" any changes to this regulation, which the group regarded as "very fair to all interests."[74]

And as a parting shot the TFPA warned that amending the current oil ordinance came at the cost of underestimating the dangers of using fuel oil. The association cited the fire of February 18, 1902, that destroyed the Louisville Bolt and Iron Company—a $100,000 loss—owing to an oil-feed pipe leading to a furnace that exploded, scattering burning oil all around and reducing the two-and-one-half-acre facility to ashes.[75]

Even so, Brown was not finished. On March 4, 1902, he told the *Houston Chronicle* that fuel-oil ordinances were "of supreme importance" to the city and should be written to "cause the general use of the Beaumont product locally." Unquestionably, Brown argued, fuel oil was safe and a key to Houston's future. But Houston, he complained, would be the last place to take advantage of this wonderfully cheap product because of the "brick wall that has been thrown around this city by indifferent legislation." Brown concluded that the council should follow Alderman Thomas's earlier suggestion to have a committee of citizens and councilmen amend the present oil-tank ordinance.[76]

Both Brown and Thomas believed the City Council had acted arbitrarily, allowing the insurance interests to shape its decisions about storing oil. The solution was negotiation, which in effect was a plea for coordinated "decision making in policy planning." Rational decision-making can either be "substantive," focusing on "what to choose," or it can be "procedural," that is, "how to choose." The "how" that Brown and Thomas preferred was face-to-face group negotiations. Further, asking stakeholders to discuss the implications of fuel-oil regulations suggested that public "[p]olicy formulation" involved reconciling various municipal factions, whose needs might be quite different from one another. The outcome of this process would be an ordinance "acceptable . . . for almost all parties concerned."[77]

The City Council had spent several weeks looking into charges that the city water company had illegally stored fuel oil in a substandard tank and did not immediately call a conference.[78] However, Brown and Alderman Thomas could not be ignored. On March 17, insurance envoys Jerome B. Cochran and Alfred K. Taylor met with Houston Business League representative Henry F. MacGregor; F. N. Gray, delegate from the Manufacturers' Association; aldermen M. D. Bennett and H. L. Robinson, and Mayor J. D. Woolford. The goal was to modify the current oil-tank-storage ordinance, which Brown, the *Houston Chronicle,* and other oil-distribution sympathizers regarded as excessively restrictive. Despite a low-key campaign by the *Houston Chronicle* and its rival, the *Houston Daily Post,* to prove that Beaumont oil was not highly flammable, the delegates were unable to reach an agreement.[79]

Even so, Houston residents kept the oil-storage controversy alive. Downtown "Property owners & Merchants" petitioned the City Council on April 3, 1902, asking them to block the "erection of [an] oil tank under [the] sidewalk of the Binz Building." They argued that the proposed tank was "not in accordance with the Oil-Tank Ordinance" and were unwilling to allow an obvious fire hazard to jeopardize their lives and property. The council sent the petition to the ordinance committee and the city attorney with instructions to "prepare [an] Ordinance repealing [the] Amendment to Oil-Tank Ordinance," that is, removing the change to section 5 which allowed oil storage under sidewalks. At a meeting of the council, April 8, aldermen reinstated the ban on below-the-sidewalk storage. Insurance spokesman Jerome B. Cochran said that disallowing subsurface-sidewalk storage was "entirely satisfactory" to the insurance fraternity. According to the *Houston Chronicle,* the council approved the ordinance "upon the strength" of Cochran's statement.[80]

But there is more to learn about this ordinance controversy. Details of the political and regulatory maneuvering behind the city's risk-management agenda are available in legal proceedings that arose over the sidewalk storage of fuel oil. According to district and appellate court records, Jacob Binz signed a contract, March 11, 1902, with a Chicago engineer, Elmer Billow. In behalf of the NSC he agreed to "furnish all material and labor" and to construct "a complete fuel oil plant" which would power two boilers in the basement of the Binz Building. This included a supply tank to be erected in the basement extension under the sidewalk of Texas Avenue. Billow had devised a complete supply system for oil-burning equipment, a good example of the dissemination and application of technical knowledge and experience related to hazard mitigation in a package designed to meet insurance and municipal rules on oil consumption.[81]

Further, Billow related in a 1905 deposition that in late 1901 he examined the insurance underwriters' form used in Texas regarding burning fuel oil and realized that, should Binz "decide to adopt oil as fuel," he would not be able to comply with these rules. To solve this problem, Billow proposed an oil-burning plant fed by an under-the-sidewalk supply tank. Jacob Binz claimed that the local representatives of a dozen fire-insurance companies—foreign and domestic—as well as Houston officials and the City

Council, approved this iron reservoir. Further, this happened, Binz noted, while the January 1902 oil ordinance that banned sidewalk-oil storage was in effect. According to Binz, the insurance men determined that "there was no increased risk or danger of fire" from the proposed storage tank, a risk assessment shared by city officials who also found no extra "hazard or risk of fire."[82]

To smooth the regulatory path, the insurance companies "induced" the city, according to Binz, to pass a second fuel-oil ordinance (effective February 19), allowing him to have an oil-storage tank under the sidewalk. The next day insurance and city officials gave Binz permits to install the oil plant and supply tank, reiterating their endorsement of the equipment, tank, and appurtenances. With risk management policies in place, Binz signed the NSC contract, one provision of which was a brick wall "separating" the oil-supply tank "from the basement."[83]

In April 1902, the City Council passed a third oil ordinance that reinstated the ban on storing fuel oil under sidewalks. Binz blamed this on pressure from the insurance companies. They allegedly warned him that, if NSC installed oil-burning equipment or a supply tank in his building, they would cancel and never renew his fire-insurance policies. Consequently, Jacob Binz refused to allow Billow to execute their contract or to pay NSC, which precipitated a lawsuit that dragged on until 1908.[84]

In addition, Elmer Billow testified that in 1902 there was a push to prohibit oil storage "under sidewalks or in alleyways" to make "the pipe line ordinance effective." This makes understandable the City Council's revocation of several early pipeline franchises, clearing the way to grant new ones on March 10, 1902, as well as the excitement over the rumors that the Hogg-Swayne syndicate or a potential London buyer planned to build a giant pipeline system from Beaumont through Houston and on to Dallas and Fort Worth. It also explains the complete reversal of the fire-insurance interests on sidewalk oil storage, acting out of their own risk concerns and, they presumed, in the city's interests.[85]

Billow also claimed that after consummating the March 11 contract this "proposed legislation" (the so-called third fuel-oil ordinance subsequently passed in April) was under consideration. He "personally appeared in the Council Chamber and talked to many aldermen" to head it off. Further, Billow asserted that the "deliberations were freely discussed with Mr. Binz," and when Binz signed the NSC contract, Billow warned Binz that "he might be called upon to defend himself against an unfriendly City Ordinance." This was why, Billow pointed out, he designed the "furnace arrangement" so it could be "quickly interchanged with coal or wood."[86]

Although city newspapers covered the evolution of Houston's risk-management strategies—expressed in an oil ordinance that was subsequently amended and then passed again with its original specifications—they did not have access to business letters or conversations between a consumer and an engineer-equipment supplier. Elmer Billow disclosed the backroom lobbying that surfaced in dealings with Jacob Binz. He did not know if Binz petitioned the City Council "to repeal the ordinance" (effective April 11) reinstating the sidewalk-alleyway storage of oil, but could state unequivocally that Binz said that "if the ordinance was passed he [Binz] was strong enough to have it repealed."[87]

Several petitions came to the City Council after it settled upon the fuel-tank ordinance in April 1902. Jacob Binz petitioned the city once again in May 1902 for permission to put up an oil-storage tank in his downtown building, but following an unfavorable report in early June 1902 by the council fire committee and the city attorney, council members rejected the petition.[88] Late that month, the GCSFR repeated a request to place an oil tank on the block enclosed by Preston, Prairie Avenue, St. Emanuel, and Hutchins streets. The City Council on July 8, 1902, again denied permission for the tank, because the council committee responsible for fire-related matters had determined that the proposal was "not in accordance with the requirements of the oil-tank ordinance."[89]

Thus, the city had fashioned a procedure to control fuel-oil storage that in light of contemporary risk-knowledge and experiences satisfied many critics of unregulated oil use. Mobile and New Orleans required oil consumers to obtain permits from their city engineers to erect a supply tank and specified how installers arranged the equipment, but unlike Houston, these port cities adopted rules that confined massive, bulk-oil storage to specific districts defined by certain streets. Houston relied primarily on tank capacity (ten thousand barrels), minimum distance between tanks (2,000 feet), and minimum distance (250 feet) from Buffalo Bayou to determine if bulk storage tanks belonged inside or outside the city limits.

A challenge to this regulatory regime came in 1904 when a dangerous fire at the tank yard of the Waters-Pierce Oil Company on Washington Street destroyed several large tanks—6,000 to 8,000 gallons capacity—filled with gasoline, turpentine, neatfoot (made from cattle hooves to dress leather), and linseed oils, as well as kerosene. The City Council instructed the city engineer not to grant the company a permit to rebuild or repair any damaged buildings on site.[90]

The ruinous fire, stoked by flammable liquids, came at a time when the city had imposed hazard-mitigation practices upon Houston theaters and amusement halls. The deadly Iroquois Theater fire in Chicago in late December 1903 precipitated this campaign, leading Mayor Holt to appoint a special committee to investigate local conditions and document dangerous construction and operations, which the City Council matched by authorizing an investigation by its fire committee. Two days after the Waters-Pierce disaster, Fire Chief Jack P. Arto petitioned the City Council to mitigate the fire hazards posed by the Sweeney & Coombs opera house (wood construction), pointing out that earlier inspections of the opera building revealed that there were no exits in the theater, and that there was a canvas stage curtain in place, a deadly accoutrement in entertainment venues. Further, people in rented rooms in another part of the same building burned oil for cooking and heating. Arto asked the council to pass an ordinance requiring major changes in the building, such as installing a fire alarm connected to the city's alarm system.[91]

Residents around the Waters-Pierce Oil Company were also deeply concerned about flammable liquids, specifically the oil reserves kept in the company's distribution yard. They successfully petitioned the City Council to prohibit storage at the firm's Washington Street site. In addition, Chief Arto petitioned the council to condemn the Waters-Pierce storehouse and to pass an "ordinance prohibiting the storage of oil in such quantities in the city limits." The hazardous "combustible liquids" stored on site, he argued, could start a conflagration that would destroy the area around Waters-Pierce's oil yard. He complained that this was the second petition he had sent to the council on forcing all oil warehouses to be beyond the city limits. His position was a predictable and valid risk assessment from a professional fireman.[92]

Ironically, Waters-Pierce had already made plans before the warehouse fire to move their oil plant beyond the city limits. A week after Arto asked for the prohibition, the City Council referred a request from Waters-Pierce for more time to "complete their store house outside of the city limits" to a committee.[93]

Hereafter, the affairs of the Waters-Pierce Oil Company became complicated, assuming of course that they could get any worse in light of the company's legal battles with Texas officials over antitrust issues that dated to 1894. In August 1904 the Gulf Refining Company applied for a permit to build a "refining and pumping station and warehouses," also in the Fifth Ward on Brooks Street near the Houston East and West Texas Railway tracks. Gulf planned to invest a great deal of money in an array of big pumps and a "large battery of storage tanks." In its permit application, Gulf promised to create a central distribution point for all of Texas. Yet, their petition, scheduled to be considered by the City Council four days later, abounded with uncertainty, because immediately after the Waters-Pierce fire, the City Council had resolved that there would be no more "oil warehouses within two miles of the court house." And no one knew how this would affect the Gulf project.[94]

Weeks later the mayor, the council's fire committee, and Fifth Ward aldermen reported favorably on the Waters-Pierce request to construct a "brick warehouse" and other facilities near the Peters tract, and on the Gulf Refining Company's plan to build a distribution station in the Noble tract (both in the Fifth Ward inside the city limits but some distance from the courthouse). In a subsequent meeting the full council considered both requests and debated an ordinance prescribing how the appellants would handle oil. Two members of the council's ordinance committee interjected their reports, asking that the proposed oil ordinance be amended to limit the companies to 1,500 barrels of oil on site, with a minority dissent that recommended only 1,000 gallons. A tumultuous debate ensued over the oil-storage limits, which at one point produced an amendment allowing the companies to "carry 2000 barrels of each kind of oil" that they wanted to sell, but this was reduced to a total of 1,500 barrels of oil "of all kinds . . . in stock at one time." Participants also sparred over requiring a brick wall around the stations, but Judge J. W. Parker, representing the Gulf Refining Company, said his client was not going to refine oil—just store and distribute it—and would not spend $7,000 to $8,000 putting up a wall on leased ground. After the council eliminated this precondition, the meeting collapsed under the weight of trying to decide whether the proposed ordinance forced the two oil companies to operate two miles from the courthouse, or whether it allowed all retail and wholesale dealers in the city to have no more than 1,500 barrels of oil on hand for sale "without a permit."[95]

At a specially called session of the City Council, this "heated and extended discussion" resumed with the oil ordinance and proposed amendments to satisfy Waters-Pierce and Gulf Refining at center stage. However, several key Fifth Ward aldermen failed to appear, forcing the council to lay the measures over to the next regular meeting. What had been accomplished? It was clear that Waters-Pierce and Gulf Refining wanted to settle in the Fifth Ward, but the City Council could not agree on the conditions, embodied in an ordinance, that would set down how the companies could carry on business.[96]

Around mid-September 1904 the politics of risk management produced a surprising answer when the needed ordinance was called up. Immediately there was trouble, because the amended version had been altered, forcing the companies to be not one but two miles from the courthouse. An argument over this ended when the council secretary replaced "one" with "two," but members then tabled the amended version. Having agreed to reconsider the oil-company petitions, the council approved the original petition from Fifth Ward residents that called for prohibiting oil operations in their ward, leaving Gulf Refining to look elsewhere and Waters-Pierce at its Washington Street location where it had been for twenty-five years. Although this was not what some expected based on earlier council deliberations, all was not lost for the oil companies.[97]

By January 1905, Gulf Refining was ensconced in Houston Heights beyond the city limits. Waters-Pierce stayed in the battle through May when it submitted a petition for the right to build huge oil tanks, brick buildings, and sheds on property in the Fifth Ward on Noble Street, basking in the glow of an affirming petition from three hundred local residents supporting the company's request. The company agreed to a thirty-five-foot setback from the boundary of the property to the perimeter of the oil tanks. The City Council amended its special permit allowing Waters-Pierce to construct an oil plant, making these hazard-mitigation measures formal requirements. The next day the oil company announced that it would shortly let a contract for a $51,000 station to be built on 3.5 acres in the Fifth Ward, consisting of "six brick buildings and immense storage tanks surrounded by safety devices and fire preventatives."[98]

In response to the Waters-Pierce and Gulf Refining permit requests, the city of Houston invoked its police powers to insure public safety, in step with existing restrictions. Pushed by residents alarmed by the dangers of huge oil tanks, the City Council banned their use and construction in the city limits, but later issued Waters-Pierce a special permit to install such tanks in the Fifth Ward far from the downtown business center on the condition of compliance with existing large-oil-tank storage that summed up the city's risk management of petroleum.[99]

In subsequent ordinances there were changes in various details such as allowable tank volume, but generally these later regulations reflected the same concerns addressed in 1901–2. Complications arose but only in step with advancing petroleum technologies and new markets for hydrocarbon distillates. For example, in May 1911 the city passed an ordinance that controlled the handling of gasoline and other flammable liquids. The measure addressed placing tanks outside buildings, burying them belowground, or covering them with earth or concrete if put in the basement of a building, and it specified standards for tank-metal gauge, riveting, caulking, pipe connections, openings, venting, suction pumps, and check valves. Gasoline or other flammable liquids could not be stored in portable tanks larger than fifty gallons in capacity. Flammable liquids inside buildings but not in supply tanks had to be stored in five-gallon safety containers with a limit of one container per building.[100]

There was, however, one notable exception to the similarity between the 1901–2 oil-storage ordinance and subsequent enactments. Comparing the 1904 edition of code revisions to those of 1914 and 1922 shows that the city of Houston once again reversed its position on under-the-sidewalk storage.[101] In May 1913, Mayor Ben Campbell submitted to the City Council an ordinance to alter chapter 38, article 904, of the 1904 code compilation, granting permission to have oil-supply tanks "outside of buildings or under sidewalks." The council passed this amendment the same day.[102]

The change did not represent unjustifiable vacillation, nor was it the result of poor civic leadership. As time passed, some substances proved to be less dangerous than previously believed. Hydrocarbon distillates such as gasoline, benzine, and naphtha, that had made poorly refined kerosene (illuminating oil) extremely dangerous in the nineteenth century, unquestionably remained serious fire hazards after 1900 owing to effortless vaporization at room temperature. In particular, large quantities of gasoline received regulatory attention, engaging the city of Houston, as fuel oil had in 1901–2, in mediating the relationship between a source of energy and demand for it. Regulating gasoline took its place in fire codes and became a central focus of standards in fire-prevention ordinances under combustible and flammable liquids. Risk assessment involved determining the likelihood of a hazard-initiated event. As turn-of-the-century businessmen, fire-insurance underwriters, city leaders, and residents became more familiar with petroleum consumption and storage, taking advantage of accumulated experience in its use—both their own and that of professional and technical associations or groups—they employed increasingly reliable and sophisticated technology to deal with hydrocarbons. They also became, on the basis of petroleum's growing track record as a hazard, better equipped to estimate the risks posed by flammable liquids. Accordingly, Houston leaders modified or adjusted municipal regulations, reflecting lessons learned in how to avoid accidents and damage to property.

From 1901 to 1914, in the context of the oil boom on the Texas Gulf Coast, Houston played a role in creating a rational interface between a new energy source and anticipated energy demand. Houston city officials, businessmen, insurance agents, and concerned citizens took part in the politics of risk management, negotiating the terms under which suppliers and consumers transported, stored, and used fuel oil in the city limits. Starting in earnest in August 1901, the Houston City Council devised, revised, and passed an oil-storage ordinance that eventually satisfied various parties and constituencies. Despite a legal dispute over banning oil storage under sidewalks, the council settled on an ordinance in April 1902 satisfying insurance underwriters and under which businesses could work effectively to utilize the new resource. Houston's risk management of fire hazards through its police powers endured throughout the twentieth century, preserving the built environment and keeping human capital out of harm's way.[103]

4

COMPETITION AND CONFLAGRATION

Exploiting and Regulating Fuel Oil in New Orleans, 1901–1943

PART I

The Beaumont oil field in 1901 presented New Orleans with enormous amounts of crude petroleum sufficiently cost-effective to replace coal but laden with undetermined risks. With so much wealth at stake, this mineral godsend set in motion a fierce debate within the ranks of city leadership over the terms under which the municipal consumer could use this new energy source.

The Crescent City's experience with this emerging energy market focused on devising policies for managing risks associated with petroleum distribution, storage, and consumption. Risk assessment, a fundamental tool of this campaign to promote community safety as well as prosperity, identified potential safety concerns, giving direction to the city's search for pragmatic regulatory solutions to fuel-oil hazards. Risk management—strategies for the mitigation of risk—was in accord with an ongoing effort dating from roughly 1880 to improve public welfare. Community betterment required effective coordination of city resources and supervision of increasingly intricate systems to meet "rising expectations of services" whether it involved water filtration, sewage disposal, levee construction, creating a comprehensive building code, or identifying the hazards associated with crude petroleum.[1]

New Orleans was a major insurance city where many marine and fire underwriters issued contracts on vessels that called at the port as well as their cargoes, particularly cotton stored on wharves. By 1898 these companies had written $207,640,618 in fire risks in Louisiana, a figure bolstered by the huge business in the Crescent City. New Orleans was also the home of one of three resident secretaries (Clarence F. Low) of Liverpool and London and Globe Insurance Company in the United States, in addition to the resident manager in New York City. This huge enterprise along with numerous competitors provided local agents and fire underwriters access to the risk knowledge of the NBFU, the NFPA, and the networks of general and special agents, inspectors, surveyors, technicians, and engineers upon whom the insurance industry relied to identify and characterize hazards and to devise permit rules to mitigate dangerous events, substances, or conditions.[2]

As was the case with Mobile and Houston, New Orleans possessed broad police powers granted by the state legislature through acts of incorporation or statutory amendments of same to protect public health and well-being from a range of threats, nuisances, and reckless behaviors. While the city was sometimes under judicial scrutiny, most notably in litigation over the Slaughterhouse Act (1869), its police powers gave it control over streets, bridges, canals, public markets, hazardous substances, building construction, and fire limits. Armed with the authority to grant franchises, permits, and licenses, municipal superintendence elicited many battles over street railways, clean water, drainage, abattoirs, levee encroachment, and wharves/landings ownership. In this context, elements of which spanned much of the nineteenth century, controlling fuel oil was one more episode in the community's transition from a lively and loosely controlled river town and commodity outlet to a modestly ordered municipality on the fringes of an "emerging managerial society" rooted in the Northeast.[3]

Police powers gave political life to public policy in New Orleans, where fuel-oil regulation took place in the context of progressive initiatives to strengthen the community that drew upon long-standing republican antipathy to monopoly and corruption. Ending official corruption, applying professional knowledge to city problems, implementing efficient, businesslike methods in government, and improving municipal services were rudiments of this campaign, the shortage of unimpeachable politicians notwithstanding. For its adherents, "forwarding the great public interests of the city" was a fundamental sentiment, and in the case of petroleum issues, there was a developing cache of knowledge produced by fire-insurance companies, technical-standards groups, and insurance associations such as the NFPA, in which engineering and scientific expertise played a fundamental, economically progressive role.[4]

Public discussions about how to deal with fuel oil contributed to shaping public policy, which in New Orleans disclosed several informal coalitions that pursued seemingly incongruous solutions. On one side were oil promoters, city boosters, factory owners, regional machinery suppliers, and capitalists who would benefit in some fashion from unrestrained or lightly regulated fuel-oil marketing and distribution. They faced risk-averse, safety-minded property owners in various city wards, cotton factors, wholesale dealers, anxious City Council members who felt responsible for fire hazards, and vocal city insurance agents, inspectors, journalists, and underwriters. This last group had ties to out-of-town, East Coast or Anglo-American insurance companies who underwrote much of the North American-fire-insurance business or reinsured big American underwriters.[5] Through these sources New Orleans underwriters, fearful that burning fuel oil might destroy the cotton, tobacco, rice, and lumber piled on city wharves, had access to engineering, technical, and company knowledge of petroleum hazards and risks.[6]

But economic risks were not new to New Orleans. All elements of political and business leadership knew that the city had lost commercial ground to Chicago starting in the late 1840s and continuing after 1860 owing to a vast railroad network that tied the Great Lakes to New York and Philadelphia. The heavy costs of the Civil War also eroded New Orleans's competitive position in relation to St. Louis and Chicago.[7]

As important as the Mississippi River was to commerce, Crescent City residents did not want the river in their streets, as happened in 1849 when a levee failure above the city spread mud and debris through neighborhoods, driving twelve thousand people out of their houses. Flood relief and keeping the river unobstructed all the way to the South Pass engaged the city leadership, who were otherwise preoccupied after 1874 with perpetuating sentimental Civil War memories, controlling or eliminating the African American vote, debating how to get the city out of its long-term fiscal morass, or scheming to attract adequate railroad service.[8]

By 1900, city leaders, just like their Houston counterparts, firmly believed that they had to attract capital to lure industrial concerns away from the Northeast and Midwest, to finance industrial expansion, and to develop indigenous manufacturing and fabrication. Compared to most of the post–Civil War South, New Orleans had a vigorous banking system, but its capital financed commodity production in outlying rural areas—"resource processing"—or transshipment of exports to Europe, rather than large-scale production of durable goods, consumer items, or machine tools. Operating in a region that had poorly integrated capital mobilization, New

Orleans commercial spokesmen subscribed to Henry W. Grady's recipe for southern industrial growth that coupled "southern labor and resources with northern capital and know-how."[9]

Further, like many southern cities (except Richmond or Petersburg, Virginia) which were "handicapped by a dearth of cheap energy sources," New Orleans needed a reliable source of relatively inexpensive fuel. Discovering and developing oil fields on the Gulf Coast could satisfy this need. As early as 1888, New Orleans oil companies, including the Petroleum Prospecting Company, drilled wells in and around Oil Springs, Texas, fourteen miles southeast of the town of Nacogdoches.[10]

It should also be noted that the same year the Louisiana Electric Light and Power Company under General Manager Morris (Maurice) J. Hart fired its boilers with petroleum supplied by SOC. Stored in the facility's two large, underground tanks, the fuel came to New Orleans by railcars that shuttled back and forth on the plant's switch track, making it unnecessary to move coal from barges in wheelbarrows and carry off ashes in the same manner. The oil came from distant sources, requiring long delivery times and uncertain arrivals, but at the time did not elicit any safety concerns.[11]

In common, however, with Mobile and Houston, the 1901 Spindletop discovery set in motion new business schemes to find and market oil, placing squarely before New Orleans officials the need to formulate public policy on fuel oil, a task that divided the City Council over not only petroleum risks but also the legal form of ordinances giving companies authority to store and distribute fuel oil within city limits. Public figures in New Orleans crafted workable risk management policies on the issue of storing and distributing fuel oil in three sequential stages, corresponding to major ordinances which respectively established technical standards for small oil tanks and locations for large storage tanks, 805 N.C.S. (New Council Series); rules for obtaining a permit to set up oil pipelines in the city, 918 N.C.S.; and a revised policy on storing oil under sidewalks, 1159 N.C.S.

Devising a "Tank" Ordinance: 805 N.C.S.

Entrepreneurial interest in possible oil reservoirs in West Louisiana and East Texas, a curiosity which the Lucas well transformed into a commercial and economic obsession, made uncertain New Orleans's regulatory power to control the transmission, storage, and consumption of fuel oil. Officials resolved this problem by passing Ordinance 805 N.C.S. in August

1901, commonly called the *tank ordinance* because the City Council charged a special three-man committee in late June 1901 to investigate oil storage within city limits and to "design a suitable and standard tank for such oils." Companies wanted to burn cheap Beaumont oil under boilers fed by small on-site supply tanks under sidewalks. To issue permits for these tank installations the city needed uniform technical standards with which companies would have to comply. The special oil committee carried out its mandate in this respect, employing specifications already in place in cities accustomed to using petroleum fuel, but under risk-management pressures it also had to create zones for constructing large oil tanks in the range of 37,500 to 50,000 barrels to keep the small tanks filled.[12]

One firm with a deep interest in resolving fuel-oil issues was the Lone Star & Crescent Oil Company (LSCOC). It evolved from a partnership that owned land near the Spindletop well into a company financed by Beaumont and New Orleans capital and headed by beer brewer E. G. Schlieder (1853–1948). LSCOC had ambitious plans to transport Texas oil to New Orleans for industrial and domestic consumption by barge or tanker to supply commercial, processing, and manufacturing firms. By early August 1901 the company envisioned a proprietary pipeline network, fed from river vessels, that would send oil across and through wharves, levees, and city streets.[13]

This strategy was rife with potential legal complications. By charter, New Orleans had absolute control over its streets. Any construction, such as a pumping station, required a permit from the city engineer. Under New Orleans's police powers, any property owner, individuals or companies, that constructed, reconstructed, or repaired buildings could proceed only with and under city permission and supervision. And businesses were familiar with asking the City Council for ordinance permission to run pipes to or from the river to dump sewage, gain a supply of water for boilers or processing, or release water used in manufacturing. Whether the city engineer could issue permits for oil plants was not clear.[14]

J. M. Guffey Company, owner of the Lucas well, was an early entrant into the New Orleans oil market. In March 1901 Guffey applied to the directors of the New Basin Canal "for space to erect a tank for the storage of fuel oil." The purpose was to transport oil to "various brick kilns, lumber mills and plants of various kinds on the lake and rivers in the vicinity."[15] In May 1901 J. M. Guffey reorganized the company and consolidated its holdings in a Texas corporation capitalized at fifteen million dollars, making it "one of the largest oil companies in the world." In addition to owning five highly productive wells at Beaumont, a pipeline from Spindletop to Port Arthur, railroad tank cars, car-loading racks, and 300,000 barrels of iron tankage with another 700,000 planned or under construction, J. M. Guffey Petroleum Company controlled one million acres of leased land in Louisiana and East Texas. Implementing a broad strategy based on producing and transporting petroleum, the company intended to market bulk oil very aggressively in the Gulf South, the Caribbean (especially Cuba), coastwise to Philadelphia and New York, and to Great Britain and Europe.[16]

Within this scheme, New Orleans was a major retail outlet, a distribution point for inland sales, and a way station for coastal and transatlantic petroleum traffic. The New Basin Canal was an important commercial route for goods coming to New Orleans and a strategic location for servicing the fuel needs of small- and medium-sized industrial enterprises and commodity processing facilities on and about Lake Pontchartrain, canals, and the swampy, vacant lands north of the city's business district.[17]

Higgins Oil & Fuel Company also drew up a program to bring oil to New Orleans. By May 1901 the company operated two producing wells at Beaumont and had under construction four large oil barges, the purpose of which was to take crude oil, piped from the Spindletop field to the Neches River, through Sabine Pass, and out into the Gulf for passage to New Orleans and other markets. Higgins intended to form a distribution company in Louisiana, secure riverfront property and wharfage, and build storage tanks connected to several miles of pipelines to move oil to "large customers." Phil R. Rice, company agent in New Orleans, acknowledged that this all required City Council permission but did not anticipate any objections. "Fuel oil," he observed, "is bound to take the place of coal in the south," and the city will have to accommodate a demand for it.[18]

Once it became clear that the Beaumont discoveries were not a fluke and that well-financed companies such as J. M. Guffey were investing money in transportation and distribution, New Orleans businesses stepped forward to get permission from the city to set up oil-burning equipment. In May and June 1901 several companies, including Louisiana Distillery Company,[19] Jackson Brewery,[20] and Crescent City Ice Company,[21] as well as the American Brewing Company (ABC)[22] and C. H. Lawrence in July,[23] petitioned the City Council for permission to install oil-storage tanks under adjoining sidewalks or on their property.[24] The council recommitted or held over these ordinances in committees, postponing indefinitely Louisiana Distillery, Jackson Brewery, and ABC's superfluous requests in early September.[25]

To clear up legal and policy uncertainties, Councilman A. T. Moss on June 25 delayed a vote on such measures and successfully moved to create a special committee to "investigate the conditions incident to the storage of oil within the city limits" and to produce "in consultation with the city engineer" an ordinance embodying its findings that described a suitable "tank for the containing of the oils." His hazard-management strategy summarized the council's institutional responsibility to craft suitable and effective policies, which reflected an inescapable obligation arising from community leadership to mitigate fire hazards and reduce property losses. The action also exemplified how the convergence of a new energy market, risk management of the built environment, and municipal police powers produced public policy.[26]

The City Council's first special committee on oil storage invited oil-tank petitioners to attend a July 27 meeting. It was a significant event because fire-insurance attendees understood the oil problem in a framework of hazards, location, and property risks that foreshadowed late-twentieth-century deliberations on hazard mitigation and economic development. At odds with major oil distributors over the consequences of using a potentially dangerous but valuable energy source, fire-insurance representatives at this gathering set in motion a risk-management agenda for New Orleans to regulate fuel-oil supplies that satisfied the insurance industry's guidelines for safe storage.[27]

Risk management was clearly on the mind of committee chairman A. T. Moss, who announced that the objective of the discussion was to devise a "way for the storage of oils for fuel purposes in the city limits without endangering property." John G. Pepper (1864–1934), secretary of the New Orleans Board of Fire Underwriters, assistant manager of the Southern Department of Liverpool and London and Globe Insurance Company (L&L&G), a major underwriter of fire insurance in the United States, went straight to the central issue. He asked the committee to "not allow the oil barges to land in front of the city or to allow the storage of oil on the river front." If companies got permission to install storage tanks on their property or near the levee, distributors would draw oil from barges or steamers at city wharves or "in front of the city" to supply them. Fires at any landing point above the city, Pepper warned, would be catastrophic, sending burning oil downriver to destroy steamboats and wharves.[28]

At a second meeting on July 30, Moss's special oil-tank committee presented a draft ordinance to fire-insurance and business interests and then introduced it to the City Council as a measure "regulating the storage" of "fuel and illuminating oils."[29] In scope and intent it was the precursor of much-amended Ordinance 805 N.C.S. The draft limited large-scale oil storage to a back-of-the-town region of small settlements, sparsely inhabited marshlands, swamps, and pine-tree thickets, enclosed by South Claiborne, Broad, Calliope, and Napoleon Avenue.[30]

The supply-tank requirements, drawn from rules "prescribed by all insurance companies" (including L&L&G), were in use well before ordinance deliberations in New Orleans were underway, illustrating how local insurance men took advantage of risk knowledge available to them from eastern companies and technical associations.[31] Companies had to apply to the city engineer for a permit to install "individual oil tanks" accompanied by plans in compliance with fabrication and installation standards. Tanks made of iron at least three-sixteenths of an inch thick had to be set underground lower than the ignition point of the oil, inside a masonry vault (unless installed fifty feet or more from the ignition point) and outside all buildings. The ordinance also stipulated that pumping oil from rail tank cars, or ships or barges moored on the river to "general storage tanks" and distributing it throughout the city had to take place with the "greatest care" for public safety.[32]

Solicitude for public safety was an obligation that fell far short of an absolute ban on moving oil across city wharves that insurance spokesman Pepper had demanded. Uncertainty over what constituted proper care opened the door for sharp debates on mitigation and risk assessment of oil hazards, especially in light of ambitious plans to enter the New Orleans energy market. Pepper complained to Moss's special committee that J. M. Guffey Petroleum was "erecting oil tanks at Gretna" and should be "compelled to pipe their oil below the city" before distributing it to manufacturers. The city, Pepper argued, should not allow "oil-carrying ships or barges" to land inside city limits and discharge oil.[33]

Pepper had grounds for concern. Guffey's vice-president, Jesse McDowell, outlined his company's plans to distribute fuel oil "in the manufacturing districts of the city" to the Moss committee on July 30. He proposed a pipeline from Clouet Street several miles upriver to Lane Cotton Mills, along the path of Front, South Peters and Tchoupitoulas streets, with "short laterals to the various consumers on the route." Moored barges would feed this system, instead of shunting railroad tank cars in and out of the city, to avoid the "accumulation of oil" on the East Bank.[34] A month earlier, McDowell had acquired a franchise for J. M. Guffey to "lay, establish, conduct, operate . . . a pipeline" in Jefferson Parish.[35] The arrangement included an oil distribution center and a 37,500-barrel tank to store oil for city and upriver markets. He did not, however, tell the special committee that New Orleans restrictions, noted by Texas Company founder Joseph Cullinan, on large oil tank sites did not apply to Jefferson Parish and that taxes on the West Bank were lower than in the Crescent City.[36]

But this paled in comparison to the firestorm generated by a pipeline request from LSCOC that through successive amendments became intertwined with the evolution of the special committee's draft ordinance culminating in Ordinance 805 N.C.S. LSCOC viewed New Orleans as a prime market for fuel oil and a distribution point for the lower Mississippi River Valley. It asked the City Council for permission to store oil in the neighborhood around the intersection of Roman and Erato streets (inside the permissible area specified in the special committee's draft oil ordinance). The company wanted to pump oil through a pipeline from the river at Julia Street, where railcars or barges could bring oil, and to construct another pipeline on the "most direct route" from St. Louis Street to Jackson Avenue with branches leading off to various industrial consumers along the way.[37]

However, the most provocative aspect of LSCOC's proposed ordinance was its legal form (to be distinguished from so-called insurance "forms" that distilled specifications for an insurable activity). Opponents of this "pipe grant" viewed it a franchise request, an unfavorable development for LSCOC since franchise acts had for decades precipitated political squabbling and council fighting. And at least two—telephone service[38] and Illinois Central Railroad's wharf construction—were currently in dispute, not counting the drawn-out battles over sewerage and drainage and the water works.[39] The core questions were: did LSCOC's request to run pipelines under city streets provide the company with the exclusive right or privilege to engage in an activity normally denied to individual citizens (franchise), or was it merely an appeal for a license or a permit to perform an activity under municipal supervision?[40]

Squabbling over the legal nature of the LSCOC pipeline ordinance held up the company's participation in the city's energy market and made passing a general pipeline ordinance more compelling as the controversy intensified. Deliberations underscored deep opposition to monopolies and exclusive privileges, reflecting state policy as well as a burgeoning national debate over municipal utility franchises that continued into the 1920s.

A controversy about the Badger Gas Ordinance played a role in the LSCOC dispute. A. S. Badger wanted to operate a fuel-illuminating gas plant in New Orleans and pipe its products through city streets. Opponents marshaled familiar legal caveats: the ordinance was a franchise, not a mere grant or permit, to be auctioned off in public; as a franchise, it was, unless otherwise specifically excluded, a monopoly not in the best interests of the city; and its tenure—for only twenty-five, not fifty, years—revocable for failure to perform the conditions. In June 1901, supporters withdrew the request in the face of stiff opposition.[41]

Councilmen noticed its similarities to LSCOC's request. They concluded that, if the Badger measure was truly a franchise, so was the LSCOC request. As one councilman reasoned, the city was going to allow LSCOC to "tear up and occupy the public thoroughfares" and use them for the "company's convenience and profit." Since Louisiana law required that all "street-railroad" franchises and all public property for sale or lease to be put up for public auction, critics reasoned that the privileges LSCOC wanted should not be a gift.[42]

When the Streets and Landings Committee (S&LC) met August 2, it tried to pacify these critics by reporting the LSCOC pipeline ordinance favorably only after City Attorney Samuel L. Gilmore attested that "it was not a franchise to be advertised and sold" in the "sense implied by the city charter." Gilmore pointed out that LSCOC had merely asked to use city streets, and anyone else could legitimately apply for the same privilege.[43]

As for the special committee's general oil-storage ordinance, S&LC and its Fire and Lighting (F&LC) counterpart jointly amended it to satisfy fire underwriters and the Maritime Association. The committees banned indiscriminate movement of oil across city wharves and stipulated that no oil barge could land anywhere on the riverfront above Egania Street, a Third District thoroughfare that ended at the river roughly across from the lower limit of the naval station. The Egania area was downriver from the business district and sparsely inhabited by poor African Americans as well as Irish and German immigrants. The amendment provided space for oil companies to land barges near the foot of Egania and build very large storage tanks for distribution through a pipeline system.[44]

When the City Council met on August 6 it took up the altered general oil ordinance and the LSCOC pipeline ordinance, rewritten to conform to S&LC/F&L's new restrictions on off-loading oil from the river. LSCOC also asked for permission to run a pipeline from the lower parish line along the river all the way to Peters Avenue, a few blocks short of Nashville, to supply contiguous industrial users.[45] As events subsequently demonstrated, this convinced opponents that LSCOC wanted a monopoly on using city streets to transmit oil, and unless the City Council could assuage their fears, they would unceasingly challenge LSCOC petitions.[46] In the end the City Council approved the LSCOC request.[47]

The City Council approved the general oil ordinance, recorded on August 8, 1901, as Ordinance 805 N.C.S. It established the risk-management groundwork for handling fuel oil inside New Orleans, making space and location instruments of hazard mitigation. Section 1 of the five-section act set the boundaries in the rear of the city where one could store large amounts of "fuel and illuminating oil" and added the Egania Street–St. Bernard Parish line–Burgundy–Florida Walk rectangle. Section 3 specified the technical details for small oil-supply tanks used in businesses to fuel boilers. These detailed conditions matched closely those in the so-called "Texas" and "Chicago" forms already in use in the East and Midwest, as well as rules adopted in 1892 by the Chicago Fire Underwriters' Association. This enabled New Orleans to institute permit requirements known to satisfy fire-insurance companies and founded upon a city ordinance. Section 4 prohibited supplying large oil-storage tanks from "railroad cars parked along the river front" or from "ships or barges moored along the river front" starting at the Jefferson–Orleans Parish line downriver to Egania Street, and downriver on the right bank to the "lower limits of the U.S. Naval Reservation."[48]

Ordinance 805 N.C.S. was a milestone in the controversy over managing New Orleans fuel-oil risks, because it identified petroleum as a mitigable hazard and anchored risk management of it in legal contrivances—space and location—that were contingent on economic, social, and political processes. Its purpose was risk management—policies to reduce the likelihood of a catastrophic event through technical and spatial restrictions on oil storage and consumption.

Phase Two: In Search of a Pipeline Ordinance

Ordinance 805 N.C.S. settled the issue of tank storage, but not distribution. Subsequent controversies in the City Council over LSCOC's pipeline ordinance, engineered by a tenacious anti-monopoly faction, pointed to the need to normalize risk management of distribution procedures. Mayor Paul Capdevielle twice vetoed the ordinance. He thought it was an illegal franchise request indistinguishable from the rejected Badger fuel-gas ordinance, and that the state constitution required "adequate compensation" for this kind of grant. Subsequently, LSCOC agreed to pay a one-half-cent-per-barrel tariff on oil pumped through city streets, start work in sixty days, and maintain streets "dug up for pipes."[49] LSCOC tried to placate the mayor through amendments that included stipulating that its pipeline privilege was "revocable at the pleasure of the council."[50]

LSCOC's ordinance campaign elicited serious charges such as collusion with City Council members to gain special favors, one of which was an ill-conceived pipeline ordinance that gave LSCOC a lucrative and exclusive franchise that was not in the city's best interest.[51] Set in the context of numerous contested grants and franchises in New Orleans, and a rising national tide of anti-monopoly and antitrust sentiments, these accusations reflected frustration over perceived market inequities and an inexplicable, chronic failure by city government to manage risk in the best interest of New Orleans.[52]

Two politically sensitive matters involving LSCOC upset Mayor Capdevielle and played into the hands of company critics. One was the purchase of the Auxiliary Sanitation Association (ASA), a nonprofit corporation founded in 1879 that had constructed two pumping stations respectively at the foot of Nuns and Toulouse streets to service its five-mile pipeline along the riverfront from Louisiana Avenue to Lafayette Avenue. LSCOC agreed to buy this property in early September 1901 for eight thousand dollars.[53] The *Daily Picayune* condemned the purchase because it gave LSCOC "possession of a vast pumping and oil plant on the river front, with the privileges of pipes in any and all the streets at its pleasure." The mayor was so agitated that he ordered the city attorney to get a district judge to block the sale and give the city a chance to intervene based on a drawn-out legal battle over the liquidation of ASA property.[54]

A second issue was the City Council's treatment of a rival proposal from two New Orleans businessmen—a wholesale drug dealer, Lucien N.

Brunswig, and A. A. Maginnis, owner of Maginnis Cotton Mills. Acting as lead investors in behalf of the Sour Lake and Franco-American Petroleum Company, a closed corporation made up of New Orleans investors, Maginnis and Brunswig sought a pipeline ordinance for company operations. In an interview with the NODP Brunswig demanded "equal rights with any other corporation" to pipeline rights-of-way. Franco-American owned land outside the city with eight hundred feet of river frontage, where the company planned to build oil storage tanks and "pipe . . . oil into the city."[55] Yet, one S&LC member in a late August meeting insisted that the Maginnis-Brunswig effort was incomplete (hereafter M-B). M-B did not specify where their company planned to build storage tanks. Skeptics later said they would not approve the M-B request until Franco-American had oil and pipes on hand like LSCOC had.[56]

Despite this foot-dragging, S&LC reported the M-B ordinance favorably to the City Council on September 3. While the City Council approved LSCOC's amended request, the M-B ordinance immediately ran into allegations that its investors had been disingenuous. Their real strategy was to rent out the sought-after pipeline. In such a case, Councilman Dickson reasoned, "the permit should be refused and the privilege sold as a franchise." This was an ironic argument, since anti-monopolists had pounded LSCOC for allegedly wanting an exclusive franchise to distribute oil, and now a second company, ostensibly needed because "the people were entitled to . . . competition," foundered on the same legal rock. On vote the council decided to reconsider the proposal without final passage.[57]

Advancing equal treatment of pipeline applicants had become a major sticking point for LSCOC opponents. Editorials in the *Daily Picayune* were first optimistic and then angry. The M-B and LSCOC ordinances were franchises—making the issue of monopoly irrelevant—that should be approved with "necessary safeguards." But when the council did not pass the M-B request, the paper denounced this maneuver, claiming that, by design, it created a "monopoly in favor of a particular party," echoing earlier complaints about favoritism.[58] Mayor Capdevielle was also concerned about fair treatment of the Franco-American investors. He would not commit to signing the LSCOC ordinance unless the Franco-America corporation and "all other oil companies who are to benefit the city . . . have an equal privilege of putting in pipelines."[59]

For the second time, Mayor Capdevielle in early September vetoed an amended LSCOC ordinance, after which Councilman O'Connor, who wrote the original ordinance, introduced a ten-section version that addressed all objections raised to date, including project commencement and termination, and indemnifying the city against claims arising from carrying out the LSCOC pipeline privilege. The *Daily Picayune* thought it had been "amended to make provision for the public safety" to such an extent that it might "go promptly through the Council." But the paper remained unflinching over it being, nonetheless, a franchise to be disposed of by "sale to the highest and best bidder." Councilman Schabel was equally resolute, arguing that the council should treat LSCOC's request the same way it dealt with the Badger gas ordinance. Both were franchises to be advertised and sold. Despite predictions of passage, the City Council at O'Connor's request laid the ordinance over.[60]

By mid-October it was evident that the City Council had reached an impasse over pipeline requests that could only be broken by another general ordinance governing all fuel-oil pipeline activities in the city. The City Council repeatedly laid over, recommitted, or demanded amendments to LSCOC's pipeline ordinance, suggesting that, instead of ordinances granting individual company privileges, the appropriate risk-management solution was another general ordinance that rationalized permit applications for fuel-oil pipelines.[61]

In the face of this political-regulatory reality, LSCOC asked on October 15 for permission to "erect a 50,000-barrel tank for fuel oil," which meant under the terms of Ordinance 805 N.C.S. construction in the Egania Street area (approved October 22).[62] Councilman Schabel submitted a general pipeline ordinance "allowing the construction, maintenance and operation of pipe lines in the streets and public places of the city for the distribution of fuel oil" that reflected concerns of and objections made by fire underwriters, the city engineer, and "those who have followed the business of installing pipe lines." It also by default made LSCOC's long-sought pipeline ordinance redundant.[63]

The City Council passed Schabel's general ordinance as 918 N.C.S. on October 22. Its contents underlined the interactions among the LSCOC pipe ordinance and Ordinances 805 and 918 N.C.S., the last two offering regulatory uniformity as a legal barricade against LSCOC's alleged flaws of exclusivity and monopoly.[64] City Engineer Hardee wanted uniformity and fairness to be in place when oil pipeline privileges were awarded. Applicants would, in his view, "be governed by the general ordinance, and therefore require nothing but a permit."[65]

Further, Ordinance 918 N.C.S. used space and location as fundamental parameters for managing the distribution of fuel oil. Section 1 limited its legal jurisdiction to the "streets of the City of New Orleans." It also restricted "main or trunk pipelines" to "the following *spaces*" (emphasis mine) which were carefully defined by a list of various streets and "the Port Commission's territory" along the riverfront. It allowed for pipeline connections to storage tanks erected in "that section of the City of New Orleans" prescribed by Ordinance 805 N.C.S. The tanks could supply or receive fuel oil from pipes starting at the Mississippi River between Egania Street and the lower parish line. The law also required the city engineer to approve lateral lines which could not extend more than one thousand feet away from the main pipe "in any direction."[66]

New Orleans's turbulent history of granting franchises and privileges to street railway, railroads, street-paving contractors, waterworks, and sewerage/drainage companies left the city mired in a morass of expensive litigation.[67] It was the legal context of LSCOC's struggle to obtain storage and pipeline distribution of fuel oil. Risk management of petroleum drove the long fight over its fuel-oil requests which, coupled with confrontations with anti-monopoly forces possessed by deep-seated fears of municipal connivance with moneyed interests and an equally serious commitment to competition, illustrated the intersection of energy, urban space, risk, and public policy, the latter manifested in protracted efforts to distinguish between the two legal forms of privilege and franchise. Deeply enmeshed in a historical context redolent of nineteenth-century republicanism and fanatical opposition to corruption and privilege, LSCOC's ordinance requests and the passage of Ordinance 918 N.C.S. closed out the second phase of the city's efforts from August to October 1901 to regulate the distribution of fuel.[68]

LSCOC, however, did not walk away from the City Council empty-handed. The mayor not only signed the Schabel ordinance (no. 918), but also he endorsed Ordinance 959 N.C.S., which allowed Lone Star to build a fifty-thousand-barrel oil-storage tank on a square bound by Villere, Urquhart, Flood, and Andry streets, as well as lay down a pipeline from the tank along two routes that included Egania Street. The line tied into the company's pumping station at the corner of Egania and North Peters streets next to the river (see map 3), from which LSCOC could pump oil to various

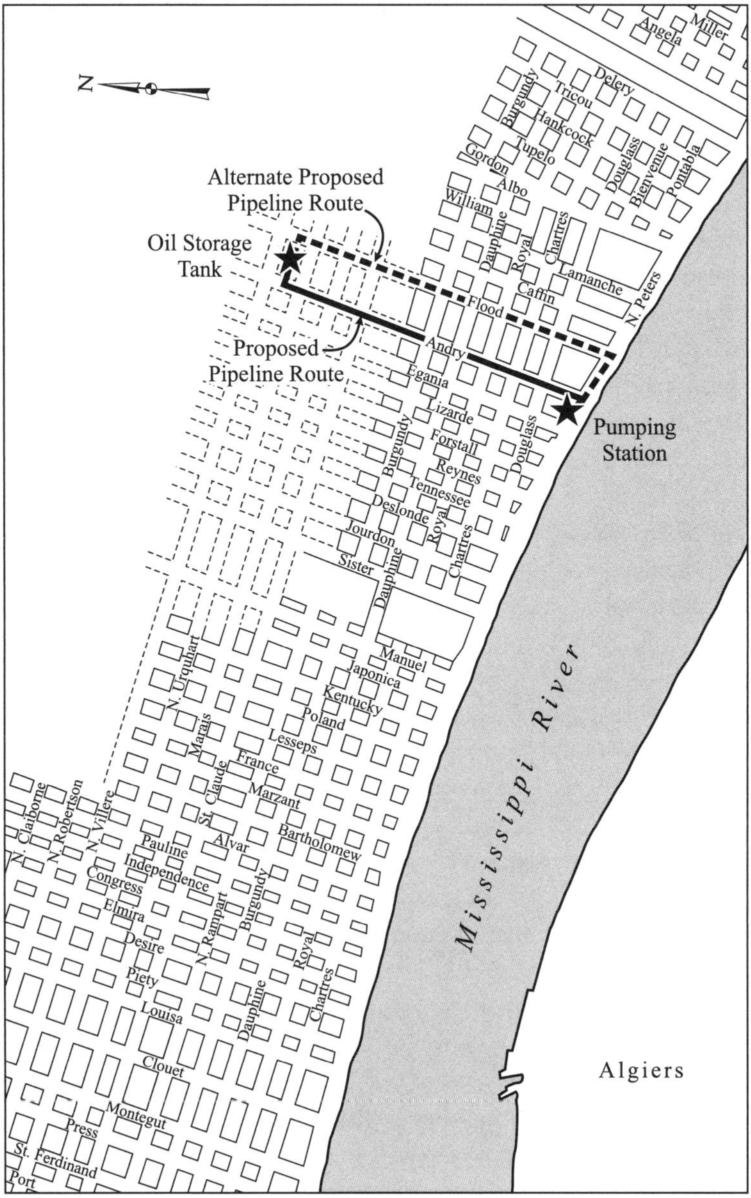

Map 3. New Orleans: Lone Star & Crescent Oil Co.'s 50,000-barrel oil-storage tank; two possible pipeline routes from the tank to company pumping station at the foot of Egania Street, permission for which was granted in October 1901 in Ordinance 959 N.C.S.

places in the city as well as off-load crude petroleum from barges and other vessels moored to a wharf directly across North Peters under provisions in Ordinances 805 and 918 N.C.S.[69]

In addition, Lone Star subsequently was the beneficiary of the city's political generosity on at least three more occasions, obtaining permission December 5, 1901, through Ordinance 1011 N.C.S. to put down a switch track on St. Claude Street to get tank-car access to the Louisiana Southern Railroad, as well as a pipeline from its storage tank to loading appliances for the required tank cars, all contingent upon Lone Star depositing a five-hundred-dollar bond, gaining the city engineer's approval for plans and specifications, and conducting the work so that it did not obstruct city streets.[70]

In March 1902 the city approved of a pipeline connection from LSCOC's main pipeline along the river to the Maginnis Cotton Mill in Ordinance 1168 N.C.S.; another pipeline near Tchoupitoulas as described in Ordinance 1204 N.C.S., April 1902, to various "steam institutions" or businesses powered by boilers; and finally in September 1902, Ordinance 1392 N.C.S. to construct a pipeline to the St. Charles Hotel.[71]

PART II

Phase Three: Making the Sidewalks Safe

The third phase of the controversy over storing and distributing fuel in New Orleans lasted from November 1901 to March 1902. In early July 1901, at the genesis of the debates that produced Ordinance 805 N.C.S., requests came before the City Council from manufacturers for permission to "place oil tanks on their properties or on the sidewalks adjoining their factories." Five months later, similar petitions to the City Council and Mayor Capdevielle set off a new round of familiar claims and counterclaims about the flammability of Beaumont oil stored under sidewalks or in enclosed oil tanks set close to buildings as well as the risks associated with pumping oil from vessels into company storage tanks.[72] Implicit in these requests was the need to overturn Ordinances 805 and 918 or to revise crucial aspects of their risk management schemes.

"[V]arious consumers of fuel oil" petitioned the City Council and Mayor Capdevielle in late October 1901 to amend Ordinance 805 to permit oil storage tanks under sidewalks. The petitioners claimed they could not comply with current insurance regulations about the distance from tank to ignition point because their property was often "covered with buildings." They argued that they should be allowed to "erect structures for storage purposes under the sidewalk abutting [our] property."[73]

Meeting on November 4, the F&LC ignored insurance representatives and favorably reported an amending ordinance that allowed fuel-oil storage under sidewalks or in "brick vaults abutting the property." A *Daily Picayune* editorial adopted the perspective of risk assessment, namely identifying hazards, determining risk levels, and instituting mitigation. Employing the rhetoric of hazard mitigation, it urged the city to leave Ordinance 805 intact because storing oil under sidewalks posed an "extreme danger from fire" accompanied by greatly increased insurance costs. Beaumont oil, the paper warned, contained "volatile constituents" such as highly flammable benzine and gasoline. The paper proclaimed that sidewalk storage was a "danger . . . too great to be risked."[74]

Determining tolerable risk levels was the task of the City Council, though it did not have credible statistical information about petroleum fires or explosions; and the politics of risk management required compromise. When the council met November 12, Councilmen Dickson and Zacharie persuaded it to recommit the amending ordinance. Dickson complained that the "insurance people" had not been heard, which explained why Zacharie had received a protest from them signed by "nearly every fire insurance firm in the city."[75]

As a prelude to political quarreling over subsurface storage, two experienced engineers offered disparate risk assessments that centered on hazard identification. Professor C. F. Z. Caracristi was an opinionated, well-connected, at times troublesome, engineer and minerals consultant, whom many observers considered an authority on Texas and Louisiana oil fields. He argued that granting "permits for underground tanks" was a "menace to property." Beaumont oil had a "small percentage of volatile content" and was "extremely liable to explosion when stored in tanks underground" from electrical sources such as "street car tracks."[76]

Clayton Billow disagreed. He designed the storage vault and pumping station that supplied oil for the boiler plant at the Chicago's World Fair in 1893, and reputedly wrote Chicago's "fire insurance regulations and . . . city ordinance" for oil-burning appliances. Billow believed New Orleans should acknowledge that burning oil was an admissible risk and could be done safely, a position that illustrated risk assessment in which "safety judgments" rested upon "measuring the risk and deciding whether it is acceptable." Since Louisiana oil was not very volatile, Billow explained, the gases in a partially filled oil tank under the sidewalk had to be mixed with oxygen to explode, a fact that proved wrong Caracristi's belief that a streetcar spark could set off an explosion.[77]

Divergent perceptions of risks associated with subsurface oil storage came to the front the same evening, November 25, when the F&LC took up the ordinance devised by Councilman Moss (chairman of the special oil committee that drew up what became Ordinance 805 N.C.S.) that granted "permission to certain steam users to have fuel oil tanks under sidewalks." Much of the published testimony came from W. A. Gordon, owner of an alcohol plant, who sought an alternative to expensive coal but was unable to move to an outlying area that allowed oil-tank placement the distance from the ignition point required by Ordinance 805 N.C.S. His answer was to place a twelve-thousand-gallon oil tank "under the sidewalk." Councilman Schabel countered that, if others demanded the same privileges, the "whole business section will be honeycombed" with oil reservoirs that will be a "constant menace to lives and property."[78]

In response to Gordon, insurance spokesman John Pepper compared the risk-management strategies of local insurance men to their counterparts in major cities. Chicago did not allow an oil tank within fifty feet of a business-district building;[79] New York, Boston, and Pittsburgh did not allow oil tanks under sidewalks.[80] Pepper warned that sidewalk storage presented a constant danger of explosion owing to gas accumulation, a menace that would increase if all businesses on a block had such systems. Addressing Schabel's point about crowded tank placements, Pepper noted that the insurance interests had conceded to oil users the installation of underground tanks, if placed on vacant lots in sparsely populated areas at least fifty feet from property on nearby blocks. From such sites pipelines could safely convey oil to multiple consumers. However, underground tanks could not be placed on blocks tightly packed with businesses, even if they were fifty feet from adjacent structures. The F&LC reported unfavorably the amending ordinance, which the City Council recommitted, November 26, to allow the Board of Trade (hereafter BOT) to investigate fuel oil.[81]

BOT president Udolpho Wolfe launched the probe in response to queries about New Orleans fuel costs, naming SOC special-agent Pendleton S. Morris, hotel executive A. R. Blakely, engineer John H. Murphy (Murphy Iron Works), James W. Porch, Progressive Union member, and machinery manufacturer Ernest M. Loeb to the Special Committee on Cheap Fuel (SCCF). He directed them to examine fuel-oil prices, study the laws of various cities governing fuel-oil storage, and determine how these factors affected insurance prices, an investigation that dominated the fuel-oil controversy from late November 1901 through February 1902. Acting in behalf of influential merchants and commercial men, the SCCF gave coherence and professional credibility to oil consumers' opposition to strict oil regulations.[82]

At the committee's first session, December 19, Clarence F. Low, manager of the Southern division of L&L&G; Albert P. Noll, president of Teutonia Insurance Company; Lucien A. Monrose, well-known insurance figure; and William A. Brand represented local fire underwriters who had recently organized an association as a "means of concerted action" in oil matters that "bear upon the fire hazard of New Orleans." Emphasizing cooperation and consensus in pursuit of risk management, Chairman James W. Porch asked them what restrictions the "insurance interests" wanted on the use of fuel oil. He invoked a well-established formula that linked the city's economic prospects to petroleum, suggesting that if attendants could find "common ground" this would allow the city to take prompt advantage of a "great supply of cheap fuel" and become a "manufacturing center."[83]

Speaking from an outlook that connected risk management with tank locations, insurance spokesman Clarence F. Low applauded conciliation, but made it clear that the chief issue for insurance officials was hazard mitigation: namely, "minimizing the dangers attendant upon the use of fuel oil generally in the [heart] of the business center." To reduce municipal risk, the underwriters had confined "oil depots below the city," blocked indiscriminate handling of fuel oil on the riverfront, and rejected hazardous storage of fuel oil under sidewalks in the business districts.[84]

Two days later, the BOT's special fuel committee held a second meeting to hear from oil producers, distributors, agents, and engineers about "the best methods of distributing and using fuel oil." They emphasized the feasibility of various distribution systems and were certain that proper tank design guaranteed safe oil storage inside the city. For example, F. M. Bullock, sales agent for J. M. Guffey Petroleum Company in Beaumont, declared that pipelines running "to stated points along the river front" or barges moored along the river were the safest and cheapest ways to supply manufacturing plants equipped with storage tanks. Experience has shown oil men, Bullock argued, "that its [oil's] use may be so safeguarded as to make it nearly impossible for an accident to occur *as it is to guard against any danger that may arise in the ordinary course of human events*" (emphasis mine).[85]

C. L. Wallis of Higgins Oil and Fuel Company underscored the importance of accurate knowledge of the properties of fuel oil—flash and ignition points—and of reliance on informed experience in assessing risks associated with its use. Oil could be safely distributed in New Orleans through general pipelines, tank cars, or by barges with pipeline connections to a consumer's underground storage tank. If the tank had been fabricated with heavy-gauge steel and proper ventilation, Wallis contended, "there is no more danger from oil than from coal." Mechanical engineer J. D. Hackstaff of the Rockwell Engineering Company cited extensive use of oil on New York City's waterfront. Some plants received oil from barges at the docks for storage in tanks inside their compounds. Hackstaff was confident that New Orleans manufacturers could safely receive fuel oil across levies or through a "common carrier" pipeline.[86]

During the SCCF's third public session, December 27, representatives of companies who consumed very large amounts of oil outlined their fuel concerns. Principal speaker Sigmund Odenheimer was a trained engineer, civic leader, and longtime manager-owner of Lane Cotton Mills. He had installed two underground-supply tanks with no vault or brick work about ninety feet from the plant boilers. Since November 1, the textile facility had burned fuel oil, relying on railroad deliveries to replenish the supply tanks. Odenheimer favored barges to supply manufacturers on the river, while industries "in the rear of the city" would get oil through pipelines tied to "central storage tanks." As for the risks, he observed that "there was danger in everything" which summed up the hazard-risk perceptions of oil distributors and consumers. Since risk was ubiquitous, one should manage operating hazards and seize the opportunity to enjoy the benefits fuel oil. Odenheimer concluded that the city should repeal ordinances that prohibited "tank steamers from landing in front of the city" or tank installations "within fifty feet of a building."[87]

Lewis H. Ashbey of the Armour Packing Company was as blunt as Odenheimer was diplomatic. Ashbey's employer wanted to use oil in a new plant, but neither the city nor the insurance interests would consent to it. He accused the underwriters of being "cautious through ignorance." That is, their habit of falling back on the precautionary principle rested not on logic or facts but on erroneous beliefs about oil safety, a stringent assessment out of step with claims that the sessions sought harmonious discussion and agreement on a fuel ordinance.[88]

Armed with such information and opinions—some of which were incorrect or unreliable—the SCCF met privately three days later with a similar group from the Board of Commissioners of the Port of New Orleans (Dock Board) in an executive session. Since the board controlled handling oil across wharves beyond the reach of the City Council, it needed to know what the BOT had uncovered about the "water transportation and delivery of oil," illustrating the importance of agency cooperation under circumstances of overlapping jurisdictions.[89]

The Dock Board in turn held a hearing a week later (January 1902) on handling oil across city wharves, a meeting which the SCCF joined, led by Chairman Porch, who squared off with his insurance nemesis, J. G. Pepper. Porch insisted that New Orleans could become a great manufacturing center if its leaders grasped the opportunity presented by fuel oil, and that would only happen if all parties acted in harmony. Porch contended that barges, moored on the riverfront, were the best means to bring oil to the city, making it imperative to "know what conditions the dock board would impose" on handling it. Pepper replied that the insurance interests did not oppose using fuel oil in New Orleans, but they were against "promiscuous landing of oil barges in front of the city" because of the enormous fire risk to the harbor. Nor would they agree to storing oil under sidewalks or streets. He implored the Dock Board to disallow landing "oil barges at the city front or above the city."[90]

Bringing oil in barges to manufacturers led logically to the issue of pumping oil through proprietary pipelines from the river to on-premises tanks, which reopened debate on an issue presumably settled definitively in Ordinance 805 N.C.S. This emerged during the SCCF sessions, as oil consumers—held back, as Clarence Low had explained in the first session, from handling oil on the riverfront—reaffirmed the advantages of water transportation to supply their plants. For example, the New Orleans Brewing Company petitioned the City Council to run a pipeline from the riverfront to its plant on Philip Street near Tchoupitoulas. The S&CL met three times, November 29 (1901), January 3 and 10 (1902), but finally laid the ordinance over for future discussion.[91]

SCCF Chairman Porch subsequently reassured the public that his committee did not advocate increasing petroleum hazards. The insurance interests had become alarmed, he thought, over something that "they are largely guessing at." Porch preferred a "system" that would provide "low-priced oil" through competitive bidding rather than relying on a monopoly to handle fuel oil (an allusion to LSCOC's pipeline ordinance). In other words, the committee wanted a rational and open market in which buyers and sellers operated freely without restraints based upon unfounded safety concerns or lack of competition among producers.[92]

In the wake of this rational, open-market, and anti-monopoly rhetoric, various agencies that had a stake in the wharves along the business district chose to "minimize the chance of worst possible outcome" or to "maximize the chance of the best possible outcome." In practice this meant eliminating oil hazards or dramatically reducing attendant risks.[93] Delegates from the Maritime and Merchants' Exchange, the recently formed New Orleans Fire Underwriters' Association,[94] Board of Marine Underwriters, the Steamboat Exchange, and the Captains' and Owners' Exchange met and adopted a position statement on the campaign to relax current oil regulations. The group opposed BOT calls to moor "oil-laden vessels. . . . above and in front of the city" which, considering the wooden wharves filled with cotton, imperiled town and harbor. Delegates also protested the apparent acquiescence of the Dock Board in the BOT campaign, having reportedly approved five places for one company to handle oil on the riverfront.[95]

In late July 1901, engineer McDowell described J. M. Guffey's plans to construct a riverfront pipeline with lateral connections to manufacturers. In a *Daily Picayune* interview in January 1902 he cited his extensive experiences dating to 1888 with SOC's Chicago pipeline followed by many fuel-oil projects in the East. McDowell thought New Orleans should call upon "practical oilmen" who were thoroughly familiar with fuel oil, reflecting a nineteenth-century belief that men of experience and accomplishment were superior to theoreticians and effete intellectuals, whose knowledge was based on books and speculation.[96] If the oilmen prevailed, McDowell promised that the city would have a "system of protection" and reap the "full advantage" of petroleum discoveries, envisioning, however incompletely, a rational system with a "formal structure" (municipal government) and specific goals (manufacturing and prosperity) ultimately validated pragmatically.[97]

The same evening the City Council, in response to divisive issues raised in the SCCF hearings, appointed its second special committee on fuel oil (the first was Moss's June 1901 tank committee), charging it with devising a new ordinance covering the "use and storage of fuel oil within the limits of the port of New Orleans." Because the oil men wanted to dismantle existing regulations in favor of a more liberal hazard-mitigation regime, Councilman Shields resurrected the protracted oil disputes of August–October 1901, introducing two ordinances, one repealing Ordinance 918 (pipelines) and another amending and reenacting Ordinance 805 (storage and handling of oil). Recognizing the renewal of political conflict over petroleum risks, the *Daily Picayune* reminded readers that reconciling the insurance and oil interests would be difficult but "not beyond" human ingenuity. Participants could rely on experience and "modern science" to agree on the "means of making oil available" without sacrificing the city's safety.[98]

In rebuttal to McDowell and other critics in an appeal to the Dock Board, a committee of the New Orleans Fire Underwriters' Association sent a letter to the *Daily Picayune* that cited storing oil under sidewalks and handling oil from barges across city wharves as significant issues. The insurance men acknowledged that satisfying fuel needs required river transportation, but to avoid a massive, harbor-front conflagration they proposed expanding the zone for handling oil over wharves upriver from Egania Street along undeveloped frontage to Poland Street (near the modern Inner Harbor Navigation Canal). This major compromise reaffirmed the risk-management strategy of Ordinance 805 N.C.S. by retaining a boundary line for large-scale storage and handling of oil, but it was moved upriver without reversing course on the prohibition of harbor-side oil activities.[99]

With arguments and rebuttals mounting on both sides, the City Council's second special committee on fuel oil had, according to Chairman J. S. Zacharie, the task to come up with "reasonable legislation" that would allow cheap fuel oil to attract "manufacturing enterprises to the city."[100] As it turned out, the committee did not meet until three weeks later on February 13, giving the oil-handling interests additional time to marshal their forces and trumpet their conviction, especially through published interviews, that fuel oil was the key to New Orleans's industrial future.

A central feature of risk analysis is assigning statistically based probabilities to events, such as a petroleum fire in a harbor, but, without data or having data of unknown reliability, oil-controversy participants had to make, in an atmosphere of uncertainty, risk estimations (degree of belief) about the likelihood of an event based on personal experiences, which are classic ingredients of subjective risk assessment.[101] In a newspaper interview, J. J. Paquette, who had testified before the BOT's SCCF in December 1901, drew upon twenty-nine years of experience in making various gases from crude oil, giving him the practical knowledge in which Jesse McDowell had expressed such confidence. Paquette described anecdotally the safety of gases under certain conditions, cautioning readers that fuel oil would explode or ignite in confined storage when sparks or flame ignited its accumulated explosive gases. He insisted that, if crude oil was stored in a "properly constructed tank," it could be used safely.[102]

The quest for compromise caught the attention of Chubb & Son of New York, a prominent underwriter of marine risks and agent for many European companies. Henry Salter, inspector general, and Max Grundner, a Chubb partner, visited New Orleans to investigate the city's new role as "an oil port." In response to questions about raising insurance rates, they said rate changes might not be necessary if the city was to adopt appropriate precautions for handling "oil barges, oil steamers and oil tanks."[103]

This running dispute came to a head February 13, 1902, when the City Council's new fuel-oil committee held its first open meeting. Chairman J. S. Zacharie stayed with the familiar themes of expectation and rapprochement. The "fuel oil question" held the city's "future development of manufacturing" in the balance, he declared, making it imperative to reconcile the various interests through "wise legislation." Goods worth $300,000,000, including a vast amount of cotton, flowed through the port, on which foreign insurance companies wrote many policies. Ill-conceived legislation could push up rates, to the advantage of rival ports. The city, Zacharie said, must find middle ground between this prospect and the advantages of "industrial development."[104]

Setting aside several proposed oil ordinances, revisions, and amendments, the committee heard from industrial oil users who unfortunately offered no practicable recommendations to resolve the fuel-oil stalemate. Boilermaker John H. Murphy complained that there was not enough room at his present location to set up supply tanks, unless he could put them under the sidewalk or inside a building, something the insurance interests had repeatedly rejected. He also objected to vaults, claiming that oil-fume accretion posed fire and health hazards. J. J. Paquette followed up on his recently made, prolix, and hair-raising case for properly constructed oil tanks with plans for tank covers and domes, allegedly making safe under-the-sidewalk oil storage.[105]

In contrast, the insurance interests through spokesman J. G. Pepper offered several specific concessions—some already made public—in an apparent effort to find a workable compromise. The primary one was permitting oil barges to be serviced as far up the river as Poland Street. Acknowledging that vaults might accumulate dangerous gases, they allowed tanks, ten or more feet from a building, to be buried underground without a vault. Nor did they object to moderately sized community tanks to supply multiple users, an idea Pepper had floated before the S&LC in late November 1901, having recognized that many businesses "had little or no yardroom" for a storage tank. While placing oil tanks in or on top of buildings was completely out of the question, the insurance men consented to "lateral pipeline[s] from the shore to the edge of the wharf to load a vessel's bunkers." The audience considered this an important capitulation, even though it could be seen as little more than a restatement of the terms of Ordinance 918 N.C.S., section 3, by which the insurance men accepted laterals that would run a short distance from the river side of the main pipeline to the water's edge. The underwriters submitted an ordinance that embodied their compromises, leaving unchanged the prohibition on storing oil under sidewalks.[106]

But not everyone was happy with these allowances. Important insurance allies such as the Sugar and Cotton exchanges and the Maritime and Merchants' Exchange made it clear during the special fuel-oil committee's February 19 meeting that mooring dockside oil barges in the harbor constituted a real or "objective risk" rather than a cultural perception.[107]

W. P. Ross, president of the Maritime and Merchants' Exchange, argued that "it was suicidal to attempt to load ships with oil amid the shipping; there was the risk of explosion, the risk of collision of the oil barge[s]." John A. Wogan, president of the Sugar Exchange, took the same position. His exchange, he stated, was against piping oil to vessels "at the water side in the shipping limits," as well as "landing . . . fuel oil barges in the river beside the shipping." Samuel Weiss, representing the Cotton Exchange, said his exchange was also opposed to unloading oil barges "in the river amid the shipping" as well as laying lateral pipelines from the main pipeline down to the water "for loading oil to vessels."[108]

The representatives' unshaken convictions about the dangers of handling oil around ships at anchorage suggested a "sufficiently high risk of . . . harm" warranting "reversal of the burden of proof." Oil distributors would have to supply "proof . . . of harmlessness" of riverside fuel-oil transfers if they wanted the insurance forces to cooperate in regulatory permissiveness.[109]

The next day, February 20, the committee reconvened and heard Charles G. Coyle, a major coal dealer in New Orleans, and Captain J. H. McFarlene, United Fruit Company's port captain, respond to the insurance case against dockside bunkering of vessels. Opportunity, practicality, and running costs highlighted their testimony. Coyle argued that efficiently supplying fuel oil to the port of New Orleans required lateral pipelines to the waterfront, saving time and money for companies that would otherwise have to bunker their ships downriver below Egania Street. McFarlene substantiated this, noting that United Fruit could annually save fifty thousand dollars if it could avoid the inconvenience and expense of sending ships "below the city." Further, he saw no dangers in refueling ships "using a lateral pipe over the wharf."[110]

Two important outlets remained where both sides could marshal their strongest arguments about handling fuel oil. The first was the presentation of the report of the SCCF to BOT directors, February 22, and the subsequent resolutions approved in a special meeting four days later. BOT directors resolved that the "contiguity" of New Orleans to the Beaumont oil fields meant that burning crude petroleum would be an "important economical factor" in developing manufacturing in the city, making it supremely important that city ordinances encourage its "wide-spread use." The city must avoid "exclusive pipe line privileges" for supplying oil and only place "limitations" on its use that "experience has shown are necessary . . . to prevent conflagration and danger to life and property." The BOT accepted the oft-repeated charge that the insurance interests, lacking field knowledge of petroleum risks, sought excessive and unnecessary regulations.[111]

Further, the directors flatly asserted that, if the Dock Commission were to designate "safe and isolated places" away from wharves used for "general commerce," oil barges operating under "proper rules and regulations" could discharge "fuel oil in bulk . . . into tanks or pipes, and . . . steamers." Since the fire-insurance underwriters had hotly contested this issue, this was unlikely to foster compromise. The directors appointed a committee to present the SCCF's report to the City Council's new fuel-oil committee for its "serious consideration."[112]

Hearings held by this body became the second venue for debating oil use. The presentation explored four issues: the cost of burning fuel oil compared to coal; the viability of the Beaumont field; the safe use of fuel oil; and the most economical way to introduce fuel oil to New Orleans.[113]

After demonstrating the savings oil offered and the reliability of its supply, the BOT report tackled the "question of safety in its use under proper rules and regulations." On this "pivotal point" of risk management the report relied heavily on the testimony (sometimes lengthy and repetitious) of Jesse McDowell, who had already publicly aired strong opinions in favor of lax oil regulation, and A. D. Childress, secretary of the Beaumont Oil Exchange and Board of Trade. McDowell recounted in a letter to the SCCF his central role in developing a Chicago pipeline and distribution system in the late 1880s and in setting up oil-burning equipment in that city. To emphasize that Beaumont crude was safe "under all ordinary conditions," McDowell disingenuously claimed that burning "enormous amount[s] of fuel oil" for fifteen years had not produced "one serious conflagration that can be traced to fuel oil." He condemned the fire-insurance interests' inexcusable ignorance and "unreasonable fear" of oil consumption, a disparaging remark that ignored many widely reported, catastrophic petroleum fires.[114]

Childress introduced several letters supporting McDowell's safety claims. Rockwell engineer J. D. Hackstaff, who had already testified before the SCCF, wrote that New England factories safely used fuel oil to raise steam and to anneal and harden metal. From the perspective of mitigating hazards through technology, Hackstaff contended that Rockwell's oil-burning systems, featuring underground piping and storage, controlled fire dangers, offering "absolutely no reason" to believe that using fuel oil increases hazard levels. Exemplifying the intersection of risk and technology, all of the correspondents regarded risk as a "property of . . . technical systems and their components" which, if properly designed and executed, excluded or minimized performance failures (hazards).[115]

The BOT report determined that the "most effective method" for "introducing fuel oil" to the city was mooring barges on the river at posts set some distance from shore. Constructed for maximum vapor ventilation to remove the possibility of an explosion, the barges would deliver oil through flexible conduits at designated hours to onshore pipelines "owned by clusters of manufacturers." This excluded "indiscriminate or promiscuous handling of oil along the river front." To expedite this scheme and circumvent monopolists (LSCOC) the SCCF had already asked the Dock Board to identify "safe and isolated places" where barges could off-load their cargoes in the manner described above.[116]

Convinced that this solution was incontrovertible, Porch charged that the insurance people could not "maintain their position with convincing reason" and had to resort to threats and intimidation, understandably unaware that by 1906 eight fire insurance companies would withdraw from the city and twenty-eight more were poised to follow, in part because of "carelessness in the storage of inflammables and explosives." Porch urged passage of an ordinance that reflected the committee findings, and to that end met with the City Council's special fuel-oil committee and reportedly made "one of his forcible and impressive speeches," quoting freely from the BOT report, skipping over some sections and reading others word for word. He emphasized the "possibility of safely handling" fuel oil, while at the same time denying that suppliers and consumers "favored the indiscriminate or promiscuous use of oil."[117]

Local exchange representatives, who had a vested interest in wharf safety through hazard mitigation, remained convinced that the oil men's proposed techniques or procedures for handling oil at dockside were ill-advised. W. P. Ross of the Maritime Exchange told the City Council oil committee that he was against any plan that would "jeopardize the cargoes and shipping in the harbor." That included handling oil above Poland Street and piping oil across wharves to vessels through connections that were not "absolutely safe." A. J. McDowell, representing the American Institute of Marine Underwriters, endorsed the risk-management regulations favored by New Orleans insurance men. Reliance Marine Insurance also supported the beleaguered local underwriters, warning that in the face of a "conflagration hazard" insurance rates might go up, familiar positions unchanged by the oil consumers' full-blown, aggressive campaign for access to city wharves.[118]

J. G. Pepper ended the hearing by assuring the council committee that the insurance interests recognized the huge economic benefits that petroleum offered New Orleans. Nevertheless, he cited examples of major fires and explosions caused by "deadly vapors" from oil and concluded that the so-called "barge privilege" would fill the river with the equivalent of floating oil tanks. Pepper thought that Ordinances 805 and 918 N.C.S. were adequate, though he wanted to strike out a clause in 918 that seemingly allowed lateral pipelines to "go over the wharves."[119]

Having granted the public several opportunities to express an opinion about handling fuel oil, the City Council's special fuel-oil committee met on March 3 and accepted three documents: a petition from steam producers insisting upon convenient, uninterrupted access to fuel oil through "water routes"; an "elaborate reply of the underwriters" to the BOT report; and a letter from the Cotton Exchange opposing oil pipes across wharves.[120]

The steam producers contended that manufacturing development demanded that only regulations found to be "absolutely necessary" should be imposed on oil distribution, surmising that some risks had "to be accepted in order to obtain benefits that would otherwise be inaccessible." They proposed stations—probably supplied by the Dock Board—as far up river as Napoleon Avenue, where "barges or tank steamers" could "discharge . . . oil through a pipe connected with all adjacent steam plants" two days each week under the supervision of city or Dock Board inspectors. The steam producers contended that these safeguards reduced any existing risk "to a minimum."[121]

In a detailed response to the BOT report, the underwriters' oil committee reiterated its determination "to advance the interests of the city," a goal that included warning the community about the "dangers incident to the handling" of oil. The committee objected "absolutely and unequivocally" to mooring "oil-laden vessels" above Poland Street or "along the city front." Precautions notwithstanding, the committee warned, oil barges put the shipping and valuable goods stored on the riverfront at risk. To the insurance men, effective risk management of oil remained paramount, even though the politics of its implementation had become heated.[122]

Jesse McDowell's derisive comments elicited a point-by-point rejoinder. Contrary to his claim that New York, Boston, and Philadelphia allowed oil to be landed in the center of their business districts, the underwriters' committee pointed out that petroleum operations took place in those cities in "outlying districts," reflecting the prevailing risk-management measures required by the insurance companies. As for oil-tank cars moving in and out of densely populated cities, there was, the committee protested, "no analogy between tank cars in transit and oil-laden barges" moored to wharves heavy with commodities. Finally, in response to McDowell's claim that oil had never been implicated in any large explosion or fire, the committee cited numerous instances of terrible fires started by mishandling fuel oil.[123]

The underwriters' oil committee also responded to testimony, which the BOT had solicited for its report, in favor of lax oil regulation, relying upon members' risk experience and the accumulated risk knowledge of fire-insurance companies and municipalities. The committee reminded everyone that the fire-insurance industry, in the course of dealing with petroleum fires, had amassed as much information on oil distribution and storage as the SCCF claimed to possess. By implementing insurance requirements, communities had reduced the number of petroleum fires, demonstrating, the committee reasoned, that they possessed extensive knowledge of oil-related fire hazards. The insurance interests, the oil committee pointed out, had throughout this controversy opposed mooring vessels carrying petroleum at city wharves and exposing ships to dangerous and massive conflagrations. The proposal to tie up ships or barges out in the river to handle oil only increased the hazard of burning oil floating on the water, as did the idea of landing oil barges at isolated places along the river.[124]

Left to make sense of this testimony, the council's oil committee amended Ordinance 805 to reflect concessions made by the insurance representatives. Tanks less than ten feet from a building had to be encased in brick rather than being put in underground vaults. Tanks ten or more feet away had to be installed at least four feet below grade. City Engineer William J. Hardee drafted the amendment, noting that twenty business concerns in the city wanted this approved as soon as possible. It allowed manufacturers to avoid expensive vaults that required yard room most companies did not have. Committee member Zacharie urged action upon matters for which the insurance men had pushed in their SCCF testimony: allowing lateral pipelines across wharves, extending wharf limits for handling oil up river to Poland Street, permitting steamers to bunker fuel oil on any wharf, and putting tanks inside buildings. The committee, however, adjourned, its members unable to agree on rewriting Ordinance 805 to accommodate these demands.[125]

Two weeks later on March 18 the City Council passed the amendment as Ordinance 1159 N.C.S., with the qualification that tanks "not enclosed in masonry" could be installed less than four feet below ground level if the city engineer determined that it was "practicable and sufficient." The new ordinance did not substantially change Ordinance 805 or 918. It did not allow oil-tank installations inside buildings or grant anyone permission to off-load petroleum at wharves above Egania Street or to pump oil from barges across city wharves to city consumers.[126]

Oil-committee members not satisfied with stop-gap measures remained undeterred. Working through mid-May 1902 they came up with a new oil ordinance to be considered on May 16 and 20, after which the results would go the City Council. However, both times there was no quorum. Most committee members wanted to keep the "present fuel oil ordinance" since it had been "working satisfactorily." The final blow came at a City Council meeting May 27. Zacharie tried to delay an ordinance granting a pipeline privilege to American Sugar Refinery so the fuel-oil committee could report "the ordinance which had been worked on for the regulation of the handling of oil." But several council members said existing oil ordinances were satisfactory, ending public disputes about how to regulate fuel oil in New Orleans."[127]

Although the city's oil ordinances were not radically changed, the crucial testimony supplied to the SCCF and rebutted by the insurance interests made it clear that the notion of risk was at the rhetorical and conceptual center of the oil controversy. It was also evident that local fire underwriters took advantage of the risk knowledge disseminated by insurance companies, technical groups, and insurance associations to bolster their position. Both sides engaged in risk assessment but disagreed about the nature of petroleum as a hazard and what risks it posed. This made determining the likelihood and estimating the possible consequences of an oil-initiated explosion or fire in New Orleans a great challenge.[128]

Some oil distributors and consumers insisted during the BOT campaign that fuel oil was not a hazard and judged the likelihood of explosion or fire as either nonexistent or extremely low.[129] But both sides in the debates had insufficient information about the annual number of oil fires and explosions and could not assess risk using traditional probability calculations. Without numerical quantification, they used phrases such as "no telling about what the disaster might be," "assuredly would not be safe," "did not think there would be any danger," "safe under all ordinary conditions," or "impressed with oil's safety," expressing risk in degrees of belief or personal judgments to deal with the unknown or uncertain likelihood of petroleum catastrophes.[130]

Relying on the history or frequency of fires or explosions, or generalizing from the experiences of people who had worked in the petroleum industry to make incomplete approximations or rough extrapolations could lead to misleading estimations of the likelihood or consequences of a disaster.[131] In the end the insurance men gained the upper hand after the SCCF hearings because they pointed to actual oil fires and asked for regulatory precautions, even if the probability of a catastrophe with major losses was low. This is not surprising. Insurance companies relied on accepting many risks, anticipating that there would be a small number of claims that could be covered by the premiums of the remaining risks, but a low-probability catastrophe, such as the 1904 Baltimore fire, could wipe out a company's assets. On the other hand, the oil interests argued from a negative: major petroleum fires were scarce or unknown, caused by improper procedures, not by any danger inherent in fuel oil. But the point made by the insurance interests was that, however unlikely oil fires might seem, when they happened—and they did happen—the result from a risk-averse insurance perspective was unacceptable capital erosion.[132]

After sorting through the contradictory evidence, self-serving opinions, and obvious human misapprehensions about hazards and disasters revealed in the exchange of opinions about the handling fuel oil, New Orleans officials did not radically change the city's culture of risk. Further, they called upon the permit system, a legally sound bureaucratic mechanism, to institutionalize enforcement of its petroleum-hazard mitigation policies, giving the city engineer an important role in this strategy. When the council formed its initial tank-ordinance committee in late July 1901, it stipulated that the committee include City Engineer W. J. Hardee in its deliberations. Hardee revised an early draft of the ordinance, and during consideration of pipeline Ordinance 918, he urged prompt action on it to coordinate its regulations with other oil privileges, so that permit approval could get underway.[133] Ordinance 805 N.C.S. required that applications for supply tanks be approved by the city engineer (section 3), and Ordinance 918 empowered the city engineer to approve and monitor construction of all large oil-storage tanks; road and sidewalk restoration above pipeline trenches; as well as pipeline maintenance (sections 2, 6, 9, 12). When the City Council's special fuel-oil committee held an open hearing in mid-February 1902 to complete a new general-oil ordinance, the fire underwriters offered their preliminary version. Section 3 called for applicants to submit detailed construction plans for oil-supply tanks to the city engineer in accordance with ordinance provisions, as well as other applicable "detail conditions."[134]

By reenacting Ordinance 805, as amended in Ordinance 1159 N.C.S., the City Council reaffirmed its faith in the judgment of its professional engineer to execute the risk management of petroleum, opening the floodgate to oil-supply-tank supplicants. It also illustrated that the city's risk strategies had a firm technical underpinning, in counterpoise to the engineering expertise that oil men had called upon to justify their regulatory complaints. The *Daily Picayune* reported in April 1902 that reconfirming the city engineer's authorization role elicited a "number of applications . . . and they are coming in . . . nearly every day." By May 19, forty-one businesses had obtained tank permits.[135]

The city engineer's response to the petitions explains how the implementation of public policy regarding petroleum risks functioned in the context of technology, hazards, risk knowledge, and insurance practices, providing details about the rush for permits, including the names of companies seeking permits, the application process, and the pitfalls to be avoided.[136]

Each request had to be accompanied by a plan or a blueprint. The plan had to have enough details to satisfy the city engineer, and if not, the city engineer might ask the applicant to come to his office to explain it. W.J. Hardee, the city engineer, told Linus W. Brown, an engineer representing ABC, that Brown's recently submitted application was inadequate and needed a drawing of the city block on which company officials wanted to put oil tanks and the distance between them and any adjacent buildings.[137]

The city engineer might reject or delay an application if its ambiguities made necessary a "close investigation and careful consideration" by the engineer. If an applicant did not follow a previously submitted plan for installing oil tanks, the city engineer, who learned of one such failure through a vigilant city building inspector, had to notify the mayor and the city attorney. Hardee thoughtfully informed the guilty party that he would like to talk about the problem before notifying anyone else to "determine if it is not practicable for you to re-arrange the installation of the tanks" to "conform to Ordinances No 805 [and] 1159." Hardee might also reject an application simply because its "proposed installation" did not conform to law.[138]

Hardee could also be quite strict with applicants who started an installation without communicating with him. He told W. T. Maginnis, a prominent cotton-mill owner, that Maginnis had to submit a plan showing the location of several already-installed oil tanks to obtain a permit. If the tanks did not conform to Ordinances 805 and 1159, Hardee wrote, "it will become my duty to require such modifications as may be necessary to make them conform to the provisions of the said ordinances."[139]

Tate, Jones & Co. persuaded Hardee to revoke an oil-tank application submitted January 27, 1902, along with an attendant permit, and to approve a substitute plan dated April 14, 1902, granting a new permit. On occasion requested changes called for Hardee merely to issue an amended permit as a supplement to one previously issued. And some diligent applicants got a permit without any trouble or challenges to their plans.[140]

Some applications contained complicated scenarios that could not be reduced to legally viable plans for a permit. One applicant wanted to install an oil tank in the "rear yard" of a drugstore on Camp Street occupied by Max Sampson, but Hardee replied that, although he had personally made a "careful investigation," he could not "without prejudice to any interests" approve the tank placement. Alonzo Church, Board of Underwriters surveyor, had advised Hardee that the board "protests against the installation of an oil tank in the yard of Mr. Max Sampson." If the tank goes in at that site, the board said, "insurance policies in the vacinity [*sic*] . . . would be seriously affected."[141]

However, some plans, although permissible according to ordinance guidelines, contained arrangements for equipment modifications that suggested actions which the city engineer was not authorized to approve. For example, Hardee told one applicant that his department did not object to "extending the vent pipe from the oil tank" already installed under a permit dated April 17, 1902, "over to the wall of the Morris Bldg." But Hardee warned he was not authorized to "privilege you to invade the premises or in any manner occupy or attach to the buildings of other persons."[142]

Other problems which surfaced in the correspondence included applications made by engineering or construction companies in behalf of a client. City Engineer Hardee warned Tate, Jones & Co. that applications for the erection of oil tanks had to be made by the parties that owned or would own and operate them. Further, Hardee pointed out that its present application was unsatisfactory. In light of the great probability that they would be called upon to prepare many future applications, he invited Tate, Jones & Co. to come to his office so he could explain his department's expectations for a successful permit submission.[143]

Even so, it should be noted that a lawyer, engineering firm, or some competent third party could present an application to the city engineer for oil tanks to be installed on behalf of a client, customer, or employer, but the letter of application had to be from those who would own or operate the tanks and who would also legally hold the permit.[144]

City fire-insurance companies were deeply interested in these permit problems. They traditionally worked with local officials to identify and mitigate fire hazards as well as aiding in regulatory enforcement. Hardee provided Alonzo Church, ex-chief inspector for the Insurance Underwriters' Association in New Orleans, with a list of companies that had obtained permits. Hardee observed that, although several companies had installed tanks without permits, he wanted "existing ordinances faithfully complied with" and asked Church to inform him of any unauthorized oil-tank installations.[145]

Finally, in most outgoing letters approving or commenting upon a pending permit, the city engineer referred to the pertinent city square or block (any quadrilateral) as defined by intersecting streets. This illustrates that space and location were elemental and indispensable components of analyzing, assessing, and managing risks associated with handling and storage of fuel oil.[146] In fact, throughout the oil controversy, geographical concepts such as place, space, and location were fundamental elements in the city's risk-management strategy. Their noteworthiness was an outgrowth of the close relationship between law and geography.

Law is a formal statement of power and its use, and therefore the basis of all regulation. Traditionally, it has been "preoccupied with jurisdiction" which implies province, range, and space, physical or otherwise. In New Orleans the ordinances that regulated and controlled hazards specified their application by place or location. The city, for example, required metal roofs on buildings in an area known as the fire limits.[147] Street boundaries defined the fire limits, suggesting that laws, rules, and instructions are formalized means for ultimate control "of some hazardous, physical process" (again implying location and space).[148]

Cultural geography examines "human endeavor" such as social processes or economic pursuits, which reflect how humans utilize their physical location. In the context of the petroleum controversy, that location was New Orleans. In this built environment humans created space and designated places to store, distribute, and consume fuel oil.[149]

Space is the realm on the Earth's surface in which humans act, the arena, however subdivided, in which human life takes place. Absolute location is a unique and "precise position on the surface of the globe" which is generally expressed in terms of a grid system such as longitude and latitude. Places are specific points on the surface of the earth which have characteristics related to how humans experience them. For humans the significance of a place can change or disappear as culture evolves.[150]

New Orleans contained various places (Jackson Square, Storyville, the Garden District) that acquired some cultural (emotions, habits, speech, expectations), historical, or political import. Streets, street names, and building numbers fixed the absolute location of these places. Naming and locating places were fundamental means by which the city set aside space for particular uses or purposes. New Orleans officials also marked off space by means of numbered squares formed by intersecting streets, or by using streets as the limits of irregularly shaped areas. The law, in turn, designated the spatial subdivisions for particular human activities—social, political, economic—which reflected the nature and goals of society and underscored the close linkage between geography and law.[151]

The question arises whether the centrality of this spatial orientation for prescriptive management of fuel-oil storage was long-lasting. The answer is found in the durability of Ordinance 805 N.C.S. as amended by Ordinance 1159 N.C.S. Ordinance 805, though altered several times, remained the fundamental statement of fuel oil and illuminating (kerosene) oil storage regulation in New Orleans through December 1943. Every change or addition modified, reconstructed, or added to the spaces in which the city allowed oil storage. Ordinance 2700 N.C.S. (1904), for example, moved the boundary of the back-of-the-city storage area from South Claiborne to Galvez, which Ordinance 2209 C.C.S. (Commission Council Series 1912–54) (1915) reversed, and which was reversed again in Ordinance 2789 C.C.S. (October 1915). Ordinance 2209 added a small area, bound by Elysian Fields, Abundance, Agriculture, and Frenchmen streets, as did Ordinance 2488 C.C.S. (1915), allowing storage in the Euphrosine–Calliope–Willow–South Claiborne sector. This was the first time the city allowed oil storage south of South Claiborne in the back-of-the-city area west of Poydras, well away from dense habitation and construction.[152]

Five years later, the City Council again amended and reenacted Ordinance 805 as previously amended by the ordinances cited above. Ordinance 5843 C.C.S. (August 1920) added three oil-storage areas. South Galvez, Calliope, Rocheblave, Broad streets, the New Basin Canal, and Napoleon Avenue defined the first in the back part of the city; Elysian Fields Avenue, Sere, Frenchmen, and Ne Plus Ultra streets the second; and Poydras, Cypress, South Miro, and Tonti streets the third.[153]

From 1920 to 1943 the City Council passed twelve more ordinances, all of which reenacted Ordinance 805 or a previous ordinance that reenacted a predecessor(s) which specifically reenacted Ordinance 805 as amended above. This included C.C.S. Ordinances 6348 (July 1921), 6526 (October 1921), 7035 (August 1922), 7378 (May 1923), 7741 (February 1924), 13179 (July 1931), 13858 (November 1932), 13886 (December 1932), 14351 (January 1936), 15259 (March 1941), 15783 (June 1943), and 15891 (December 1943). Ordinance 13886 C.C.S. replaced Rocheblave with Dorgenois. Ordinances 6348, 6526, 7035, and 15259 C.C.S. added one new site each: Lopez, Salerdo, Calliope, and Clio; Conti, Virginia, Helena, St. Louis; Miro, Lafayette, Tonti, Poydras; and Opelousa, Pelican Avenue, Powder Street, and the Mississippi River in Algiers.[154]

Exactly why the city made these changes is a difficult question. Almost all of the ordinances specified locations on the lakeside of South Claiborne with the exception of Ordinance 13179 C.C.S., which added the only square on the river above the city (Tchoupitoulas, Arabella, South Front, Joseph) and the Magnolia-Clio-Clara-Calliope square two blocks below South Claiborne. The Egania–Burgundy–Florida Walk–parish line unit, which Ordinance 805 N.C.S. set aside for storage or handling oil across city wharves, was in the Third District, inhabited by African Americans, immigrants, and workers, and by design not close to the central business district or wharves, or upriver neighborhoods. Ordinances nos. 5843, 6526, 13858, and 15891 added areas contiguous to a canal or basin. Generally, the industrial spaces along canals and basins were interspersed with residences, boarding houses, basements, or garrets of working-class people or the very poor.[155]

The City Council may have passed Ordinance 15891 so distributors could supply oil to the Higgins Aircraft Plant–Michaud. Ordinance 6526 C.C.S. apparently allowed oil in the Conti–Virginia–St. Louis–Helena streets site because the City Council only six days before had passed Ordinance 6522 granting Chalmette Oil and Refining Company permission to install five 13,000-gallon oil tanks inside nearby St. Patrick Cemetery, Masonic Cemetery, Hold Gravel Yard, Terminal Railroad tracks, and the Boys' House of Refuge.[156]

Efforts to regulate the storage and handling of fuel oil in New Orleans can be divided into three phases. In the first phase, petroleum drilling in West Louisiana and East Texas followed by distribution schemes made fuel oil accessible to New Orleans, which created uncertainty about how to regulate petroleum use. In early August 1902 city officials passed Ordinance 805 N.C.S., known as the "tank ordinance" because it specified where oil companies could erect large (37,500–50,000 barrel) storage tanks. Consumers had to obtain a permit to install small oil-supply tanks under specifications devised in the early 1890s in Chicago and embraced by many fire-insurance companies.

Phase two started in mid-August and ended in late October 1901. It featured Lone Star and Crescent Oil Company's efforts, despite charges of monopoly and City Council quarrels over the exact legal nature of the company's desired ordinance (city grant or franchise) to put down pipelines inside New Orleans. The City Council passed Ordinance 918 N.C.S., which explained how anyone who sought to distribute fuel oil in the city could obtain permission to lay pipe.

The third phase started in late October 1901 and lasted until late May 1902. The political apex of this phase came March 21, 1902, when the City Council approved Ordinance 1159 amending Ordinance 805 N.C.S. Oil men lobbied for permission to store oil under sidewalks and to handle it across city wharves. To accommodate them, a special City Council oil committee worked hard to devise a new, major ordinance to replace Ordinance 805; however, the City Council ultimately declined to change Ordinances 805 or 918, satisfied that prevailing regulations worked well. This ended the initial public controversy about fuel oil.

New Orleans, acting under its extensive police powers, used space and location to manage fuel-oil risks. The city stayed on course amending and reenacting Ordinance 805 many times through 1943 to prevent petroleum catastrophes. The rationalization of fire risk offered credibility to community aspirations for economic growth and sustained a commitment to hazard mitigation, public safety, and capital efficiency.

5

HAZARDS, RISKS, AND VULNERABILITY

Galveston, Texas, 1901–1937

The politics of risk governance in 1901–2 in Mobile, Houston, and New Orleans show that fire underwriters helped to shape their municipal risk strategies. This was also the case in Galveston. Galveston's participation in the evolution of risk knowledge, research, and dissemination stretched across the twentieth to the early twenty-first century, producing policies to control storing and using fuel oil. In addition, the city benefited from technical expertise, risk management experience, and insurance rules such as those promulgated by the Chicago Fire Underwriters' Association in 1892, and the concerted efforts of the NBFU, the SETA, and the NFPA.[1]

Engineering organizations and professional-standards groups provided scientifically informed, technical information condensed in fire codes that fire-insurance companies and their surrogates promoted. Insurance underwriters and insurance associations encouraged municipalities to adopt the codes, which summed up the industry's risk knowledge and perspectives. Gulf South communities such as Galveston embraced the risk-management agenda, illustrating a gradual, incremental participation in rule by experts, scientists, and educationally, technically accredited social managers that started in the nineteenth century and reaches to the present day.[2]

In September 1901, a level-four hurricane destroyed Galveston, Texas, and killed roughly six thousand people, making it the most lethal storm in U.S. history. It was a hazard or a "physical event" that brought "loss of life or injury, property damage," as well as "social and economic disruption." There was no official forewarning until the day before it arrived, and few people took these cautions seriously. Nor was there an organized evacuation, raising the vulnerability of the city to calamitous levels. These human actions amplified the consequences and therefore the risks of natural hazards presented by this hurricane, demonstrating that climate, near-sea-level human habitation, human error, and demographics often increase a city's vulnerability to natural hazards. The tragedy also confirmed the personal and communitarian nature of decisions made by the populace: Galvestonians assessed the risks of coastal living (estimating the likelihood of a hurricane and the magnitude of its anticipated damages) and made both pragmatic, individual judgments and collective social and political decisions, the sort of determinations communities must make about public affairs.[3]

Handling natural hazards such as hurricanes, floods, or tornadoes, or man-made hazards—industrial equipment, poorly designed bridges or theatres with no outside exits—reflects the strategies and policies of risk management that are essential features of organized societies. Galveston, for example, employed a range of practical solutions and prudent restrictions. The city raised its grade level, built a seawall, banned new or refurbished wooden structures within the city limits (fire limits), required adequate water supplies for fire suppression, prohibited oil lamps around stored cotton, and set the conditions under which merchants could store and handle flammable liquids such as fuel oil.[4]

During the nineteenth and early twentieth centuries, Galveston officials managed petroleum risks by controlling where and how one could refine, store, and consume flammable liquids such as naphtha, kerosene, and fuel oil. As the petroleum industry employed increasingly complex and hazardous technologies, the city of Galveston had to contend with the social costs inherent in their use or application in transporting, storing, or consuming fuel oil. Municipal police powers, as well as tort law and liability, determined which group—consumers or producers—would bear the risks of these technologies, as a way of dealing with the uncertainties associated with hazardous industrial processes and systems.[5]

To reduce petroleum risks, Galveston leaders in 1901 had to deal with uncertain outcomes without sufficient data to mount a rigorous quantitative risk assessment of safety issues, dangers, and potential catastrophes. They resorted to subjective (qualitative) risk assessment, which incorporates elements such as individual perception of risk, extreme risk aversion leading to precautionary policies, or community willingness to accept some levels of risk for economic gain. Qualitative assessment expresses a degree of belief in the likelihood of a hazardous event (for example, an oil fire) and the magnitude of its anticipated consequences based upon historical knowledge and personal experiences with hazards.[6]

This methodology guided policies to mitigate natural and man-made hazards such as protection from hurricanes after the 1900 storm and the proposed storage of large amounts of petroleum within the city. The latter was the source of the tension between fuel-oil disputants, one side of which featured oil men who claimed that their viewpoint, based upon their industrial, oil-field, and commercial experiences, rested on *objective facts,* not opinion; while to the contrary, their insurance critics, it was claimed, offered subjective opinions not founded upon facts, a charge the fire-insurance interests vigorously denied. Fire-insurance companies viewed natural hazards as well as massive oil-field and oil-tank fires as objective risks, events, and conditions, and they provided their agents and underwriters with risk knowledge that standards associations such as the NFPA or the Underwriters' Bureau of Fire Protection (later the UL), and insurance groups such as the NBFU or the SETA collected, codified, and promoted through suggested ordinances and voluntary-consensual fire codes.

The companies also benefited from fire-insurance maps in charting and recording hazardous activities, manufacturing processes, and storage of flammable substances. The Sanborn fire maps published in 1899 for Galveston were extensive and showed in detail the city's built environment, allowing companies to have ready-at-hand risk knowledge of the exposure of buildings to nearby hazardous activities or materials.[7]

Insurance maps are valuable records of accumulated risk knowledge that were adjusted for and attuned to new and changing risks. When fire-insurance agents submitted new contracts to a supervisor for approval, the package often included hand-drawn sketches of property, diagrams or rough maps of the surveyed risk. The well-known Sanborn maps were sophisticated and therefore widely used because they could provide block-by-block details and sometimes had updates of a specific area. Many other companies and entrepreneurs also developed maps specifically for the insurance business such as Pinney's atlas of Boston that showed in color the firms and plants that used hazardous and flammable substances. John Whiting, one-time surveyor for the Boston Fire Underwriters' Union, produced in 1877 a schedule or inventory of Boston buildings "expressively for the use of insurance companies" that succinctly described what sort of activities went on in them to complement fire-insurance maps.

Such materials were quite valuable for identifying hazardous activities and potential dangers to adjacent property, as well as providing information on construction and materials such as wood or brick. They shaped and informed building and fire codes, keeping surveyors, map publishers, and consumers appraised of changing field conditions and new hazards.[8]

I will examine the 1901–2 fuel oil controversy in Galveston within the framework of hazard mitigation, risk management, and community vulnerability, concepts that in practice were informed by technical and professional expertise, risk knowledge, and experiences. The political-regulatory contest over storing and transporting fuel oil on and through city streets, was contemporary with similar developments in Mobile, New Orleans, and Houston.

Dangerous weather events were familiar problems in these three communities, but unprecedented storm damage in 1900 demonstrated that Galveston was especially susceptible to the ravages of natural hazards. This raises the question whether Galveston's experiences with a catastrophic hurricane (natural hazard) disposed municipal leaders to be risk averse, willing to embrace only conservative risk-management measures, all the while relying on a repertoire of police powers, to deal with the man-made hazard of stored fuel oil.

Petroleum flooded coastal Texas as a result of the famous Lucas well that came in January 1901, less than forty miles north of Galveston. Energy issues, however, surfaced in the city and county alongside long-standing financial problems, which subsequently became an issue in the campaign for commission government. Part of the problem was that municipal revenue set aside to make interest payments on bond issues dating to 1881 had been used instead to pay off other obligations. Bondholders filed a lawsuit over the city's alleged failures to make timely redemption payments. Reallocating budget funds ran into legal snags, and the city defaulted in December 1901 on its bonded indebtedness. Galveston also struggled with its sewer system in the aftermath of a court decision that invalidated part of existing sewer bonds while awarding ownership of the sewer plant to the city, building on a controversy over municipally operated or privately owned utilities dating from 1896.[9]

In the aftermath of the 1900 hurricane, Galveston officials called upon the state to subsidize storm-damage reconstruction and recovery—grade-raising and building a seawall in particular—by remitting state taxes for several years.[10] They turned to the federal government to rebuild and restore the port. On a different front, volunteers and citizen associations distributed charitable donations, benefactions, and benevolent gifts that came from around the United States.[11]

Private rebuilding and construction efforts did not face as many political and legal hurdles. The Southern Pacific Railroad (SPR), for example, announced in 1901 that it would move ahead with building a planned set of wharves in the port, and it would rebuild its storm-demolished depot in the West End.[12]

These issues did not distract Galvestonians, who saw once-in-a-lifetime financial opportunities in the Beaumont oil field. With oil-wealth mania in full bloom by March 1901, many Galvestonians formed, directed, or invested in new oil companies. A notable example was D. R. Beatty, a Galveston real-estate agent who obtained oil leases on ten acres near the Lucas well. Galveston investors thronged to his projects, which included several spectacular gushers brought in under the aegis of the Texas Western Oil Company. Beatty's feats raised oil fever in Galveston to new heights and led his admirers to claim that "nothing is too big for this remarkable man."[13]

Several other established Galveston figures participated in the oil fervor. John Sealy, whose family owned the Galveston Wharf Company (GWC), was a director of Texas Western, and with several investors, including George Sealy, gained city approval to set up an oil distribution system. William L. Moody Jr., an important Galveston financier, played a financial role in oil affairs as a director of several oil companies such as High Island Land & Oil and the National Oil and Pipe Line.[14] Prominent councilman Ira H. Kempner and three other Galveston businessmen were on the Board of Directors of Roby Oil Company.[15]

Landowners in Galveston County benefited from oil fever. County records from May 1901 tracked the ongoing rush to lease land for oil prospecting, despite machinery and skilled-labor shortages, by entrepreneurs—risk-takers—who believed that "oil abounds in paying quantities in Galveston County and on the island." They were wrong but did exemplify oil-field risk-takers from Houston to Beaumont and Port Arthur.[16]

Manufacturers and businessmen also had a great interest in coastal oil developments. During the spring and summer of 1901, several Galveston companies, convinced that the developing Beaumont field was a reliable source of oil that justified conversion expenses, switched from coal to fuel oil for power. Brush Electric Light and Power, Galveston Red Snapper (ice plant), Texas Ice and Cold Storage, the Merchants & Planters Compress and

Warehouse, and Galveston Brewing companies applied to the City Council for permission to install iron tanks and pipelines to burn fuel oil. Details of their tank installations were similar to the insurance guidelines adopted in New Orleans that governed oil-burning equipment, which in turn reflected basic concepts and the practical application of technologies evident in the rules devised and circulated by the Chicago Fire Underwriters' Association in 1891–92. By 1901, large insurance companies such as L&L&G prescribed rules for petroleum use comparable to the Chicago standards. There was also a connection to major insurance associations deeply involved in creating technical standards and codification. L&L&G's president, Henry W. Eaton, became president of the NBFU in May 1897, putting his firm at the nexus of the NBFU and NFPA efforts to establish oil-tank rules and standards for safe use.[17]

Brush Electric Light installed a large tank about five blocks from its plant, filled it from a railcar, pumped the oil to a supply tank positioned behind its power house, and set a small backup tank in its yard. In late August 1901, Merchants and Planters Compress fired up its boilers with fuel oil pumped from a large, brick-enclosed tank on its Thirtieth Street lot via a pipeline to its plant across the street. Suppliers used railcars parked temporarily on Postoffice Street to refill the twelve-thousand-gallon tank.[18]

The Texas Star Flour Mill wanted to burn oil for steam as early as 1899 but postponed the change owing to doubts that the small Corsicana field could provide the needed petroleum, a problem which the Beaumont discoveries solved. By May 1901 the mill had in place a complete oil-burning apparatus, fed by a twelve-thousand-gallon underground tank set about fifty feet from the power house, replete with requisite gauges, standpipe, and ventilation. The Galveston Red Snapper Company petitioned the city for permission to install a "tubular tank" under the sidewalk in front of its building on Twenty-Second Street and Avenue A and shortly thereafter sought approval for a pipeline from a wharf-side tank at Twenty-Second Street to the boiler room of its ice plant. The Texas Ice & Cold Storage Company followed suit, gaining permission to put an oil tank "under the pavement" on the west side of Twentieth Street between avenues A and B. By December 1901, eleven "steam power plants" were burning fuel oil in Galveston, and plans had been made to burn fuel oil in the city's waterworks.[19]

With commercial oil consumption established, Galveston could benefit from a pipeline connection to the Beaumont field, increasing the community's revenues, buttressing Galveston's aspirations to be the premier Gulf Coast shipping port, and providing cheap fuel to manufacturers. Early in April 1901 the *Galveston Daily News* disclosed that Galveston men were considering building a pipeline from Beaumont to Bolivar Point on the eastern side of the Galveston channel, setting off public discussions about raising the necessary funds. Observers believed that a pipeline would attract shipping to Galveston. After bunkering fuel at Bolivar Point, steamships could put in at the nearby Galveston wharves, taking on or discharging cargoes while saving time and money. Barges could shuttle oil-filled tank cars from a proposed Bolivar Point terminal to its existing counterpart in the port of Galveston to bunker vessels or for city distribution.[20]

The plans to bring sufficient oil to Galveston businesses in tank cars relied on railroad connections to support all fuel-oil consumers. One account described the formation of a company to supply industrial users such as the Brush Electric Light Company, including plans to construct "reservoirs" in Galveston supplied by tank cars or barges for under-the-streets pipeline distribution tied to on-site supply tanks. Rumors surfaced in April 1901 that another company was being organized to import oil for small consumers within the city.[21]

Creating an efficient transportation link from the oil fields to Galveston, using railcars, pipelines, or barges, entailed the uncertainties and risks of handling petroleum in an urban setting, underscoring the need for effective risk management. The city approached this issue armed with uncontested legal control of its streets and wharves. It had routinely approved of water pipelines under the streets, but fuel oil was an obvious fire hazard that introduced additional risk in the minds of insurance underwriters and nervous citizens.

Concerns over burning oil notwithstanding, managing the risks that accompanied the growing use of oil in Galveston complemented a recovery-through-growth agenda to rebuild and rehabilitate the city and port, and foster manufacturing and commercial growth. These goals required the City Council to determine if it had the specific authority to regulate fuel-oil use—equipment, distribution, location—and, if it had such powers, how should they be exercised? Galveston had pre-1901 legal resources to deal with risks surrounding "any inflammable oils of any kind or nature." The city had created a "Petroleum Oil and Inflammable Commodities" ordinance as early as 1871 that covered among other things the storage of "any product by distillation of coal, petroleum or rock oil." It also specified that anyone who stored, sold, refined, or manufactured more than two barrels of the "fluids, products, oils" including naphtha, benzine, or kerosene had to obtain certificates from the chief engineer of the Fire Department and the mayor that confirmed compliance with the hazard-management stipulations in the ordinance. These requirements included iron storage tanks surrounded by levees or earth embankments, and use of premises that did not endanger contiguous structures. The elements of this routine exercise of municipal police powers was still in place in 1893 and constituted the regulatory antecedents of the 1901–2 fuel-oil regulations.[22]

Burning oil for steam and concomitant distribution schemes progressed from the summer of 1901 into 1902. The Galveston City Railway, Galveston Brewing Company, the GCSFR, Elevator A, and the city waterworks laid plans and sought permission to install fuel-oil equipment. Two groups petitioned the City Council in 1901 for oil-distribution privileges, one led by George and John Sealy, and the other by Captain Charles Clarke, a successful maritime operator and dredging contractor, who had announced plans to organize a company to move oil coastwise in three large tank barges. Using the Neches River, Clarke intended to load the barges directly from field wells, allowing cheap and quick transit to Galveston to supply railroads with motive fuel. Political infighting within the City Council began over awarding Clarke a free franchise, a conflict similar to though not as bitter as the drawn-out fights in New Orleans over oil-pipeline franchises, fears of monopolies, and municipal interests.[23]

Clarke's request moved smoothly through the early stages of City Council deliberation—petition, favorable committee report, and approval—but the council ultimately rejected an amended version that would have allowed him to construct piers to handle oil on and east of Seventh Street, which fronted the ship channel. Seventh Street intersected waterfront property owned by the GCSFR, prompting Alderman W. F. Stewart to warn that the railroad might expand wharf operations along this area if the port grew as expected. He thought that land east of Fourth Street would do just as well, if Clarke wanted to handle oil across city wharves for his company's steamers and barges without having to pay wharfage to Sealy's GWC. It was also noted that the distribution franchise that Clarke wanted was quite valuable—George Sealy put forward twenty-five thousand dollars for it; the GCSFR offered even more money—which required the council to consider all offers.[24] Unable to get approval for wharf operations, Clarke changed his plans to Pier 37 and later received permission from the Board of Commissioners (BOC), January 30, 1902, to construct an oil tank and related buildings at that site.[25]

Sealy's 1901 bid for the franchise initially did not fare well. Since his proposal originally went to the council's ordinance committee as a petition, members ignored it until Sealy resubmitted it as an ordinance, similar to Clarke's approved measure, but embellished with two perquisites: the city would receive 7.5 rather than 5.0 percent of the stock of any company subsequently formed by Sealy and his investors; and work outlined in the ordinance would commence within eight months, not twelve as Clarke had pledged.[26]

It is unknown whether this beneficence influenced anyone. Alderman A. Bornefeld, a Sealy apologist, pleaded for equal treatment, but complained in a council meeting a week later, May 27, 1901, that the ordinance committee had reported out the Clarke ordinance in an adjourned meeting but declined to consider the Sealy measure in the same circumstances. At a meeting on June 3, Alderman F. M. Gilbough joined Bornefeld in advancing the Sealy project. Detecting a ploy to subvert Sealy's legitimate interests and professing a concern for the "city's best interest," Gilbough opposed allowing Charles Clarke to construct piers on the waterfront, a gesture, he insisted, the state would have to ratify. Bornefeld added that the Clarke franchise was worth "hundreds of thousands of dollars" and should undergo legal review.[27]

Alderman C. H. McMaster refocused deliberations on the Sealy ordinance—just reported favorably by the ordinance committee—and suggested that it should also be reviewed. Although Bornefeld insisted that "justice and fairness" demanded its immediate passage, McMaster countered that he did not believe in granting privileges to one company and then giving the same thing to a second company, unless the first grantee "failed to avail itself of the privileges contained in the grant." He viewed Sealy's ordinance as an attempt to upstage Clarke, acting in the interests of the GWC, from which Sealy, as president, would get special wharfage rates. Clarke on the other hand needed an "independent place" to land oil, McMaster pointed out, which the GWC might try to block by asking for "prohibitive" tolls. In the end the ordinances went to the city attorney for legal analysis.[28]

Although the council reviewed City Attorney James B. Stubbs's opinions at another meeting, June 17, it took no action on the Clarke-Sealy requests. Stubbs believed that Clarke's request to construct piers at the end of Seventh Street and streets eastward to the bayfront was legally conditional on an existing ordinance. In March 1898 the city had granted permission to a ferry company to build a landing at the end of Seventh Street, making Clarke's project subordinate to the earlier commitment. By charter the city had the right to construct and regulate wharves and to control wharfage rates on city-owned property, giving it the authority to approve Clarke's request as long as his privileges did not impinge upon public use of the streets in question."[29]

As for the Sealy ordinance, Stubbs cited the city's chartered power to control streets, direct "laying and repair of gas pipes . . . in the streets," and manage city property that allowed the council to pass or reject the ordinance as it saw fit. He advised the City Council to reserve the right to "control the location and construction of storage tanks," thereby insuring "public convenience and safety," a clear statement of Galveston's responsibility to undertake risk management of petroleum.[30]

Aldermen Bornefeld, Stewart, and McMaster, joined by drilling-contractor and Clarke associate Alfred S. Newson, and City Attorney Stubbs, resumed debate over the ordinances the next evening at a called meeting but accomplished very little despite several votes and motions, deferring action on the Sealy-Clarke requests until the next council meeting. One of Sealy's collaborators, businessman and steamship agent Charles Fowler of Fowler & McVitie, in an attempt to expose what he thought was arbitrary opposition, maintained in a written statement that certain councilmen who held grudges against the GWC believed that Sealy and his business partners "practically owned" the city, making it "high time to emancipate the community from the clutches of such octopi." Fowler countered that George Sealy's "courage and enterprise" had made the city a huge cotton and grain port with the "finest equipped system of wharves and terminals in the United States." And Fowler added, after the Lucas well came in, Sealy decided to branch out from his coal business and distribute the newly available petroleum, a decision that initially prompted the ordinance request.[31]

In early July the GCSFR, a railroad which George Sealy had reorganized and then sold to the Atchison, Topeka, and Santa Fe, offered the city $27,500 to relinquish municipal claims to the railroad's land between Fourth and Tenth streets north of Avenue A and the ship channel, claims based on the allegation that the right-of-way of city streets went through this land. The railway argued that Fifth and Sixth streets did not extend through the property in question, and it was unlikely that Seventh through Ninth streets would ever traverse it. The railroad wanted to "quiet" its title—to have control of the entire area without interference or disturbance—preventing later subdivision of the premises into parcels by others (such as Charles Clarke), which would make the land unfit for "wharf and terminal purposes."[32]

The offer came on the heels of charges by Sealy proponents that the amended Clarke ordinance was a bad deal because the proposed piers, wharves, and buildings would encroach upon GCSFR property and block future commercial development in the area. The offer also resurrected demands that the council offer a land-use franchise to the highest bidder.

That resolution did not pass, but the Sealy ordinance finally gained approval, as a grant of privileges tainted by George Sealy's entrenched interests along the wharf and in the GCSFR but unscathed by frequent concerns about distributing valuable privileges to private interests without considering Galveston's municipal interests. Technically, the Sealy proposal did not include using city wharves; a favorable arrangement with the GWC would allow access to these wharves. This move discredited the amended Clarke ordinance, removed a competitor, and gave the Sealy project credibility with no apparent liabilities.[33]

However, the proposal and its associated ordinance had unanticipated consequences, as it highlighted issues that embroiled city leaders in a struggle over how to manage petroleum risks. This conflict placed the GCSFR in the center of a struggle between oil consumers and fire-insurance companies similar to fuel-oil battles in Houston and New Orleans. The ordinance's flaws were evident to Mayor Walter C. Jones, who vetoed the Sealy ordinance—the right to lay pipes and build oil tanks in Galveston—on the grounds that it did not explicitly reserve to the city the right to "control or direct the location and construction of the tanks for the storage of oil." "Public safety," Jones argued, required an affirmation of the city's power to manage the risks associated with "inflammable material." Jones clearly thought that specifying locations for oil storage "some distance from buildings" was effective hazard-management, mitigating the "danger from accidental fire." The city, Jones observed, must assert its right to "designate the place where and the manner in which fuel and other oil shall be stored and kept." He added that Galveston residents "have an interest" in the placement and regulation of oil-storage tanks to "prevent the spread of fire or the occurrence of any disaster." His arguments convinced the council to sustain the mayor's veto.[34]

Expressing his concern about the lack of risk-management provisions in the Sealy ordinance, Jones referred to an "ordinance of a similar nature" that was "recently passed" in Galveston, making the city engineer responsible for supervising the "location and construction" of oil-storage tanks. The "similar" ordinance was the original Clarke ordinance, passed May 9, which put the construction, location, and features of "buildings, tanks, reservoirs, machinery and equipment" for the "safe use and distribution" of fuel oil and gas within Galveston under the supervision of the city engineer. He had a charter-based responsibility to oversee the repair and construction of streets, including excavations for pipelines or oil tanks. This duty supported the city attorney, who routinely received oil-tank petitions from the City Council and was responsible for turning them into legitimate ordinances. On June 17 and then July 1, the council added the city engineer when it became clear that the Sealy petitioners wanted to install oil tanks under the sidewalks, enterprises that involved adherence to technical specifications, defined locations, and yard grades.[35]

Mayor Jones may have vetoed the Sealy ordinance to remind George Sealy, a member of the Deep Water Committee, that his efforts to discredit Jones to advance the case for commission government had unseen costs. But certainly Jones recognized that the dispersal of petroleum appurtenances (small supply tanks and large storage tanks) throughout an urban, built environment raised the specter of disastrous fires or explosions and demanded centralized risk management.[36]

The initial attempt to mitigate these concerns was a response to large-tank storage and engineering reviews of oil-distribution ventures. They converged in an ordinance approved August 19 that allowed the GCSFR to construct two oil tanks in Galveston, of 20,000- and 1,000,000-gallons capacities. The railroad had embarked on a major project to adopt fuel oil for its locomotives. The plan called for eleven large tanks scattered along the GCSFR line, one of which in Somerville, Texas—1,500,000 gallons capacity—was almost finished. The proposed design placed the large tank inside a reservoir, surrounded by an embankment that would occupy twenty-seven feet of sidewalk and street on block 583. The plans for the reservoir had "to be approved by the City Engineer." The GCSFR–West Galveston tanks dramatically magnified oil risks, which in Houston and New Orleans elicited intense opposition from the fire-insurance interests, a reaction to be duplicated soon in Galveston.[37]

Following the path of various Galveston businesses in adopting fuel oil for power production, Austin Bertrand Wood, owner of the American, Model, and O.K. laundries, petitioned the City Commission in early October for permission to install three oil tanks adjacent to the laundries "according to the regulations of the board of underwriters," underscoring the semi-governmental role of fire-insurance companies who constrained the behavior of their clients through technical conventions. The insurance companies also worked closely with and lobbied governments for laws or ordinances that codified their understanding of risk and hazard-mitigation procedures.[38]

A week later, October 8, the city got a preview of coming debates over identifying fuel oil as a hazard and determining what policies should follow. Commissioner I. H. Kempner objected when Fire Chief John H. Gernand, former land surveyor for the Santa Fe Railroad and fire chief since 1891, asked for additional time to study the request, expressing a risk-averse and cautious approach to using fuel oil. He argued that some tanks were explosion hazards, having been placed too close to buildings. Kempner dismissed these claims: the tanks had City Council approval, making a delay pointless; "insurance experts" and other officials saw no danger in the tanks' placements; there had been no rate increases; and pumping oil from the tanks to supply boilers was widely practiced in Texas and not considered hazardous. The commission agreed to an extension, but at its next meeting, October 15, Wood's request never got out of committee. The proposal violated a subsection, article 239, of the *Revised Ordinances,* apparently because at least one of the laundries was in or alongside a residence.[39]

However, a related, concurrent discussion within the City Commission overshadowed Wood's unsuccessful petition. The commission faced a new challenge born out of insurance and maritime-business concerns over the recent activities of oil consumers: months spent obtaining city permission for tank installations, pipelines and oil-burning plants, installing equipment, arranging for petroleum supplies, and producing power, as required under terms provided by the city's charter and revised ordinances from

1893, as well as established rules devised by major fire-insurance companies. Cotton dealers, steamship agents, the Galveston Maritime Association, and the Galveston Cotton Exchange petitioned the City Commission to "adopt an ordinance regulating the handling and storing of crude and illuminating oil." This was necessary, they argued, to "guard against the great hazard" involved in using oil when the "season of commercial activity" in which the city receives and ships cotton, grain, and other commodities began. If there were wharf fires, shippers would reroute their consignments to rival ports. The draft ordinance attached to their petition, they believed, contained "all that is necessary to secure protection," mirroring similar sentiments in Mobile and New Orleans.[40]

It proposed that city-approved supply or storage tanks for storing fuel or illuminating oils (burning fluid, kerosene) must be constructed and used under the supervision and approval of the city engineer. Such tanks must conform, regardless of capacity, to familiar standards: minimum tank-plate gauge; underground installation inside a buried and ventilated, masonry vault; positioned so that any release or overflow would not stream into the bay or threaten property or shipping. Future petitions must include "settings and protection" of the proposed tank(s) and be approved by the city engineer.[41]

All subsequent proposals—revocable and subject to amendment—would fall under the city engineer's supervision and the specifications noted above. Several controversial details raised concerns among those who wanted to distribute or consume petroleum. Aboveground tanks would be placed at least one hundred feet from all buildings on an unobstructed line to their oil plants. In addition, railcars containing fuel or illuminating oil could not be left overnight on "any dock, pier, bulkhead or wharf." Tanks or "ships, barges or lighters" could not be supplied with fuel oil from moored vessels, nor could vessels tied up at a dock store or keep fuel or illuminating oils as cargo. No one could handle oil dockside or over wharves, except at docks or piers beyond the city limits (east of First Street or west of Fifty-Seventh Street).[42]

Disquiet surfaced quickly. Commissioner Herman C. Lange called his colleagues to consult with "oil experts and insurance men" to frame a more permissive ordinance. The required risk analysis entailed serious consideration of the "danger from fire and other accidents" presented by petroleum, followed by, Lange believed, hazard mitigation, removing sources of risk such as oil storage under sidewalks or streets. In terms of risk management, their initial task would be the "identification of potential risks and their consequences," which led to regulatory balance: they must encourage the adoption of this necessary fuel while regulating oil-burning plants without "imposing unreasonable hardships." The challenge was to create a rational interface between oil and consumers: use credible engineering and technical knowledge filtered through professional experiences and political judgment to create rules understood by all parties, allowing the exploitation of nature without harming humans or the things they value.[43]

Critics complained that the ordinance was ambiguous—too indefinite for the GCSFR which stopped construction on its huge oil tank at Forty-Fourth Street, an unnerving action for city officials. The potential changes required by the proposed ordinance had stymied a major railroad project and cast a shadow over the city's economic viability. Businessmen had expected that an intelligible and coherent ordinance would sanction the building of large tanks on the "outskirts of town" while permitting small businesses to store oil safely in the city center. The ordinance also confused oil men because it repeatedly used the term "supply" tank, which in their parlance was not the same as "storage" tanks, leaving them uncertain whether the regulations for supply tanks covered all types of oil tanks, including large storage tanks some distance from a manufacturer, even though the regulations seemed to pertain to small supply tanks in the yard or rear of their facilities. Similar confusion swirled around sections that dealt with underground tanks but subsequently referred to tanks "wholly or in part above ground" (existing tanks built prior to the proposed ordinance).[44]

The *GDN* observed that the commissioners did not read carefully the insurance men's proposed ordinance. This document, not surprisingly, followed the "conditions and provisions included in the fire-insurance oil-fuel clause attached to policies," a reference to stipulations governing tank fabrication, installation, and location in an oil-burning plant that closely resembled rules outlined in insurance-permit forms, particularly the so-called "Texas form" mentioned above.[45]

Commission President Judge William T. Austin echoed Lange's call for a special commission meeting to confer with "expert oil men, insurance men and men versed in the technical features of installing an oil plant and its operation." Austin wanted to encourage Galveston businessmen to use fuel oil by a publicly approved "sensible law" that provided little more than the minimum "protection [necessary] against accidents and fire." Risk mitigation through commonsensical regulations was desirable, as GCSFR Superintendent William C. Nixon made clear in his declaration that the railroad would comply "with all reasonable regulations in the equipment and storage plants." Ironically, the oil interests hinged their case on professional expertise and experience-based engineering, while their insurance opponents drew up risk knowledge from companies and professional associations that offered standards and methods certified by technically derived, hazard-mitigation methods.[46]

Lange and Austin urged the city to pursue such regulatory controls, demonstrating that they understood risk management in a way consonant with our current perspectives. Crafting a rational, regulatory balance rested on achieving socially acceptable levels of risk, in which guarding public well-being and meeting the needs of stakeholders such as city residents, oil men, and insurance underwriters, were fundamental, risk-acceptance criteria. This trade-off between reasonable regulation at acceptable levels of risk and the costs the city could bear to mitigate petroleum risks required prudent allocation of Galveston's limited resources, making cost-distribution of risks and hazards a political activity.[47]

To accomplish this, they wanted a less rigorous, minimalist approach to oil regulation, a strategy associated with tolerable risk and what is now called risk-informed or performance regulation. Risk-informed regulation focuses on the realities of a system, such as large-scale petroleum storage, highlighting known areas of hazards, rather than devising prescriptive standards that constitute a "defense-in-depth" against hazards, asking operators for a massive, comprehensive risk-management program designed to eliminate "temporary, reversible" hazards or conditions as well as very low-probability, high-consequence events.[48] If Galveston citizens would accept minimal regulations, tolerating the occasional accident or small fire, it was not necessary, from Commissioner Valery E. Austin's viewpoint, to expend resources on risk protection against incidental, low-consequence, or extraordinary events. The alternative would be to act irrationally, initiating severe regulations that handicapped Galveston manufacturers, discouraged small business activities, or blocked a major railroad from operating in the city.[49]

At the special meeting, October 23, attended by utility, railroad, and commercial oil consumers, Commissioner Kempner acknowledged that many critics objected to his previously submitted oil ordinance, a document that various shipping and cotton companies had "compiled from ordinances adopted by other large American ports."[50] After this acknowledgment, Kempner offered a substitute ordinance that rested upon the petroleum-oil and flammable-commodities rules articulated in Galveston's 1893 *Revised Ordinances*. The substitute ordinance's maligned predecessor had specified the technical details of tank fabrication and supply-tank, oil-burning apparatuses, the placement of and intervals between large storage-tanks, and prohibitions on storing or moving oil across wharves. The substitute instead started with two sections that were almost identical to the opening articles of the *Revised Ordinances,* adding the phrase "illuminating or fuel oil" to a list of dangerous liquids.[51]

The initial insurance-inspired ordinance prescribed petitioning the city government for council/commission approval if the application details conformed to ordinance specifications as determined by the city engineer, who was also responsible to see that these specifications were carried out. It was explicitly aimed at individuals, copartnerships, or corporations, tempting one to see it as a shot across the bow, aimed at the Santa Fe railroad's planned construction of several gigantic oil tanks to supply locomotives. But Kempner's substitute ordinance dropped any requirements for oil-burning plants, presumably in the belief that this was covered by long-standing insurance rules. Instead, in the same vein as its 1893 template, the newer draft ordinance established the city's legal authority over "person or persons" who wanted to handle various flammable liquids, and required their operation or business to be inspected by the fire chief. He would certify that their premises contained the requisite "tanks, levees, excavations or embankments" to protect adjacent properties. After the BOC president approved this certification, he would then issue a certificate allowing storage, manufacturing, or refinement of the liquids at issue. This update of the 1893 *Revised Ordinances* controlled fuel oil through the streamlined authority of commission government.[52]

Sections 6 through 10 of Kempner's substitute ordinance set forth permit restrictions on storing kerosene, "dangerous, inflammable, or burning" fluids, or the products of the distillation of "coal, petroleum or rock oil" such as "naphtha, kerosene, benzine or benzole." Above or partially belowground tanks had to be one hundred feet from all buildings and surrounded by embankments or excavations, while entirely belowground tanks could be less than one hundred feet from nearby buildings but contain no more than 15,000 gallons. Tanks more than one hundred feet away from buildings could contain 100,000 gallons or less, and any two so constructed had to be five hundred feet apart. Tanks set down on city streets or alleys could hold a maximum of 625 gallons, had to be entirely underground, at least ten feet from property lines, and capable of supporting, without risk to traffic, overarching pavements or sidewalks.[53]

During a subsequent discussion, tank fabricant William H. Penman of Chicago, charged with constructing the huge Santa Fe tanks in Galveston, announced that the railway had suspended work until the oil-ordinance matter "was settled." He said that he had built tanks throughout the country and, with one exception in Chicago, had never encountered "any restrictions." While agreeing that tank "restrictions in thickly populated business districts" were appropriate, he cited SOC's large refineries—built on the fringes of various cities—at which there had been many fires but none that were uncontrollable. J. H. Hill, general manager of the Galveston, Houston and Henderson Railway (hereinafter GHH), pointed out that 100,000-gallon tanks could not provide his firm with a three-day supply of fuel, without which the railway would have to send locomotives several miles out of town for oil, implying that such circumstances might hamper or threaten railroad service to the city.[54]

Insurance representative Charles R. Brown responded with a flurry of information on the flammable character of Beaumont oil, a dangerous hazard as illustrated by restrictions on handling and storing oil in Chicago, Philadelphia, and elsewhere. He defended the initial oil ordinance as a safety precaution, which insurance men believed provided oil consumers with the "proper safeguards" against risks. Additionally, the ordinance was a "virtual reproduction of the New York ordinance." When questioned by Penman about the Chicago restrictions, Brown insisted he had read from a city ordinance, not insurance rules. In contrast, Charles Clarke & Co. glumly advised that, if the "proposed ordinance" gained approval with its ban intact on handling petroleum products across city wharves, it would end Clarke's current barge-line oil business, which brought oil in near the Eighteenth Street wharfs at prices competitive with tank-car rates at Houston, a market the company planned to exploit by barge.[55]

Kempner's substitute ordinance did not mollify the oil men, who thought it would "seriously cripple the oil fuel industry in Galveston." Its passage was, therefore, in doubt, because no one thought the city could "frame a law" that would win universal approval. But in fact the City Commission favored "reasonable and liberal" restrictions that encouraged fuel-oil usage while safeguarding "life and property." This position incorporated the central purpose of municipal police powers in all circumstances: a resolution of competing interests that was at the heart of the politics of risk management.[56]

The oil interests had several functional concerns. The ban on handling oil "along the wharf front" east of Fourth and west of Fifty-Sixth streets, they contended, isolated their operations from Galveston's shipping activities. More importantly, this prohibition would end Galveston's competitive oil-barge traffic that kept down railroad rates for oil transportation, forcing city consumers to pay whatever prices the railway companies demanded. The ban also threatened Galveston's chances of bunkering oil for steamships bound for coastal and Caribbean ports. They wanted the city to designate "certain places along the water front" for moving oil, leaving storage to be placed elsewhere in Galveston.[57]

Another issue was the 100,000-gallon limit on the capacity of large storage tanks, which the railroads claimed was insufficient to maintain rail traffic. The Santa Fe, for example, needed large supplies on hand to keep motive power going, without sending locomotives miles out of town to get oil, burning coal only in Galveston, or losing the services of several locomotives that would have to ferry oil to multiple small city tanks to keep them filled. Further, railroads serving Galveston insisted that it was ridiculous to think they would spend huge amounts of money building large oil-storage tanks but not equip them with safety features or embankments to prevent spills or fires.[58]

Adding insult to injury, C. L. Wallis, the general manager of Higgins Oil & Fuel Company, sent the BOC and Commissioner I. H. Kempner two letters—printed in the *Galveston Daily News*—accusing the "insurance and shipping interests" of trying to shut Galveston "off from handling oil by water," robbing the city of its "natural water competition," and putting city manufacturers in a poor competitive position. Preferring "reasonable and rational methods" rather than "building a brick wall around" the city, Wallis insisted that Beaumont oil was safe to use, making it pointless—from a risk-informed perspective—to require the construction of masonry walls around underground tanks because this did not diminish fire risks. He was surprised by the "ignorance" displayed by many insurance men, who failed, he claimed, to distinguish between "stable" crude and flammable refined oil, a shortcoming Wallis condescendingly attributed to a lack of "thought and investigation which the business requires." Wallis invited the commissioners to visit Beaumont and see the existing "oil field, storage tanks," and pipelines.[59]

At its next meeting and with President William T. Austin in the lead, commissioners amended Kempner's substitute ordinance and sent a revised draft to the city attorney. Expanding the permit-license system, the draft required oil users whose stock exceeded 80 gallons to obtain a certificate of inspection and an approved license issued by the BOC president that demonstrated compliance with the ordinance's conditions and regulations. The BOC deleted section 9 that limited large-tank capacities to 100,000 gallons and modified section 8, which had limited tanks less than one hundred feet from any building to 15,000 gallons, allowing owners to install a tank in excess of that amount if commissioners issued them a permit, a qualification designed to placate the Santa Fe Railroad. Aboveground tanks still had to be one hundred feet from any building, but underground or partly underground tanks could be as close as ten feet from structures. Commissioners also changed section 10: tanks holding up to 625 gallons could be put under streets and alleys but had to be removed if ensuing construction set buildings less than ten feet from the tank. And finally, the commissioners removed section 13, which banned handling oil over city wharves, heeding Charles Clarke's warning that this prohibition could end marine transportation of petroleum to the city. They replaced it with section 5 that allowed handling and storing oil on wharves but only under a special BOC permit.[60]

Opponents were unmoved. McClure Kelly, insurance underwriter and member of the Texas Fire Prevention Association (TFPA)—one of the founding member organizations of the NFPA—declared that the insurance and shipping interests preferred the original ordinance they had submitted to the BOC, which Kelly noted was identical to the oil ordinance New Orleans had adopted.[61] He read a letter from J. G. Pepper, the forthright and plainspoken agent for L&L&G in New Orleans, warning about the dangers of "immense" oil tanks such as Galveston's Santa Fe tank project. Commissioner Kempner dismissed the letter, asserting it was common knowledge that the New Orleans oil ordinance passed because of pressure exerted by the J. M. Guffey Petroleum Company and LSCOC. Moreover, it became clear that allowing the Santa Fe to finish its tanks was solely a "matter in the discretion of the board."[62]

The BOC ignored the complaints, including a last-minute letter from Charles R. Brown in which he strenuously protested the amended ordinance. Brown and McClure Kelly spoke for the TFPA, which battled municipalities over hazard mitigation and risk management, including inadequately funded fire departments, oil storage, and allegedly lax oil ordinances. The TFPA was particularly concerned about the GCSFR's construction of very large oil-storage tanks along its line, as well as similar tanks proposed by the Houston & Texas Central Railway in Fort Worth, Houston, Denison and various other points, making it clear that Galveston and other Texas cities faced policy cancellations if they were to proceed.[63]

Charles Brown informed the Galveston BOC that the TFPA strenuously objected to tank placement on or under streets, sidewalks, and alleys; construction of large storage tanks aboveground within the city; or allowing the fire-department chief or the president of the BOC "as a matter of public policy" to approve the storage or handling of crude oil "in any building in the city limits." Nonetheless, the commissioners unanimously approved the oil ordinance on November 5 during a three-hour meeting.[64]

The ordinance favored those who wanted to store large amounts of oil in Galveston but also opened the door for more modest proposals that the BOC could decide on a case-by-case basis, particular with regard to transportation plans that would integrate all elements of a petroleum system for Galveston businesses, including production, transportation, and "marketing or consumption."[65] Charles Clarke & Company, for example, had modified its dredges and tugs to burn oil, operated several oil-tank barges with more under construction for its numerous harbor contracts, and eventually obtained, a special BOC permit to handle oil on wharves.[66]

On the other end of the scale, the International Development Corporation (IDC) announced plans to operate oil tankers between Galveston and Mexico, to build a pipeline from the Beaumont oil fields to Bolivar Point, and to construct as much as 20,000,000 barrels of tank or earthen storage with the "necessary terminals and wharves" at Bolivar to supply tank steamers. General Manager H. D. Mepham,[67] a St. Louis railway promoter, observed that IDC, having picked Galveston over New Orleans and Port Arthur, expected city officials to shield its ambitious plans from harm by "preventing unwarranted restrictions" and by refraining from taking "unfair advantage" of the company in the future. Considering that I. H. Kempner and Thomas Taylor, city commissioners and powerful cotton merchants, were the principal Galveston stockholders in IDC, corporate promoters clearly expected the city to resolve the political and regulatory uncertainties about oil in their favor. The new oil ordinance of November 5 accomplished this resolution with regard to the construction of large storage tanks while making room for moving oil from Bolivar Point to the wharves for shipment to Mexico under a special permit. As the *GDN* pointed out, the Gulf & Interstate Railway had regularly moved standard tank cars on barges from Bolivar Point to Galveston wharves, a technique easily applied to petroleum and compliant with the then-pending oil ordinance.[68]

Although the new oil ordinance encouraged fuel-oil adoption in Galveston, an eight-month shortage of tank cars had almost crippled oil deliveries, leading to lawsuits and rulings from the Texas Railroad Commission over whether the railroads had deliberately withheld or misdirected rolling stock. Some suppliers, seeing little prospects for relief, refused to take any more contracts. Even so, officials, unable to turn their backs on potential fuel-cost savings, planned to construct two oil tanks at the city waterworks' pumping station large enough to hold a thirty-day supply.[69]

While the GCSFR had its own fleet of tank cars to supply its locomotives and did not have to worry about availability—apart from angry customers and the Texas Railroad Commission—it did have to apply under the terms of the new oil ordinance for a special permit to resume construction on its large oil tanks, since they were over the 15,000-gallon-capacity threshold and required earthen embankments. General Superintendent Nixon asked the BOC, November 18, for the permit, providing blueprints showing the location and details of a massive tank—ninety-five feet in diameter, thirty-feet high, holding 37,500 barrels or 1,500,000 gallons of oil—as well as a 20,000-gallon supply tank, a crane to hoist a pipe between the aforementioned tank and a locomotive tender, and pipelines running through city streets.[70]

Even though Nixon urged the commissioners to act promptly, his application went to a committee, made up of City Engineer Sias, Fire Chief Gernard, and Commissioner Valery Austin, that recommended granting Nixon's request if the GCSFR reversed tank positions to meet spatial requirements between tanks and contiguous properties or oil tanks belonging to other businesses, putting the large tank on Block 524 rather than 538 and the smaller tank on Block 538. Commissioner Kempner asked for a delay to allow all railroads an "opportunity to be heard" before final passage. Although Commissioner Austin pushed for immediate approval of Nixon's request, the board's president, W.T. Austin, postponed the committee report until the next meeting to give the GCSFR time to consider its position.[71]

A week later the board heard Valery E. Austin read another letter from Nixon, who announced that GCSFR had put a similar big tank at Cleburne, Texas (37,500 barrels), "in service" in an area more densely populated than the location of the Galveston tank. He reviewed similar circumstances about GCSFR's large tanks in Brownwood, Temple, and Alvin, Texas. Board President W. T. Austin ignored the implication that these large tanks were safe and responded that the Galveston site was acceptable if the "proper safeguards" were in place, reaffirming the need for hazard mitigation. Fire Chief Gernand supported this position by defining the required mitigation in practical terms: the "embankments about the tanks" had to be high enough to contain oil released from a tank failure. The board referred the petition back to committee to avoid more misunderstandings about what the city expected from GCSFR, and it requested a meeting with railroad officials so they could "ascertain just what was desired" by the city and "arrange for the proper safeguards connected with the construction of the tanks."[72]

By the end of the year the GCSFR, as well as the GHH, a venerable line hard hit by the 1900 hurricane that destroyed its bridge across Galveston Bay, had ordinance proposals ready for the commissioners. Both lines operated out of the western part of the city away from the business district and associated valuable real estate. Their detailed ordinances incorporated risk-management strategies for mitigating hazards. The GHH proposal included the construction of two large tanks, each sixty by thirty feet, and a 25,000-gallon delivery tank in the railroad's Galveston yards, along with a crane and connecting pipes under the streets. Earth embankments surrounded the tanks, and they were fitted with pipes to smother fires, should they occur. The Santa Fe asked to build two tanks, their locations reversed from its initial petition to coincide with the BOC's committee recommendations, with emergency reservoirs and pipelines for storing and supply locomotives with fuel oil.[73]

After several conferences among GCSFR representatives, J. H. Hill of the GHH and the BOC committee that had scrutinized the two railway ordinances, participants came to a "definite understanding" of their common goals based upon what was politically possible. The improved ordinances gained BOC approval in January 1902, and the GCSFR promptly resumed tank construction the next morning.[74]

Various retail and commercial firms converted to burning fuel oil in 1901 under fire-underwriter rules for oil-burning equipment. When the new oil ordinance passed in 1902, other businesses moved ahead with their plans to meet Galveston's regulatory requirements for oil-storage tanks. Implementation of these plans relied upon licenses or permits that were familiar legal instruments of municipal police powers. Awarding these special oil permits displayed a developing, bureaucratic rationalization of risk marked by the evolving interactions among urban built environments, energy markets, and risk management in the late nineteenth and early twentieth centuries. The focused use of special permits to control oil interests remained a permanent fixture in Galveston's risk management of petroleum and other flammable liquids until World War II.

Adjudication of oil-permit applications confirmed that the relentless use of instrumental reasoning drove risk-management efforts to find solutions for all hazardous conditions or events. It also demonstrated the BOC's broad authority to carry out the city's risk-management policies, as it applied relevant ordinances, tempered by common-sense pragmatism when encountering unanticipated or unusual circumstances. When necessary, the BOC was willing to shift from a rigidly prescriptive to a somewhat flexible regulatory philosophy. When required to deal with unique, infrequent, or unconventionally hazardous circumstances "without specific direction" from prescriptive rules, the BOC devised commonsensical solutions to maintain the city's hazard-mitigation and risk-management requirements.[75]

With detailed policies and a new ordinance in place, the BOC efficiently processed unambiguous applications. For example, the owner of the Model Laundry, having made a failed bid in October 1901 for a fuel-tank permit, successfully petitioned the BOC to install a 625-gallon tank—the maximum allowable capacity—under the sidewalk next to a building on Block 503. Clarke & Courts, a successful stationery-printing operation that occupied a five-story building at Twenty-Fourth Street and Ships Mechanic Row, asked for an identical under-the-sidewalk tank. Seeking to "embark in the petroleum business," W. R. White sought permission to construct a warehouse and oil tanks at Forty-Fifth Street and Avenue R, while George H.

Henchman, a cement supplier and roofing contractor, wanted to install an oil tank on Block 571. President Austin agreed if Henchman were to install the tank on a sidewalk location chosen by the city engineer "in strict compliance" with the new oil ordinance.[76]

With a clear oil policy in place, the BOC was also able to correct or reject deficient petitions. Wisrodt Grain Company, a large feed dealer, wanted to build tanks for "lamp oil" (coal oil or kerosene) on Block 570, which President Austin initially rejected, noting that the city's oil ordinance did not "discriminate between the different kinds of oil except as to quantity" but did require a stated tank size which the petition did not include. Subsequently, Fire Chief Gernand reported that the petition was flawed because Wisrodt wanted to store the oil in a wooden building. The lawyer for the Little Sisters of Charity made the same mistake. He petitioned the BOC for an oil tank at St. Mary's Infirmary, but no approval was forthcoming because the petition did not specify the size of the tank. It was a technical and legal rather than a theological misstep, for which the Sisters' lawyer could be forgiven.[77]

Using the new ordinance to deal with specific petitions, the BOC provided several modifications to resolve apparent difficulties. John P. McDonough, owner of McDonough Iron Works, a large boiler and foundry establishment, wanted to install a 1,700-gallon tank under the street on Avenue A between Twenty-Sixth and Twenty-Seventh streets adjacent to his works, whose tank exceeded the 625-gallon limit. Having initially rejected the petition, the BOC amended the city oil ordinance (enshrined as Article 279 in the city's 1902 *Revised Ordinances*) by abolishing the 625-gallon limit in favor of leaving the capacity of tanks placed under streets and sidewalks to administrative discretion, as long as the decision took into consideration "the surroundings of the particular locality." President Austin explained that a special BOC committee had recommended allowing large tanks on streets, alleys, and sidewalks "in certain parts of town where there would be no objection to them." In "such isolated places," large-capacity tanks would not jeopardize nearby property.[78]

In January 1902, when the BOC gave the GHH permission to construct storage tanks and pipelines, General Manager J. H. Hill asked if the railway could extend its pipelines to the railroad-shop boilers, believing on the basis of legal counsel that the BOC could "add or abridge" GHH's tank ordinance. However, on the advice of City Attorney J. Z. H. Scott, Hill decided to draw up a "separate ordinance to embody the amendment." Two months later, the BOC reversed its legal formulae, and GHH applied to move its already approved oil-supply tank from Block 584 to 582 to be near the road's roundhouse. Commissioners agreed but insisted on passing an amended, rather than a separate, version of the original GHH tank ordinance. With large-scale oil operations concentrated on the city's west side (GHH's tanks were adjacent to the GCSFR tanks), the BOC had to avoid contradictions and ambiguities in the legal instruments that described its risk-management policies.[79]

The risk policies set forth in the new city oil ordinance required tank owners to obtain a certificate of inspection, leading Fire Chief John H. Gernand in late September 1903 to visit an SPR tank site at the foot of Forty-First Street on the SPR docks. The SPR and its subsidiary, the Southern Pacific Terminal Company, operated freight and wharf activities including storing fuel oil for operations. Gernand informed the keeper of the railway's power house that he needed a permit to use the tank. Gernand reported this notification to the BOC, which empowered him to "see that the law is complied with" and to examine the "situation and condition of the tank and surroundings." Subsequently, Gernand informed the BOC that he had "actually and thoroughly examined" the premises of the 10,000-barrel tank, noting that it was within a catch basin, more than one hundred feet from any building and not a threat to any "adjoining property." SPR was granted the necessary permit.[80]

In 1904–5, W. A. Guyton & Co., controlled by W. A. Guyton and Charles H. Moore, sought the city's permission to run pipelines from Avenue A to various points, one of which was Eighth Street, the beginning of the grade-raising canal where fuel-oil-burning dredges and barges entered the city with sand. The pipelines would feed loading racks for fueling grade-raising machinery along their routes. Guyton also asked the city if it could build oil tanks "upon piling in the grade raising canal in a manner which is not in compliance" with the oil rules. President Austin and Fire Chief Gernand recommended granting a permit for these tanks with a total capacity of 14,000 gallons, "on the edge of the canal," believing that "there will not be any danger from fire . . . if it can be done under existing ordinances." The BOC issued a temporary permit to construct "tanks along the canal right of way for the purposes desired and as the President may deem proper." Here again was a presidential exercise of discretionary power, as allowed in the oil ordinance, to apply the general principles of risk management to special circumstances, which President Austin noted in this case involved a tank not covered in the city's current oil ordinance.[81]

During the first four decades of the twentieth century, petitions, ordinances, and interpretations of the oil ordinances, as well as 1940s efforts to pass a new building code with fire-prevention provisions for "storage of oil or oil products," demonstrated that Galveston's commitment to the risk management of flammable liquids on the basis of general principles of hazard mitigation set down in the November 1902 oil ordinance continued through World War II. Risk management through tank licenses during World War II point to the city's resolve to manage risks even amid wartime conditions. C. W. Karbach in late August 1942 asked the city commissioners for a nine-dollar refund "on his oil tank license" issued in the previous May for his now defunct business on Broadway.[82] City officials also continued to use fire-insurance-company rules to regulate the technical details of oil-burning systems, which had surfaced in 1901–2 in the "Texas form" and L&L&G demands. This symbiotic relationship with the fire-insurance industry perpetuated the companies' semi-official role of governance in local affairs. When Nehi Bottling Company asked the city in 1928 for permission to install an underground oil tank on its premises, the city stipulated that the city fire marshal and the building inspector supervise and approve the installation, which had to be "in accordance with the Underwriters Specifications." They gave the same reply to a request to install a similar tank at the Jewish Synagogue on Avenue I.[83]

Galveston's *Revised Ordinances* of 1902 and 1917 framed the city's ongoing management of petroleum risks. The 1917 *Revised Ordinances* follow closely articles 273–81 of the 1902 *Revised Ordinances* with some added details. Article 278 (1902) provided that aboveground tanks had to be one hundred feet from the nearest building and tanks "wholly or partly under ground" could not be "less than ten feet" from the same. Article 244, the 1917 counterpart to 278, replaced these distances with a three-column table for aboveground tanks that showed capacities in the first column, starting with three hundred gallons or less up to unlimited capacity, with matching columns showing allowable minimum distances to the adjoining properties and to nearby oil tanks. There was no change in the ten-feet-from-a-building rule for tanks partially or completely underground.[84]

The 1917 tank-capacity-distance schema played an important role in a

1929 hazard-mitigation incident in the GCSFR yard on Galveston's west side. The company had a large oil-storage tank on block 524 (Forty-Third, Forty-Fourth, Market, Postoffice streets) that had been part of the 1902 oil-ordinance debates. Superintendent A. P. Hall considered trading the tank for a newer one on block 587 or exchanging block 524 for block 587, but nothing was done, in part because the GCSFR would have had to build a new pipeline and move its oil-pump house. Subsequently, officials scheduled repair of the tank's steel shell and roof, and overhaul of an oil line used to fill the tank, as well as a new paint job.[85]

On May 17, rain filled the tank's catch basin and floated it "thirty to forty feet" to one of the containment dikes, leading Superintendent Hall to suggest using rope and a block-and-tackle to "swing the tank back to position as the water goes down." When that proved to be impractical, he told Division Engineer Z. A. Green and W. H. Kluever to relocate the container without dislodging its concrete bottom or allowing it to disintegrate. Refloating the tank failed, and cracks in the bottom ruled out refilling the tank with oil.[86]

To determine the legality of leaving the tank in its new position, W. A. Hudson, the GCSFR's right-of-way agent, investigated Galveston's ordinances at the request of the company's engineer, K. B. Duncan, "with respect to distance from other structures." This was a fundamental issue during the 1901–2 oil debates in Galveston and New Orleans. Galveston addressed tank locations in its 1902 *Revised Ordinances* for aboveground and belowground tanks. Hudson reported in a detailed letter, accompanied by a typescript copy of the city's tank-distance rules, that there was "not room enough" on block 524 to "comply with the revised ordinance of 1917." The displaced tank was too close to other structures, violating the inter-tank distances set forth in the city's 1917 oil ordinance.[87]

Hudson advised in his report that going before the Galveston "City Council" [*sic*] or leaving the tank in its new location might "wake up a sleeping dog" and force the railroad to move it. Green agreed, advising the superintendent that repair work should be suspended until it could be determined that the tank "could be safely left in its present location from a legal standpoint." GCSFR's chief engineer, F. Merritt, believed that moving and repairing the tank should be done to "avoid any controversy with the City or adjacent property owners." Hall also asked G. G. Moore, general manager of the GHH, to find out if "your Insurance people" would complain about leaving the tank at its present position near a GHH tank. Two months later, Moore replied that he had "taken this matter up with our insurance Underwriters," and they have "very serious objection[s]" to leaving the tank in its new location, promising to "cancel insurance on our structures" that are in "close proximity to the tank." Moore asked Hall to move the tank back to its original location so that "our insurance will not be disturbed." Subsequently, the railroad repaired the base and moved the tank back to its proper location.[88]

This event confirms the importance of the risk-management politics of 1901–2. It set in motion hazard-mitigation strategies employed by Galveston throughout the twentieth century. City officials and fire-insurance underwriters set down ordinances in 1901–2 that reflected the growth and dissemination of risk knowledge. A network of fire-insurance companies, research laboratories, engineering and technical associations, insurance alliances, and cooperative ventures contributed to the regulation of the storage and handling of fuel oil in Galveston. They created rules, regulations, and standards that promoted new or improved methods and procedures in hazard mitigation. The NBFU, NFPA, and the SETA assisted Galveston fire-insurance agents and underwriters, enabling them to use scientific and technical studies in the struggle to persuade city officials to adopt ordinances that reflected the latest insights into petroleum hazards and risks as well as mitigation procedures and strategies. This grew into a general concern over hazard mitigation of flammable liquids, the numbers of which increased geometrically as the twentieth century passed. Thus, the battle over regulating fuel oil in Galveston drew the city into the rationalization of risk management, reliance upon science and technology, and awareness of the connection among risk, technology, and city governance.

CONCLUSION

These four Gulf South cities created policies to manage risks associated with storing, transporting, and consuming petroleum for fuel. The commercial and mercantile elites of the cities responded very quickly to the discovery and development of the Beaumont oil field in the first six months of 1901 and made plans to bring petroleum to their communities to foster manufacturing.

When the Corsicana field opened in 1892, it provided oil to Houston and Galveston through companies such as the SOC's invisible surrogate, Waters-Pierce Oil Company. Kerosene was its dominant product, but the company also handled fuel oil, which was moved by railcars to consumers. Chicago Fire Underwriters' Association, responding in 1892 to the growing use of fuel oil, particularly to power the Columbian Exposition, devised a set of permit rules that controlled how oil plants should be constructed, installed, and operated to burn fuel oil. Underwriters in Louisville and Philadelphia adjusted their regulations as well, and major fire-insurance companies such as L&L&G required a similar regimen to supply the technical details and operating rules of oil-burning plants that had antecedents in insurance rules devised in the late 1880s.

In turn, the TFPA, representing fire-insurance representatives and agents in the state, followed this path, recommending two versions of the rules, one known as the "Chicago form" and the other the "Texas form," differing chiefly in how far away a supply tank could be from an oil-burning appliance in a building. The "Chicago form" suited dense building construction, while the "Texas form" was more liberal and appropriate for low levels of population and an uncrowded built environment.

The four cities had in common a strong sense of risk and disaster, in particular petroleum explosions and fires, that could threaten business districts, and in Mobile, New Orleans, and Galveston, huge stocks of cotton stored dockside. Merchants and manufacturers in New Orleans and Galveston set plans in motion to burn fuel oil before the municipalities had time to come up with ordinances that would incorporate the existing insurance rules on oil consumption, adding important local details involving space and location that controlled how and where huge oil-storage tanks and pipelines could be built, as well as sites for barge distribution, as hazard-management techniques.

In the four cities, two groups—the oil producers-distributors-consumers and their engineering and business entourage, facing off against risk-averse residents as well as maritime and fire-insurance interests—negotiated the terms of managing petroleum hazards, colliding repeatedly in City Council sessions and committee meetings over the actual risks generated by petroleum use.

With a long history of devastating fires coupled with companies planning to convert to fuel oil, Mobile passed strict rules about its use in December 1901, followed by a successful insurance campaign to block J. M. Guffey Petroleum from building a massive tank close to the central business district. The city withdrew its approval of the tank and passed a new oil ordinance in May 1902 that put the Guffey site just outside the boundaries where such tanks could be constructed. Much of the Mobile campaign for safe oil storage was a battle with a single, major oil company. Small businesses and manufacturers complied with city ordinances on storing and handling fuel oil, since many of them were familiar with the fire-insurance rules on which the ordinances were based.

Houston received applications for pipeline franchises and oil consumption early in 1901, which also fell under rules in place for oil-burning equipment provided by the TFPA and required by major fire-insurance companies such as the L&L&G. As oil use increased, the insurance interests asked for an ordinance regulating storage and consumption, which the city approved in early January 1902 after months of debate. Complaints about its ban on storing oil under sidewalks produced a favorable amendment, but the city reversed course in April and reinstated the prohibition. This became a central issue in a protracted legal battle between Jacob Binz, owner of the Binz Building, and the National Supply Company, represented in Houston by Elmer Billow. Binz contracted with the NSC to install an oil-burning plant in the building's basement beneath the sidewalk, in the belief that the city had decided to allow such arrangements. But before Billow could complete the project, the City Council reinstated the ban, and Binz refused to pay NSC for materials and work done. The result was lengthy litigation.

Bitter fights in New Orleans over oil storage and distribution started in the late summer of 1901 to deal with a rising tide of petitions for oil tanks and pipelines. The battles fell into distinct installments. First, the city by August approved a tank ordinance to regulate the technical arrangements of oil-supply tanks used to feed boilers and machinery. It incorporated well-known specifications provided by fire-insurance companies and established locations behind the city and downriver in the Third District for large-scale storage tanks. Debate over it became intertwined with a long fight over a request from LSCOC for pipeline and oil-storage privileges, which featured angry exchanges about franchises, monopolies, and political collusion and intrigue. Second, in October 1901 the city passed a companion ordinance that dealt with pipeline, barge, and railcar distribution of oil, a risk-management tool that ultimately forced LSCOC to give up its pipeline request and follow the city's rules, although the city tried to mollify Lone Star by approving a major oil-storage tank, pipeline, and pumping station on and along Egania Street.

And third, from November 1901 to March 1902 oil men and fire underwriters clashed over amending the so-called tank ordinance to relax rules on handling oil across wharves, placing oil tanks inside buildings, and requiring masonry tank enclosures. The insurance underwriters grudgingly made limited concessions on these meticulously scrutinized details, giving each a fine-grained defense in behalf of hazard mitigation. As in Mobile and Houston, the fire-insurance interests, acting under the advice and assistance of the formidable risk-knowledge industry spearheaded by their home companies, standards associations, and engineering expertise, had the upper hand.

Galveston followed a different path owing to the broad powers of the commission government and the dominant role of Commissioner I. H. Kempner, who initially submitted an insurance-sponsored ordinance, but under criticisms that focused on the urgent need to promote petroleum consumption, promptly crafted a revised version more satisfactory to the oil interests. Just as some New Orleans officials were hostile to the LSCOC efforts to set up an oil-supply system along the riverfront, so also various councilmen-commissioners in Galveston resented the power of John Sealy and the Galveston Wharf Company. It was quite evident when the Charles Clarke Company petitioned the city for permission to set up an oil business on the wharves. The Galveston BOC eventually made concessions about handling oil across wharves and passed an ordinance that shifted municipal risk management of fuel from a prescriptive to a flexible, administrative stance, granting the president of the BOC and the city engineer considerable discretion in resolving unforeseen questions raised by applications for oil consumption and storage. These officials worked within the city's general risk-management ground rules such as location, inter-tank distances, and capacities but granted easements on a case-by-case basis.

What is the significance of these events? First, the thread that runs through all of the Gulf South struggles over petroleum is a nineteenth-century risk culture which important insurance associations such as the NBFU and the NFPA gradually embraced. As an inchoate form of fire-prevention engineering, it rested in its early stages upon professional know-how, practical experience, and eventually upon science and technical analysis.

Gulf South officials through insurance men—agency managers and special agents, surveyors, consultants, underwriters, and inspectors—crafted risk-management policies, basing their risk-assessment and hazard-mitigation strategies regarding the handling and storage of fuel oil on accreted risk knowledge from the nineteenth century. The insurance interests—local and special agents, inspectors, experienced fire-insurance-company leaders—drew upon this knowledge to rationalization insurance rules and methods for dealing with petroleum hazards, in which engineers and scientists such as C. F. Chandler, C. J. H. Woodbury, Henry Morton, and A. M. Schoen played an important role.

Second, the city-council/commission proceedings in the Gulf South cities show that the insurance interests successfully persuaded city officials to convert insurance rules into legally enforceable regulations that prescribed how and under what circumstances a business could reconfigure its energy sources—coal to petroleum or gas for example. This was a major accomplishment in the long-running, fire-insurance campaigns dating to the Civil War era to raise their voluntary insurance norms to the level of mandatory, public directives. Fights over details such as acceptable distances between oil tanks and adjacent property; confining oil activities to specific zones inside a city or to areas outside of a town; specifications for piping, pumps, venting; and designating tank locations and installations above or belowground, were generally decided in favor of the fire-insurance companies or at least resulted in compromises they could live with.

Further, through the politics of risk management, city councils adopted enforcement provisions that granted city engineers or mayors, who signed off on permits for storage tanks or oil-burning apparatuses, the discretion to make exceptions to certain details of an ordinance or to make liberal interpretations of ordinances, not clearly or completely contrary to regulatory intent, that allowed businesses or manufacturers to burn fuel oil under unanticipated or unique circumstances.

Hypothetically, this might mean approving a storage tank in an oil yard with a capacity greater than regulatory limits owing to the difficulty of delivering or parking tank cars on a railroad spur, thereby reducing the number of times the tank would have to be refilled. The administration of permits and their requirements rested upon broad, municipal police powers that gave Gulf South cities wide scope to protect and promote public safety and the well-being of their citizens, which included the mitigation of oil hazards and installations rules.

Third, the oil-regulation disputes disclose the importance of petroleum-risk management in the governance of a new energy market crucial to the industrialization of cities that needed cheap and reliable energy carriers. The battles over the storage and handling of petroleum took place at the intersection of energy, municipal regulation, urbanization, and technology in U.S. history, eliciting concerns over hazard mitigation and risk management well before the appearance of more sophisticated risk-management such as employing quantitative-statistical techniques for manufacturing-quality control in the mid-1920s to the 1950s; wartime and post–World War II studies of vacuum-tube failure rates and aircraft vulnerability; and post–World War II, probability-based QRA models applied to "complex engineering systems" as found in the 1950s through the 1970s in the nuclear-power, chemical, or aerospace industries.[1]

The oil rules were also a response to small-scale, incremental industrialization in the Gulf South, a process which has always varied in countries, regions, and locales by "timing . . . government policy, pace of change and cultural circumstances," not to mention a devastating civil war. By the 1890s, industrialization had begun in several southern locations, such as textile manufacturing in the Carolinas and Georgia. In the case of Houston it was the wave of the near future.[2]

Finally, Gulf South officials, fire-insurance representatives, and the oil interests engaged in the politics of risk management. Through rhetoric and political interactions, they produced regulations for the storage and consumption of petroleum, establishing themselves as agents or social actors in a complex, socio-technical system built around petroleum exploration, production, consumption, and regulation.

The risk management of petroleum at the beginning of the twentieth century did not precipitate technical revolutions; rather, it made good use of technical accomplishments in energy production and consumption already underway. But it is clear that in the Gulf South—Mobile, New Orleans, Houston, and Galveston—the evolution of the standardization and formalization of hazard mitigation and risk policies heralded the beginning of a fledgling modernization of a major coastal area of the South.

EPILOGUE

Risk Management of Flammable Liquids in the Twentieth and Twenty-First Centuries

From 1902 through World War II to the present the four Gulf South municipalities kept their petroleum risk-management schemes in place. With the coming of zoning measures, World War II pressures to expedite material handling, and the accumulation of statistics about oil-related disasters, the cities modified their regulations to reflect growing knowledge about petroleum safety and advances in engineering. However, their treatment of petroleum hazards maintained a core of basic mitigation techniques laid down in 1901–2 that was developed, modified, and adapted to the growth in technical knowledge and experiences in storing flammable liquids throughout the twentieth century.

Mobile relied upon spatial controls to keep large, oil-storage tanks in city districts where industrial and manufacturing activities required huge infusions of energy. The city confined large oil tanks to an area defined from 1902 to 1907 by Water, Delaware, Franklin and South Carolina streets, moving the southern edge in 1909 to Tennessee. This expanded in 1919 to include land east along Virginia Street to the river channel and much of the Choctaw Point tract south to an extension of Tennessee Street, encompassing the switching yard and pier of the Mobile, Jackson & Kansas City Railroad. Amendments in 1916 to the city's 1907 ordinances set aside a good part of Farmer's Island for storage tanks up to ten thousand barrels in capacity.[1]

In 1934 the city added two areas south of Eslava and the business district that bordered the river channel, as well as an area north of the business district around the intersection of Water and Beauregard streets along the tracks of the Mobile & Ohio Railroad. Officials also permitted large oil tanks within an area from Choctaw Point along the Garrows Bend Channel to Live Oak Street, winding through several streets north to Delaware and east to Water Street. The city retained the proviso that oil tanks built in this area with a capacity exceeding twenty-five hundred barrels had to be enclosed in levees and embankments that could retain slightly more than 100 percent of a tank's capacity.[2]

The city's 1947 code set out new areas for large oil-storage tanks near railroad operations: one small zone centered on the intersection of Savannah and Royal streets alongside the Louisville & Nashville Railroad yard; and another between One Mile Creek and Marion, and St. Joseph and Conception streets north of the business district, where there were several railroad and SOC facilities. Reflecting city growth and changing land-use patterns, officials in 1951 expanded oil-tank construction north of town starting with the L&N right-of-way and the Alabama State Docks up the Mobile River to Three-Mile Creek, which the 1955 code edition precisely described in surveying terms. In contrast, the city never changed from 1902 to 1955 the area for off-loading oil from barges and ships on the riverfront. It had to take place south of the business district between Texas and Virginia streets.[3]

Mobile officials also retained the original oil supply-tank specifications from 1902—required at that time by all fire-insurance companies—through the 1934 and post–World War II code editions of 1947 and 1955. This remarkable continuity demonstrated an ongoing participation in the technical and engineering risk-knowledge from the late-nineteenth century as modified and augmented through the twentieth century. This was based on growing bodies of rules and standards that reflected scientific and engineering refinements and experiences with handling flammable liquids. There is also evidence of the city's tentative steps, that started in 1934 and continued after World War II to 1947, into a new realm of hazard mitigation—petroleum as an environmental hazard. The city prohibited pumping or discharging petroleum, mineral, or organic oils into the Mobile River or adjacent waters from ship bilges and tanks, shore storage, or tank cars. Special care had to be taken in loading or discharging oil from ships to prevent accidental spills or leaks.[4]

In the 1960–92 period, Mobile adopted several consensual fire codes—sometimes excluding specific parts—including the 1965 and 1970 editions of the American Insurance Association's *Fire Prevention Code*. In 1978 the city adopted the *Southern Standard Fire Prevention Code* (SSFPC) issued by the Southern Building Code Congress International in Birmingham, Alabama; embraced the 1985 edition, excluding the section based on the *NFPA 101, Life Safety Code* (*LSC*), in February 1986; and in 1992 took up the 1991 version of the same code, adopting as well the NFPA's *National Fire Codes*.[5]

The Alabama state fire marshal's role in fire prevention, dating from 1917, included the statutory responsibility to devise rules and regulations on storing combustible materials and "highly inflammable materials" such as "crude petroleum, or any of its products . . . or inflammable fluids or compounds," adding a second layer of fire-risk management. This created complications for Mobile fire inspectors in the early 1980s, who had to enforce the city's AIA *Fire Prevention Code* and the fire marshal's *National Fire Codes*. The NFPA developed this comprehensive set of technical standards—written in legally enforceable language—as research-based rules that insurance companies, technical or professional associations, and municipalities could recommend, adopt, or impose upon clients or industries to mitigate hazardous conditions or events. By 1966 the codes included the *NFPA 101, LSC*, the groundwork for which an NFPA Committee on Safety to Life laid in 1913. Long known as the *Building Exits Code*, the *LSC* covered issues such as building design, exits, interior finishes, and other hazardous features. The Mobile City Commission in February 1981 declined to adopt the *National Fire Codes*, because the *LSC*, if thoroughly enforced, would have put many Mobile buildings in violation of state law.[6]

In keeping with the city's risk management of fire hazards, officials embraced model fire codes designed for voluntary, national adoption because they represented a "broad spectrum of fire prevention experience," reflecting decades of accumulated hazard-mitigation procedures and technical expertise. In 1994, three regional organizations set up the nonprofit International Code Council (ICC) to create a set of complimentary safety codes. The International Conference of Building Officials (ICBO), the Building Officials and Code Administrators (BOCA), and the SBCCI had historically issued model codes on building construction, fire, electricity, and other areas. BOCA codes were used in the Midwest, while most southern states followed the SBCCI materials. The ICC produced its first codes in 2000, followed by revisions every three years up to 2012. The *International Fire Code* addresses fire-protection requirements for existing and newly constructed buildings, as well as substances, processes, and materials stored or used in these buildings, including flammable liquids such as petroleum and its products. In the 2009 version, chapter 34 covers flammable and combustible liquids and cites ASTM and ASME standards, the *International Building Code,* the *International Mechanical Code,* and the venerable NFPA 30.[7]

Mobile adopted several ICC codes, including the fire codes (*IFC*) for 2006, 2009, and 2012, in step with the Alabama state fire marshal, who also adopted ICC model codes. This put Mobile squarely in the tradition, established in 1902, of devising rules, regulations, and techniques for managing petroleum hazards on the state and local level, sustained by the accumulation and dissemination of technical and research-based risk knowledge, risk-assessment procedures, and risk-management strategies.[8]

In both *IFC 2009* and *IFC 2012,* fuel oil is covered in chapter 6, section 603, not chapters 34 or 57, which are respectively about flammable and combustible liquids. Fuel-oil storage in outside, aboveground tanks outside of the point of consumption is addressed in chapter 6, section 603.3.1 (*IFC 2012*), which limits storage to 660 gallons, and anything above that capacity is covered in NFPA 31 (Oil-Burning Appliances). Fuel storage inside buildings can also have an aggregate, maximum capacity of 660 gallons. Aggregate capacity can increase to 3,000 gallons if stored in "protected aboveground tanks" or a single tank and located in a room in the building served if protected by an "automatic sprinkler system." Section 603.3.3 cites NFPA 31 for rules regarding fuel-oil storage in underground storage tanks. The move from chapters 34 (2009) or 57 (2012) reflects decades of experience with storage and consumption of fuel oil providing sufficient data to predict the probability of fire-initiating events using quantitative risk analysis.[9]

Houston followed a different path between 1904 and 1943. Its 1914 code of ordinances included a detailed ordinance from 1911 regulating the storage and use of "gasoline, or other volatile, inflammable liquids." The city's massive, revised code of 1922 expanded upon these rules, banning retail supply tanks, pumps, or compressed air along segments of Main Street in downtown Houston as well as new "filling" stations within the city, but without the segregated areas for large-tank storage so prevalent in Mobile's regulations. There were also several pages of detailed restrictions on the materials and specifications of tanks that stored flammable liquids such as gasoline, which tied allowable tank capacities to their distances from the lowest levels of nearby buildings, with additional rules for tanks that supplied buildings through "long-distance" pipe systems and for portable containers.[10]

Similarly, article 4 in the 1914 code on oil was identical to the 1904 code except for a 1913 amendment that allowed oil-supply tanks to be under sidewalks. Rules on oil-tank permits, construction, placement, embankments, or reservoirs for large, aboveground tanks (maximum capacity ten thousand barrels; two thousand feet separation between tanks), were unchanged. The city's 1922 and 1947 codes vary only slightly from their predecessors. The 1914 code required two thousand feet between large storage tanks, but by 1943 that had been reduced to two hundred feet.[11]

There was, however, no diminution of the size of the 1943 code (1,342 pages), reflecting city growth and new risk-related knowledge. To accommodate rule proliferation, Houston published its 1958, 1968, 1974, and 1984 codes in ring binders. Using loose-leaf page inserts, the city could routinely update practices and standards arising from new technical insights and scientific information that called for modified hazard-mitigation techniques and risk-management strategies.[12]

Houston's 1974 *Fire Code,* chapter 17 of the city's complete code, is over six hundred pages. It was one stage in the serial development of the 1958–84 codes with a common format, as well as familiar topics that evolved incrementally as a result of growing information, issue clarification, and efficacious hazard-mitigation techniques. New topics reflected state-of-the-art technical insights, newly recognized hazards, and industrial changes. A noteworthy part of chapter 17 is a four-page inventory of standards and accepted fire-prevention methods from "publications by recognized national authorities and technical or scientific organizations" that were "adopted by reference" into the code. This important directory included selected standards, practices and tests from the AIA, ANSI, API, ASME and the ASTM, as well as three pages of NFPA standards, most notably NFPA 30 Flammable Liquids Code (1973) and 31 Oil Burning Equipment (1972), the birth of which involved NBFU engineer Allen M. Schoen and his trip through the Gulf South in 1902.[13]

Unlike pre–World War II city fire codes, chapter 17 was a comprehensive guide for a great variety of fire hazards in Houston, including tire-retread plants, bowling alleys, compressed-gas operations, flammable finishes (paint-dipping tanks, automobile undercoatings), dangerous chemicals, lumber yards, ovens and industrial baking, places of public assembly, welding, stack-storage of combustibles, liquefied-petroleum gases, cryogenic liquids, schools, hospitals, and motels, as well as flammable and combustible liquids. Chapter 17's articles, nearly identical in topic and order to those in the *Southern Standard Fire Prevention Code 1974 Edition,* show that its broad scope mirrored evolving model fire codes that addressed not only technical standards but also practical problems and hazardous circumstances faced by state or municipal fire officials.[14]

Article 15 on flammable and combustible liquids stipulated that all equipment for storing and handling these substances had to conform to NFPA 30, unless stated otherwise in "laws or regulations legally in effect." Further, this section placed combustible and flammable liquids in classes I, II, IIIA, and IIIB, and classes IA–IC based respectively on flash-point temperatures. It also identified, based on class and volume, which liquids required a storage permit.[15]

It banned aboveground storage tanks in fire zone no. 2, which was downtown Houston, and, for allowable areas, stipulated minimum distances based on tank capacities from public streets or buildings as well as surrounding earthen, metal, or concrete dikes. Other rules described concrete or brick vaults for tanks placed inside buildings; and restrictions on tanks installed underground, piping apparatuses, and venting. In all areas of concern, though updated to reflect current risk knowledge and technical details of materials and tank construction, the fire code imposed the same basic constraints found in the Chicago Fire Underwriters' rules from 1892 and the specifications incorporated in Houston's 1902 oil ordinance required by fire-insurance companies.[16]

Houston eventually adopted the 2006 edition of the *IFC,* a regulatory enactment that was the culmination of more than a century of risk management of flammable liquids, particularly petroleum. This confirmed the city's commitment to ordinances and codes based on science, technical expertise, and professional experiences, showing that in terms of core issues the *IFC* was the regulatory capstone of Houston's 1902 efforts to control petroleum distribution and consumption.[17]

During World War II, Galveston, a community with a long history of hazardous natural events, had a large stock of buildings and homes that were not fire-resistant. On the eve of Pearl Harbor the city did not have an enforceable building code—it had never been published as required by law—leading the mayor to appoint a building-code committee of architects and contractors to examine the relevant ordinances that had been revised repeatedly to accommodate changing circumstances. A revised building code that established minimum construction standards acceptable to the National Board of Fire Underwriters would "insure the public health . . . [and] safety to persons and property" and hold down insurance costs.[18]

The committee's draft followed an NBFU-recommended building code and addressed fire prevention under many topics, including "storage of oil or oil products." The Texas Board of Insurance Commissioners and the NBFU contributed advice and guidance. But in October 1942 the city, having submitted the building code to the NYBFU for suggestions, had not received a response owing to uncertainty about the wartime availability of building materials. Subsequently, it was decided to keep its legally crippled code—originally adopted in December 1914—and defer rewriting it until the war ended, when "structural materials will be . . . changed . . . and new systems adopted." Galveston finally adopted a revised, detailed building code in 1946 that complemented its regulation of the storage of petroleum and flammable liquids.[19]

The city struggled to keep pace in fire service. An NFBU survey of the city's firefighting capabilities cited inadequate water supplies, water storage, insufficient pumping capacity, and out-dated equipment, a negative assessment similar to NBFU reports in 1910 and 1925. The report also recommended the adoption of a "complete fire prevention code governing hazardous materials, processes and occupancies" under supervision of the city fire marshal with the NBFU's recommended fire-prevention ordinance serving as a guide in "framing new regulations." Since the report came out in July 1946, three months after the city had adopted its new building code in March, officials agreed with the substance of the NBFU report but took no action, claiming the city did not have the money to tackle the problems cited.[20]

The requirements subsequently became a campaign issue in city elections in 1947, but not much was done. A Chamber of Commerce insurance committee strongly recommended in 1948 that city commissioners act promptly on the NBFU recommendations to "restore Galveston's good fire record credit." The board agreed to study the matter, which also included a call to strengthen the "fire prevention code" and rigidly enforce it.[21]

Notwithstanding Galveston's protracted hesitation to improve its fire-service and fire-prevention flaws, its 1946 building code did address the storage of petroleum and its products with regulations that followed the general principles devised in 1902: a permit system and inspection; tank location, orientation by grade, spatial separation; embankments or excavations for large-storage tanks; and vaults for underground supply tanks.[22]

Attention to petroleum storage and consumption surfaced in 1953, when the Galveston Insurance Board recommended that the city adopt the NBFU's recently issued *Fire Prevention Code.* But despite repeated pleas and petitions, the city relied on the regulations in its 1948 code of ordinances and did not pass a model fire code until 1966. Yet, it became clear that by 1950 code developers had subsumed the control of flammable liquids under general fire prevention. Fueled by the wartime acceleration of industrial and business paradigms that emphasized standardization, quantification, and control, scientists and engineers produced a growing stream of theoretical and technical developments, with which code developers and standards groups kept pace. Fire-prevention codes expanded to safeguard human life in many day-to-day, hazardous conditions that fire services might face. Their topical coverage widened and deepened to include building egress, occupancy restrictions, hazardous manufacturing processes, airports and aircraft fueling, high-stack storage of dangerous chemicals and combustible liquids, and bulk storage of compressed gasses. Quite clearly, fire-prevention codes took on some attributes of and intermeshed with building codes, having a common purpose, central to controlling flammable liquids, of protecting human life and property.[23]

Enforceable zoning laws and state fire-marshal activities made fire prevention more complicated than in the pre–World War II period. In March 1971 the city adopted—with some exclusions—the 1970 update of the NBFU/AIA's 1965 *Fire Prevention Code,* noting it was "more comprehensive . . . more technologically advanced and more comprehensive." Subsequently, amendments to the code defined the chief of the fire-prevention bureau as the city fire marshal, who would be the chief inspector of fire hazards, but it was noted in 1977 that this amendment did not specify the exact powers of the marshal. As the city tried to enforce fire regulations that applied to building safety, it became apparent that city codes sometimes conflicted with state or federal requirements. The Fire Department campaigned in 1978 for adoption of the SSFPC, which would complement the Southern Standard Building Code already in place. Galveston officials followed through, accepting a code that covered a broad range of real-world situations and conditions, including "flammable and combustible liquids."[24]

Galveston had also taken on the *NFPA 101, LSC* in 1978, causing controversy through the early 1980s as the city fire marshal tried to enforce it. The *LSC* addressed fire-safety issues about building egress and occupancy, requiring lighted exit signs, fire-suppression equipment in restaurants, and a minimum number of building exits and entrances. The city diligently supported fire-risk management, adopting the 1994 edition of the *Standard Fire Prevention Code,* although it is not clear if this came from the SBCCI or the AIA. With the merger of the major code publishers into the ICC, Galveston adopted the *IFC 2006* edition and later its updated 2009 version, keeping its legal system technically up-to-date and capable of managing a broad range of petroleum and hazardous flammable- and combustible-liquids risks.[25]

Starting in the nineteenth century, New Orleans addressed petroleum risks—kerosene, fuel oil, and gasoline—employing fire limits, segregation of flammable and combustible substances, as well as spatial restraints on bulk-oil and supply tanks to regulate petroleum and its products. The attention devoted to hydrocarbons was justified, but eventually mitigating petroleum hazards was subsumed under the broad rubric of fire prevention, especially when cities after World War II adopted interconnected model codes in which control of flammable liquids and petroleum storage intersected building use and occupancy standards, static-free ignition or machinery, topics that otherwise received attention in building, electrical and mechanical codes embellished with cross-references to one another.[26]

Fire prevention in New Orleans reflected a national push to end fire waste that had its origins in the nineteenth century, which one commentator contrasted with insured losses. Losses were "transferals of value," a conversion of material reality into a quantifiable economic abstraction in which fire underwriters spread risks across their clients as opposed to leaving individuals to bear the entire cost of a fire. Fire waste on the other hand was an "absolute destruction of wealth—the product of man's thought, toil and self-denial." As a solution to this permanent risk to life and property, fire prevention was a rationalization of the construction, maintenance, and protection of the human-made environment based in part on mapping and collecting detailed hazard information about insured property.[27]

As communities became more complex, codes grew in size and technical precision and became more difficult to interpret consistently across states and local communities, where definitions and specifications were stretched to fit a unique set of circumstances or address unanticipated configurations of hazards, or the codes challenged accepted ways of dealing with hazards that heretofore had been overlooked or not officially contested. This was one reason that fire prevention became a controversial topic in Louisiana and New Orleans throughout the twentieth century as governments struggled to apply developing risk-knowledge to sound building construction and fire codes. Early on, New Orleans's template was the *National Building Code* issued by the NBFU in 1905 that ran to four editions by 1915, although other organizations subsequently issued similar guides.[28]

The NBFU became embroiled in an early twentieth-century ruckus in Louisiana over the Louisiana Fire Prevention Bureau (LFPB), a state-approved insurance-company agency that conducted building and residential inspections. State legislators targeted it as a scheme to fix insurance rates whose inspections, critics maintained, should be conducted by public officials. At the same time, lawmakers created the office of state fire marshal to deal with fire prevention, an action applauded by fire-insurance interests. Legislators also insisted that the LFPB was a tool of the SETA, a founding member of the NFPA and alleged conspirator to raise insurance costs. The bureau denied that the SETA had engaged in such activities and claimed that strict inspections by their engineers and risk-informed agents under the NBFU building standards earned communities reduced fire-insurance rates; although it did not offer a resolution to complaints that city building and electrical inspectors sometimes disagreed with bureau men over code interpretations. It was a debate over uniformity in mitigating risk. The state and New Orleans sought to mollify merchants and contractors about these disagreements without surrendering their police powers to the NBFU, just as the city had rationalized petroleum-hazard mitigation through the permit system.[29]

New Orleans officials feared catastrophic fires from handling and consuming petroleum, but failed an NBFU inspection that indicted the city for obsolete firefighting equipment, inadequate water supplies, out-of-date building laws, and a lack of fire coverage in peripheral neighborhoods, all of which made the Fire Department and the city engineer look incompetent and unprepared to prevent major fires that the city had sought in the case of petroleum to eliminate.[30]

In November 1908 the city engineer, William J. Hardee, announced that a new building code was in hand, assembled with the help of an LFPB attorney and a city architectural draftsman, who adapted the NBFU's *National Building Code* to city needs. It was a hard-won accomplishment but not without controversy. Only a year later the NOBT's Municipal Affairs Committee held hearings that blasted the NBFU and the LFPB, who acted, it was claimed, in collusion to impose "exorbitant insurance rates" on the business district based on the underwriters' stinging 1906 inspection report.[31]

The city adopted a new building code in January 1910, though controversies surfaced over LFPB approval from a fire-risk-management perspective of electrical installations that subsequently proved to be operationally defective. The city also cooperated with various groups who waged countrywide fire-prevention campaigns. The New Orleans Association of Commerce, a group formed in 1913 to protect the interests of small businesses, took part in city affairs to create an industrial and business-friendly atmosphere, in the same way the NOBT tried to promote manufacturing and commerce by legitimating fuel oil in 1901–2. The association's fire-prevention committee in the 1920s worked with the National Fire Waste Council, sponsored by the U.S. Chamber of Commerce, to promote fire prevention as a way of gaining recognition for New Orleans as an efficient community with an "optimal business environment."[32]

Fire prevention acquired political credentials, reaffirming that controlling fire, whether from burning petroleum or illegally storing it in a building, made risk management of hazards a political act. Mayor T. Semmes Walmsley lobbied the Louisiana Rating and Fire Prevention Bureau (LRFPB) in 1931 for lower insurance rates, noting that the city had just adopted a new building code. At the same time he urged the public to recognize the need for improved firefighting equipment and a city department that would identify fire hazards. R. P. Strong, LRFPB manager, had already announced that he anticipated an NFBU inspection of local fire hazards by its Committee on Fire Prevention and Engineering Standards, which would provide information about appropriate rates. But staying abreast of risk-management trends was an administrative responsibility with political consequences. Harold Stone in a Kiwanis Club speech in 1934 condemned city conditions and insisted that it needed a "fire prevention code." Association of Commerce official R. B. Montgomery chimed in, calling for the city to improve its fire-alarm system, enforce the building code, and cooperate with the state fire marshal.[33]

The city closed its Fire Inspection and Prevention Bureau in 1935, bringing calls early in World War II from State Fire Marshal Campbell B. Palfrey for new safety regulations in New Orleans, residential fire-hazard inspections, strict enforcement of building and electrical codes, and mandatory fireproof means of egress from buildings, echoing concerns about a terrible boardinghouse fire on Dauphine Street in September 1942 in an overcrowded city where most buildings were put up "before any modern building code was in effect." This may have also been a response to the NBFU's inspection of New Orleans the same year.[34]

State attention and local apprehensions about a conflagration in a densely populated municipality set in motion a campaign for a new fire-prevention code at the close of World War II. A city committee on fire prevention, city attorneys, and the state fire marshal Palfrey called on the City Council to set up a fire-prevention board staffed by real-estate representatives, Association of Commerce officers, insurance underwriters, and safety engineers who would establish rigorously enforced fire regulations. Additionally, they wanted the same ordinance to address fire prevention; the storage of "explosive, combustible and inflammable materials, compounds and gases"; automatic fire alarms; guarantee safe egress from public buildings such as factories, schools, hospitals and theaters; and arson investigations.[35]

The pressure for effective fire prevention from the state level was evident in 1946 in a meeting called by the state fire marshal after consulting with Safety Commissioner Bernard J. McCloskey and the city fire chief. A. C. Hutson, assistant chief engineer and member of the NBFU's Committee on Fire Prevention and Engineering Standards—and fresh from Atlanta and the horrendous Winecoff Hotel fire—spoke on avoiding such catastrophes in New Orleans. Out of this came a draft fire-prevention code circulated citywide courtesy of the Association of Commerce. Further, the state fire marshal announced a state meeting for January 1947 in New Orleans to discuss strengthening the state building code through amendments drawn from an up-to-date version of the NBFU building code and other codes that "aimed at prevention of unnecessary fire loss of lives and property."[36]

There was also a "comprehensive study" of New Orleans government in 1946, undertaken by "nationally recognized experts" in municipal affairs at the request of the mayor-designate. W. S. Haldeman, a Chicago civil engineer, examined the regulation of buildings, elevators, boilers, and plumbing, and the NBFU looked at city fire services and the "fire prevention system," moving up their scheduled 1950 inspection to provide New Orleans its hazard-mitigation expertise.[37]

Concern over fire prevention had flourished on the federal level since World War I. Several New Orleans officials attended a conference in Washington, D.C., on fire prevention. They heard recommendations that had already been incorporated into the city's fire-prevention code, which a subgroup of a citizens' housing committee had drafted and presented to the Fire Department for City Council consideration in the summer of 1947. A revised draft, made public in late 1948 as a comprehensive building code, employed "standards and specifications" developed by the National Bureau of Standards, Underwriters' Laboratories, and the American Standards Association. New Orleans's quest to control fire, so hotly contested in 1901–2, received significant help from these and other associations, professional groups, and organizations that developed engineering and technical standards. Relying upon risk knowledge to regulate the storage of combustible and flammable materials and liquids was firmly ensconced in the city's fire-prevention strategies about the occupation of buildings and processes or activities carried on in them.[38]

Risk experts and their allies among New Orleans's ruling elements sometimes referred to model fire-prevention codes as building codes, an understandable conflation because fire prevention and building codes touched upon common and related topics. Note the apt and careful distinction explained in the 1976 edition of the AIA's *Fire Prevention Code*: "The Fire Prevention Code has limited coverage of such features of life safety in buildings as are normally considered proper for inclusion in a building code. If a suitable building code is not in effect, it is recommended that consideration be given to adopting the National Building Code recommended by the American Insurance Association as a companion document to the Fire Prevention Code" (AIA, *Fire Prevention Code: A Code Prescribing Regulations Governing Conditions Hazardous to Life and Property from Fire* [1976]).

These codes focus on public safety in a variety of circumstances but from different perspectives. Building codes established fire districts in which fire-resistant construction was required, while fire rules restricted oil storage in the same area. Further, fire officials increasingly dealt with matters such as fire-suppression equipment, alarm systems, building egress and occupancy, the last two functions of design parameters set by building codes. Building codes stipulated ventilation in industrial buildings susceptible to dust explosions, but it was the fire codes that determined what constituted adequate ventilation in terms of fire-risk management. Building codes set forth guidelines for construction of buildings in which hazardous materials might be stored or used in processes, requiring cross-references to the fire-prevention code that regulated the use of flammable and combustible solids and liquids; or pointing code inspectors to a technical standard such as NFPA 30 on flammable liquids or NFPA 31 on oil-burning equipment.[39]

New Orleans emerged in the post–World War II period from lengthy struggles to put enforceable building and fire-prevention codes in place. The city relied upon NBFU publications: the *National Building Code* and the 1946, 1955, 1960, and 1965 iterations of the *Fire Prevention Code*, both of which cited various technical standards, especially those of the NFPA. Problems surfaced when Louisiana adopted the NFPA's *Code for Safety to Life from Fire in Buildings and Structures* (1966), opening the door to conflicting interpretations among state and local inspectors.[40]

In the 1970s the city and the state addressed fire prevention. A committee commissioned in 1968 submitted a revised fire-prevention code in 1970 to Mayor Victor H. Schiro that examined overlapping code jurisdictions. The committee recommended shifting examination of "construction plans" for code compliance from the fire-prevention division of the Fire Department to the Department of Safety and Permits which enforced the building code. The mayor observed that government and private interests could cooperate in devising "standards . . . in the best interest of the community."[41]

In August and October 1973 the state legislature held hearings about reworking its fire code under pressure from the state fire marshal and many local fire chiefs to require sprinkler systems in buildings with three or more floors. The Factory Mutual companies championed automatic sprinklers as early as 1875, followed by the NBFU and the NFPA, viewing them as an efficient means of hazard mitigation in mills, factories, and buildings.[42]

Following a third hearing in November, the LRFPB declared in a 1974 report prepared at the request of a legislature subcommittee that sprinklers in high-rise buildings would dramatically reduce insurance rates. Proposed legislation made enforcement of sprinkler installations a mandatory activity for all future state fire marshals, a power thought to be necessary to make any new fire-prevention laws efficacious. This led to a prolonged fight over which fire or building codes the state should adopt. One legislative faction that included the state fire marshal supported the 1973 edition of the NFPA's *Fire Prevention Code,* while other legislators and many fire chiefs favored the 1974 edition of the SSFPC and the amended 1973 *Southern Standard Building Code* (1974).[43]

What was at stake was mandatory sprinkler systems in high-rise buildings, which gained credibility in the wake of the Howard Johnson Motor Lodge fire in July 1971, the deadly fire at the Rault Center which killed four women who jumped to their death from the fifteenth floor, and the Up Stairs Lounge fire in the Vieux Carre in June 1973. It was thought that sprinklers would have saved most of the victims, and the city scoured the Quarter and uncovered one thousand fire-code violations. One year later, seventeen bills that strengthened the state fire-safety code came out of a legislative committee, but the controversial sprinkler law stalled. After considering several rival bills, the legislature compromised and adopted the *NFPA 101, LSC* for old construction and SBCC's building code which required sprinklers or compartmentalization on all new construction over three floors in height. The squabbles and political maneuvering in the state legislature did not, however, permanently derail demands for comprehensive fire prevention statewide or in New Orleans, where NFPA 30 remained a referent standard to regulate flammable and combustible liquids. Risk management in New Orleans, despite corruption, backdoor deals, truculent property owners, historic preservation, and intermittent enforcement, moved forward.[44]

After several decades of updating New Orleans's fire-prevention code, the city by 2003 had passed through several editions of the SBCCI's *Southern Standard Fire Prevention Code,* and then adopted the *IFC* in 2006. The *ICC*'s model codes rely heavily on standards developed by the NFPA and widely accepted technical and governmental bodies. Chapter 45 of the 2006 *IFC* as amended cited the NFPA's *Uniform Fire Code* as well as seven other standards. Chapter 34 prohibited storage of flammable and combustible liquids "in above-ground tanks outside of buildings . . . within the inner fire limits as per the New Orleans Building Code," illustrating the durability, abiding importance, and currency of the risk management of petroleum dating to 1902 at the intersection of the twin components—fire and building codes—of contemporary fire prevention.[45]

What the Gulf South cities had in common was a century-long commitment to risk-management policies based on technically informed, consensus standards that undergirded codes advanced by the NBFU, NFPA, and other code bodies for voluntary federal, state, and local adoption, facilitated through sample ordinances or laws. The codes reflected the developing expertise of various professional organizations that promulgated standards for a broad spectrum of engineering specialties and technical areas, including handling petroleum. Regional code groups such as the Western Fire Chiefs Association, Building Officials and Code Administrators International (BOCA), and the Southern Building Code Conference (SBCC) developed consensus building codes specific to their regions, and over time shared many common engineering and technical specifications that came from the Bureau of National Standards, the American Society of Mechanical Engineers, Underwriters' Laboratories, and many others. The apex of consensus came with the combination of the regional codes in the *International Fire Codes* in 2006, 2009, and 2012 that moved fuel oil from the chapters on flammable and combustible liquids but still referred readers to NFPA 30 and 31.

Gulf South cities starting in 1901–2 mitigated petroleum hazards through the politics of risk management. Relying upon and staying abreast of scientific and technical advances in risk-knowledge and fire-suppression techniques, they passed ordinances and adopted consensual codes, modified as necessary to reflect local conditions and economic-political realities, up to the present day.

In so doing, these cities took part in a process of modernization that brought the identification, quantification and control of fire hazards and risks, in step with a national agenda in engineering, science and mathematics—to create models, paradigms, and abstractions of material reality. This campaign paralleled the post–Civil War national, political, and economic integration of the United States. It also reflected technological rationality, which features application of scientific knowledge to solve hazard- and risk-related problems. The goal is objective and practical—to mitigate or eliminate hazards—culminating in risk management or devising strategies or policies to achieve this end. This example of instrumental rationality calls for techniques or methods to reduce the likelihood or consequences of a hazardous event, guided by a "calculus of efficiency" in risk assessment that demands in the realm of fire-insurance underwriting a reduction or elimination of capital loss.[46]

While Americans had from colonial times taken and calculated risks through rational use of their resources, modernization in the South in 1900 was a historical process contingent upon past experiences, environment, social arrangements, and political realities. The ambit of these conditions renewed, revised, refreshed, and perpetuated the historical momentum of modernization. Within this context, risk management of petroleum hazards throughout the twentieth century in the Gulf South was part cause and part consequence of a new level of reliance upon accepted professional, scientific, and technical standards and their implementation through public policies and regulations for public benefit, despite countervailing customs, traditions, and habits in the South that never allowed modernization to be thoroughgoing but did make room for its expression.[47]

Thus, modernization in the Gulf South involved controlling petroleum risks and advancing public safety and well-being while addressing environmental, medical, industrial, commercial, and financial uncertainties. Faith in quantification expressed through identification of hazards and some form of probability determination sustained a drive to control and contain hazards that is characteristic of the modern world and the regulatory infrastructure of the twenty-first-century South.

NOTES

INTRODUCTION

1. On comments about fuel oil as a hazard, risk that embodied likelihood and consequences, and potential disasters from fires or explosions, see "Insurance and Fuel Oil," *NOTP*, January 15, 1902, 4, col. 4; "Particulars about Beaumont Oil," *NOTP*, January 23, 2, col. 2; "Exchange Heard About Fuel Oil," *NOTP*, February 20, 10, col. 1; "The Shippers and Fuel Oil," *NOTP*, February 26, 4, col. 1; and "Fuel Oil in New Orleans," *NOTP*, February 27, 4, col. 3.

2. On oil tanks and lightning hazards, see "The Protection of Oil Tanks from Lightning," *Electric World* 3, no. 26 (June 28, 1884): 208–9. Refineries inside city limits were regulated, and although there were some massive conflagrations in and around places such as New York City or Philadelphia, fire-insurance companies had different views about refinery insurability. Hazards and risks came into play because refineries require a great deal of oil-tank storage which, if a company wanted coverage, had to be configured as prescribed in insurance rules and accepted procedures. For incomplete insurance on an oil factory and adjacent buildings, see "Destruction of an Oil Factory in East Cambridge—Eight Buildings Burned," *BDA*, October 19, 1870, 1, col. 6. In 1878 a Home Insurance Company manual advised agents that "Earth Oil Refineries, Storehouses and Tanks," were prohibited risks; see C. K. Francis, *Manual Issued by the Home Insurance Company of New York* (1878), 18 (quotation). In 1902 an NFPA member advised inspectors to see that refineries of oil, lard, and other inflammable materials were "as near fireproof as possible"; see George Velten Steeb, *Agents' and Inspectors Pocket-Book of Fire Protection* (1902), 27 (quotation). However, in *Fire Insurance and How to Build* (1903), fire-insurance veteran Francis C. Moore advised readers to decline oil refineries (probably lard or cotton seed), and that few companies would write policies on petroleum/kerosene warehouses and tanks (521).

3. In this study, risk management refers to a formal, analytical process of decision-making, carried out by public institutions such as state or municipal governments, in tandem with fire-insurance companies, associations, or organizations, certain members of which were engineers, technically informed insurance men, or consultants. Risk management produces specific policies or strategies embodied in insurance rules or requirements, which may be codified in law, that identify or characterize certain hazardous processes or substances to regulate or control the same. See John Salter, "Risk Management in a Disaster Management Context," *Journal of Contingencies and Crisis Management* 5, no. 1 (March 1997): 60.

4. For Mobile's economic growth, see John S. Sledge, *The Mobile River* (2015), 183–89.

5. For a commentary on what the Ship Channel did for Houston in the twentieth century, see Joseph A. Pratt, "Letter from the Editor: The Houston Ship Channel and Its Port," *Houston History* 12, no. 1 (Fall 2014): n.p.

6. On the completion of the Ship Channel, see Marilyn McAdams Sibley, *The Port of Houston: A History* (1968), 136–41.

7. On Galveston, see David G. McComb, *Galveston: A History* (1986), 150, 169, 174–75. On the importance of Houston's civic leadership, see Joseph A. Pratt, "8F and Many More: Business and Civic Leadership in Modern Houston," *Houston Review of History and Culture* 1, no. 2 (Summer 2004): 3–6; on the rivalry with Galveston, see Sibley, *Port of Houston,* 57–58, 76, 86–87, 164–68.

8. For the populations of Galveston and Houston, see *Thirteenth Census of the United States,* vol. 3: *Population 1910* (1913), table 2, "Composition and Characteristics of the Population for Cities of 25,000 or More," 852–53; for Mobile, see *Thirteenth Census of the United States Taken in the Year 1910: Statistics for Alabama* (1913), Supplement for Alabama, table 2, "Population of Incorporated Places: 1910, 1900, and 1890," 586; for New Orleans, see *Thirteenth Census of the United States Taken in the Year 1910: Statistics for Louisiana* (1913), Supplement for Louisiana, 569, 580.

9. For worker productivity in Mobile, see *Fourteenth Census of the United States,* vol. 9: *Manufactures 1919* (1923), table 6, "Average Number of Wage Earners, Value of Products, Mobile," 25; New Orleans, 511; Houston and Galveston, 1452.

10. On the Ship Channel and petroleum, see Sibley, *Port of Houston,* 151–53.

11. For fuel consumption in Mobile, see *Fourteenth Census of the United States,* vol. 9, table 18, "Fuel Consumed, For Selected Industries and for Cities: 1919," 33; New Orleans, 518; Houston and Galveston, 1464. On risk knowledge, see Andreas Klinke and Ortwin Renn, "Prometheus Unbound: A New Approach to Risk Evaluation" (2009), 169; and Scott G. Knowles, *The Disaster Experts: Mastering Risk in Modern America* (2011), 10–12, 14–15.

12. On promotion, see Diane Barnes, "Boosterism," in *Encyclopedia of the United States in the Nineteenth Century* (2001); C. Vann Woodward, *Origins of the New South, 1877–1913* (1971), 131, 133–34, 136, 138, 144, 146–48; Paul Gaston, *The New South Creed: A Study in Southern Mythmaking* (1970), 45–79; James C. Cobb, *The Selling of the South: The Southern Crusade for Industrial Development,* 2nd ed. (1993); David Goldfield, *Black, White, and Southern: Race Relations and Southern Culture, 1940 to the Present* (1990), 112; and *Region Race and Cities: Interpreting the Urban South* (1997), 197. For a plea that Texans should take advantage of cheap fuel oil for manufacturing, rather than leaving it to outsiders, see "A Prosperous Year," *GDN,* December 29, 1901, 6, cols. 4–5. A prominent publication applauded regional spokesman Richard H. Edmonds (1857–1930), editor of the *Manufacturers' Record,* for "bringing in capital to fully develop the new oil territory" and for urging manufacturers to take advantage of the Gulf South's inexpensive petroleum. See "A Champion of the Southern Oil Industry," *OIJ* 1, no. 13 (November 15, 1902): 9, cols. 2–3.

13. On oil hazards, see "The Fire Marshall's Notice to the Public," *USIG&M* 3, no. 14 (June 1856): 99; "Destruction of an Oil Factory in East Cambridge—Eight Buildings Burned," *BDA,* October 19, 1870, 1, col. 6; "New Jersey: Destructive Fire at Weehawken," *BJ,* July 13, 1874, 4, col. 3; and "Fire Record: At Cleveland, O.—Loss $50,000," February 4, 1888, 4, col. 8. On agents, boards, and inspections, see W. F. Fox, "The Special Agent," in *Proceedings of the Twelfth*

Regular Meeting Fire Underwriters' Association of the Northwest, Chicago, Ill., September 14 and 15, 1881 (1881), 173, 175, 177, 183; and the NBFU's inspection of "all the large cities of the country, to ascertain their condition in regard to danger from fire" in the summer of 1874, in "The National Board of Fire Underwriters. Their Action in Regard to Chicago—A General Inspection of Large Cities," *NYEP*, October 3, 1874, 4, col. 2. On William Muir, an NBFU special agent and member of the Pennsylvania Provisional Committee, who took part in an investigation of coal-oil risks in tanks and pipelines in the Pennsylvania field, see "Minutes of a Meeting of the Executive Committee of the National Board of Fire Underwriters . . . January 4, 1877," 530–31; and Muir's NBFU status in "Address to the National Board of Fire Underwriters, of the United States, by Henry A. Oakley, President, at the Decennial Meeting, Held in the City of New York, April 26, 1876," 10, in Henry A. Oakley, *Addresses as President of the National Board of Fire Underwriters of the United States, on Several Occasions, by Henry A. Oakley, 1871–1876* (1876).

14. "Insurance Department," *NODS*, March 21, 1891, 3, col. 1.

15. See, for example, F. Crace Calvert, "On the Manufacture and Application of Various Products Obtained from Coal (Coal-Gas Excepted)," *Mining Magazine: Devoted to Mines, Mining Operations, Metallurgy* 4, nos. 5–6 (May–June 1855): 362–70; continued in vol. 5, no. 1 (July 1855): 28–32 (originally published in the *Journal of the Society of Arts* 3, no. 105 [November 24, 1854]: 17–20); and "Dry Distillation of Coal and Its Products," *Mining Magazine* 6, no. 2 (February 1856): 141–46; C. B. Mansfield, "Processes for Distillation of the Products of Coal," *Mining Magazine* 6, no. 3 (March): 197–201; "Processes for the Distillation and Purification of the Products of Coal," *Mining Magazine* 6, no. 4 (April): 295–300, and nos. 5–6 (May–June): 383–91; W.H.D., "Critical Examination of the Value of Naphthaline and Petroleum as Substitutes for Cannel Coal," *E&MJ* 20, no. 15 (October 9, 1875): 357–58, from *Bayerisches Industrie and Gewerbeblatt* 1 (1875): 1–5; "Apparatus for the Manufacture of Gas from Oil," *AM&IW* 33, no. 9 (August 31, 1883): 3, cols. 1–3, from *Revue Industrielle*. See also Ross Thomson, *Structures of Change in the Mechanical Age: Technological Innovation in the United States, 1790–1865* (2009), 160–89.

16. On engineers, manufacturing, and mechanization, see Thomson, *Structures of Change in the Mechanical Age*, 1–12, 15–99; Paul Israel, *From Machine Shop to Industrial Laboratory: Telegraphy and the Changing Context of American Invention, 1830–1920* (1992), 2, 13–14, 17–20; and David R. Meyer, *Networked Machinists: High-Technology Industries in Antebellum America* (2006), 1–16.

17. Monte A. Calvert, *The Mechanical Engineer in America, 1830–1910: Professional Cultures in Conflict* (1967), 3–8, 13–14, 23–24; Yhouda Shenhav, *Manufacturing Rationality: The Engineering Foundations of the Managerial Revolution* (2002), 25, 30, 104–5; and Peter Meiksins and Chris Smith, *Engineering Labour: Technical Workers in Comparative Perspective* (1996), 62–70. On tacit knowledge, see Larry Stapleton, David Smith, and Fiona Murphy, "Systems Engineering Methodologies, Tacit Knowledge and Communities of Practice," *AI and Society* 19, no. 2 (April 2005): 159–68. *Rationalize* refers to analysis, categorization, formulation of objectives, and identification of the means to obtain the same in engineering, process safety, military operations, financial strategies, and many other fields.

18. On fire testing, see "How Underwriters' Tests are Made," *InsEng* 4, no. 2 (August 1902): 99–106, and "Growth of the Underwriters' Laboratories," *InsEng* 4, no. 2: 194–95. Examples of

engineers and scientists include John M. Ordway, industrial chemist at MIT, for whom see chapter 1, note 113, below. See also "Prominent Insurance Engineers: William A. Anderson" (one of the founders of the NFPA); and "William H. Stratton," *InsEng* 4, no. 1 (July): 45–46; and no. 2: 153–55, respectively. Stratton worked successively for the AFMFIC and the Factory Insurance Association of Hartford, was a member of the ASME, and became chairman of the Executive Committee of the NFPA.

19. On Morton and acetylene rules, see NBFU, Thirty-Second Annual Meeting, "[Report on Lighting, Heating and Patents]," *AmEx&Rev* 52, no. 6 (June 1898): 262, 266 (quotations 1–3); and Alexander C. Humphreys, "Obituary: Henry Morton, Ph.D., Sc.D., LL.D. December 11th, 1836–May 9th, 1902," *Stevens Institute Indicator* 19, no. 3 (July 1902): 230 (quotation 5), 241 (quotation 6), 259 (quotation 4); and Rev. Edward Wall, "The Work and Character of President Henry Morton," *Stevens Institute Indicator* 19, no. 3 (July 1902): 270, 274. On Humphreys, see Alexander C. Humphreys, "Is a Knowledge of Business Methods of Importance to the Engineer," *Stevens Institute Indicator* 14, no. 4 (October 1897): 413–23.

20. On the development of standardization, which eventually included fire codes and regulations, see Andrew L. Russell, *Open Standards and the Digital Age: History, Ideology, and Networks* (2014), 25–57. I use the term "standards" to mean "established norms or requirements applied to technical systems." See Dong-Hee Shin, Hongbum Kim, and Junseok Hwang, "Standardization Revisited: A Critical Literature Review on Standards and Innovation," *Computer Standards & Interfaces* 38 (February 2015): 152.

21. *Annual Cyclopedia of Insurance in the United States 1915*, established by H. R. Hayden (1915), "National Board of Fire Underwriters," "National Fire Protection Association." Quotation from Everett U. Crosby, Henry A. Fiske, and H. Walter Forster, *Hand-Book of Fire Protection* (1919), 15.

22. See Edward Atkinson, *The Prevention of Loss by Fire and the System of Factory Mutual Insurance: An Address by Edward Atkinson, of Boston* (1885), 11, for comments on petroleum illuminants in mills; and "The Prevention of Loss by Fire: An Applied Science," *InsEng* 4, no. 3 (September 1902): 248–61.

23. On the NBFU, see Harry Chase Brearley, *The History of the National Board of Fire Underwriters: Fifty Years of a Civilizing Force* (1916) (hereafter Brearley, *History NBFU*), 87–92; and on its relationship to the NFPA, see *Proceedings of the Thirty-Fifth Annual Meeting of the National Board of Fire Underwriters Held at the Rooms of the New York Board of Fire Underwriters . . . 1901* (1901), 104. For the ASTM and its relationship to government agencies, municipalities, and the NFPA fire codes, see U.S. Department of Commerce, National Bureau of Standards, *Standardization Activities of National Technical and Trade Organizations*, by Robert A. Martino, NBS Misc. Pub. M169 (1941), 5–9, 10 (NFPA); *Standardization Activities in the United States: A Descriptive Directory*, by Sherman F. Booth, NBS Misc. Pub. 230 (1960), 1–2. On the development of Associated-Factory-Mutual fire standards, see Sara E. Wermiel, *The Fireproof Building: Technology and Public Safety in the Nineteenth-Century American City* (2000), 104–37. On the distribution of test reports related to fire hazards to local boards of underwriters, inspection bureaus, federal-state-municipal governments, and insurance companies, see Underwriters' Laboratories, *The Organization, Purpose and Methods of Underwriters' Laboratories* (1916), 7.

24. On the federal government and risk, see National Academy of Science, National Research Council, Commission on Life Sciences, Committee on the Institutional Means for

Assessment of Risks to Public Health, *Risk Assessment in the Federal Government: Managing the Process* (1983); U.S. Department of Commerce and Labor, National Oceanic and Atmosphere Administration, Committee on Environment and Natural Resources, *Ecological Risk Assessment in the Federal Government* (1999); National Academy of Science, National Research Council, Commission on Life Sciences, Board on Environmental Studies and Toxicology, Committee on Developmental Toxicology, *Scientific Frontiers in Developmental Toxicology and Risk Assessment* (2000), 26–27; U.S. Environmental Protection Agency, Risk Assessment Forum, *Framework for Cumulative Risk Assessment* (2003), and *An Examination of EPA Risk Assessment Principles and Practices* (2004); and 42 U.S.C. 85 §§ 7401, 7403, 7405 (2013); as well as John D. Graham, "Historical Perspective on Risk Assessment in the Federal Government," *Toxicology* 102, no. 1–2 (September 1995): 29–52; Jay O'Laughlin, "Policies for Risk Assessment in Federal Land and Resource Management Decisions," *Forest Ecology and Management* 211, no. 1–2 (June 2005): 15–27; and William Keller and Mohammad Modarres, "A Historical Overview of Probabilistic Risk Assessment Development and Its Use in the Nuclear Power Industry: A Tribute to the Late Professor Norman Carl Rasmussen," *RE&SS* 89, no. 3 (September 2005): 271–85. State laws and municipal regulations often mirrored with some exceptions the federal government's standards and policies. For international risk-management standards see a preview of the International Organisation for Standardisation (ISO) publication *ISO 31000: 2009(en) Risk Management—Principals and Guidelines,* iso.org/obp/ui/#iso:std:iso:31000:ed-/:v1/:en (accessed February 9, 2016).

25. On "ever-present" risk inherent in all systems, see Gary H. Eaton and Donald E. Little, "Risk Assessing & Mitigating to Deliver Sustainable Safety Performance," *Professional Safety* 56, no. 7 (July 2011): 35.

26. Paul K. Freeman and Howard Kunreuther, "Managing Environmental Risk Through Insurance," *International Yearbook of Environmental and Resource Economics,* ed. Henk Folmer and Tom Tietenberg (2003), 169; Ortwin Renn, "The Role of Risk Perception for Risk Management," *RE&SS* 59, no. 1 (January 1998): 51 (quotation 1); and Osama Ahmed Jannadi and Salman Almishari, "Risk Assessment in Construction," *Journal of Construction Engineering and Management* 129, no. 5 (October 1, 2003): 492–93.

27. See Francis, *Manual Issued by the Home Insurance Company of New York,* 97 (quotation 1); Jeremiah Griswold, *The Fire Underwriters' Text-Book* (1872), 7 (quotation 2), 8; and Legal Dictionary, www.legal-dictionary.thefreedictionary.com/insurance (quotation 3) and thefreedictionary.com/risk (accessed January 12, 2016).

28. On risk assessment, see Jonathan H. Klein and Robin B. Cork, "An Approach to Technical Risk Assessment," *International Journal of Project Management* 16, no. 6 (December 1998): 346–48; for defining risk from an engineering perspective, see Terje Aven, "On How to Define, Understand and Describe Risk," *RE&SS* 95, no. 6 (June 2010): 623–24; and "A Risk Concept Applicable for Both Probabilistic and Non-Probabilistic Perspectives," *Safety Science* 49 (2011): 1080–82. See also Aven's apt observation that "Risk is either the possibility/uncertainty/ chance that the activity will have some undesirable consequences, or the activity (person, gasworks) itself, that which is often also referred to as a risk source or a threat." See "The Risk Concept—Historical and Recent Development Trends," *RE&SS* 99 (2012): 36 (quotation above).

29. Graham, "Historical Perspective on Risk Assessment," 30 (quotation); for risk management, see European Spatial Planning Observation Network, "Glossary of Terms," 1, www.gtk

.fi/projects/espon/glossary.htm (accessed February 2, 2007); and Salter, "Risk Management in a Disaster Management Context," 60–61.

30. On vulnerability, see Fikret Berkes, "Understanding Uncertainty and Reducing Vulnerability: Lessons from Resilience Thinking," *Natural Hazards* 41 (2007): 284, 287; Michael H. Kottow, "Vulnerability: What Kind of Principle Is It?" *Medicine, Health Care and Philosophy* 7 (2004): 282–84; Anand Patwardhan and Upasna Sharma, "Improving the Methodology for assessing Natural Hazard Impacts," *Global and Planetary Change* 47 (2005): 253–65.

31. See Stephen Salisbury, "The Emergence of an Early Large-Scale Technical System: The American Railroad Network" (1988), 37–68; Israel, *From Machine Shop to Industrial Laboratory,* 2, 24–56, 176–83.

32. G. Melvyn Howe, *People, Environment, Disease and Death: A Medical Geography of Britain Throughout the Ages* (1997), 48; "Disease and the Environment," in *Disease and the Environment,* ed. Alun R. Rees and H. J. Purcell (1982), 5; "President Jasper M. Dresser's Address: Delivered at the twelfth annual meeting of the Fire Underwriters' Association of the Northwest at Chicago, September 14, 1881," *InsTm* 14, no. 9 (September 1881): 560, col. 2; "Traps to Catch Firebugs: They are Set at the National Convention of Fire Underwriters," *NYHRD,* May 17, 1889, 4, col. 2 (quotation 2); and D. A. Heald, "National Board of Fire Underwriters: Extracts from the Address of President Heald at the Twenty-Third Annual Meeting," *InsTm* 23, no. 6 (June 1889): 300 (quotation 1).

33. On industrialization and the effect of technology and machines on modern culture and social relationships, see Langdon Winner, *Autonomous Technology: Technics-out-of-Control as a Theme in Political Thought* (1977), 3, 47–50, 196–97, 205.

34. For storms, see "Galveston in 1900," *GDN,* January 1, 1901, 4, cols. 1–2; "Gaining By Leaps," *GDN,* January 1, 1901, 8, col. 2; "Storm Property Loss At Galveston," part 2, *GDN,* January 1, 1901, [1], cols. 2–3; "Summary of Losses and Rehabilitation," "Losses on Residence Property," *GDN,* January 1, 1901, [1], col. 6; and part 2 passim for many other references to and descriptions of the 1900 storm's effect on Galveston; Erik Larson, *Isaac's Storm* (2000); Patricia Bellis Bixel and Elizabeth Hays Turner, *Galveston and the 1900 Storm* (2000); Casey Edward Greene and Shelly Henley Kelly, eds., *Through a Night of Horrors: Voices from the 1900 Galveston Storm* (2000); Mary Jo O'Rear, *Storm Over the Bay: The People of Corpus Christi and Their Port* (2009); and Hugh W. Stephens, *The Texas City Disaster, 1947* (1997). For petroleum as an additional hazard, see U.S. Chemical Safety and Hazard Investigation Board, *Investigation Report: Refinery Explosion and Fire, BP Texas City, Texas, March 23, 2005,* Report No. 2005-04-I-TX; National Commission on the BP Deepwater Horizon Oil Spill and Offshore Drilling, *Deep Water: The Gulf Oil Disaster and the Future of Offshore Drilling* (2011); and William R. Freudenburg and Robert Gramling, *Blowout in the Gulf: The BP Oil Spill Disaster and the Future of Energy in America* (2011), an extended plea for preventative risk-management policies and thoughtful energy choices. On humans and the environment, see C. Emdad Haque, *Hazards in a Fickle Environment: Bangladesh* (1997), 4–5, 55–64; and Brian Frehner, *Finding Oil: The Nature of Petroleum Geology, 1859–1920* (2011), who points to "nature and culture" coexisting in a "tangled dialectic rather than discrete categories of analysis" (3).

35. On cataclysmic fires associated with drilling, distributing, or storing oil—otherwise harmless inorganic physical matter—see "Oil Fire in Jersey City," *Brooklyn Daily Eagle,* September 23, 1868, 3, col. 8 (Verplanck, Richards & Co. oil refinery); "An Oil Fire," *Brooklyn*

Daily Eagle, September 3, 1877, 3, col. 9 (SOC tank fire, Hunter's Point, NY); "Large Fire in Cleveland," *Brooklyn Daily Eagle,* September 7, 1885, 2, col. 4 (SOC works in Cleveland burned); "Many Burned," *Brooklyn Daily Eagle,* May 11, 1891, 6, col. 1; "Fires in an Oil Field," *Brooklyn Daily Eagle,* May 3, 1900, 1, col. 5; "Standard Oil Plant Burning at Bayonne [N.J.]," *Brooklyn Daily Eagle,* July 5, 1900, 6, cols. 1–2 (massive $1,000,000 fire); "Swan & Finch Refinery Burned," *Brooklyn Daily Eagle,* May 8, 1902, 20, col. 4 (Staten Island Sound, 1500-barrel crude-oil tanks burned). For the Beaumont field, see "In the Oil Fields," *GDN,* August 4, 1901, 3, col. 1, wherein it was reported that J. M. Guffey Petroleum Co. "had constructed levees around their tanks to prevent the spread of blazing oil in the event of fire" at a Spindletop well.

36. On the oil boom, national and international audiences, see "The Rush to the Amazing Oil Region in Texas," *Leslie's Weekly* 92, no. 2385 (May 25, 1901): 504–5. On the approximately 178 companies chartered in Texas from Jan. 15 to April 29, 1901, see "Organization of Oil Companies in Texas," *GDN,* April 29, 1901, 4, cols. 1–2; and May 1901 issues of the *NOTP* which contained an enormous number of new oil-company advertisements. On images of derricks covering Spindletop hill, see "A Busy Oil Field," *GDN,* June 9, 1901, 5, col. 1. On Gulf-South oil production, see table 2.1, "Domestic crude production by major fields: 1900–1919 (millions of barrels)," in Harold F. Williamson et al., *The American Petroleum Industry: The Age of Energy, 1899–1959* (1963), 16. In 1901, production was 3.7 million barrels, 18 million in 1902, and 36.5 million barrels in 1905. It fell thereafter to 9.7 million in 1910, rallied in 1911 to hit 11 million and then back down to 8.5 million in 1912 and 1913. Production recovered and rose to 24.3 million, 24.2 million, and 23.4 million from 1917 to 1919. On the sensational Lucas well at Spindletop, January 10, 1901, and public expectations of great oil production and wealth, see "The New Treasure Field of Texas / Its Discovery, Its Development, Its Present Status and Its Great Future. I. Texas Strikes Oil," *HC,* January 31, 1902, 4, cols. 4–5.

37. On insurance rules and requirements see chapter 1, below. On public discussions before the New Orleans City Council's fuel-oil committee, see "Streets and Landings," *NOTP,* January 25, 1902, 8, col. 3; "City and Suburban," *NOTP,* February 14, 2, col. 1; "Insurance Agents and Oil Men at Issue," *NOTD,* February 14, 1902, 5, cols. 1–2; and "Oil as Fuel Discussed in Council Committee," *NOTD,* February 27, 1902, 12, cols. 4–7. According to "Much Oil in Use," *NOTD,* January 29, 8, col. 5, the City Council's fuel-oil committee met with insurance representatives to "pass on the matter of what restrictions are necessary to insure the safe handling of oil at this port." For the role of special committees in Galveston, see "The City Council," *GDN,* August 5, 10, col. 3; and "In Special Session," *GDN,* December 13, 10, col. 2.

38. On socio-technical (STS) and large technical systems (LTS), see Frank W. Geels, "Transformations of Large Technical Systems: A Multilevel Analysis of the Dutch Highway System (1950–2000)," *Science, Technology, & Human Values* 32, no. 2 (March 2007): 123–26; Lea Fuenfschilling and Bernhard Truffer, "The Structuration of Socio-Technical Regimes—Conceptual Foundations from Institutional Theory," *Research Policy* 43, no. 4 (May 2014): 772–91; Angela Weber Righi and Tarcisio Abreu Saurin, "Complex Socio-Technical Systems: Characterization and Management Guidelines," *Applied Ergonomics* 50 (September 2015): 19–30; Florian Schöttl and Udo Lindemann, "Quantifying the Complexity of Socio-Technical Systems—A Generic, Interdisciplinary Approach," *Procedia Computer Science* 44 (2015): 1–10; Paul Pao-Yen Wu et al., "A Framework for Model Integration and Holistic Modelling of Socio-Technical Systems," *Decision Support Systems* 71 (March 2015): 14–15; and Pieter Vermaas et al., *A Phi-*

losophy of Technology from Technical Artefacts to Sociotechnical Systems* (2011), 67–70. On the nineteenth-century "building and establishment phase" of energy systems with a "diverse range of technologies and engineering practices, or technical styles," see Ronan Bolton and Timothy J. Foxon, "Infrastructure Transformation as a Socio-Technical Process—Implications for the Governance of Energy Distribution Networks in the UK," *Technological Forecasting & Social Change* 90 part B (January 2015): 540.

39. For the notions of coevolution, complex systems, and energy transition, see Francis G. N. Li, Evelina Trutnevyte, and Neil Strachan, "A Review of Socio-Technical Energy Transition (STET) Models," *Technological Forecasting & Social Change* 100 (November 2015): 290–91; on the intensification and complexity of energy transitions, see Paul Rutter and James Keirstead, "A Brief History and the Possible Future of Urban Energy Systems," *Energy Policy* 50 (November 2012): 78–79. For a concise definition of "energy transition," see Roger Fouquet and Peter J. G. Pearson, "Past and Prospective Energy Transitions: Insights from History," *Energy Policy* 50 (November 2012): 2.

CHAPTER ONE

1. For a contemporaneous explanation of "petroleum and its products," see P. H. Vander Weyde, MD, "Petroleum and Its Products," *Scientific American* 14, no. 25 (June 16, 1866): 412. Vander Weyde was a chemistry professor at Girard College in Philadelphia.

2. Gesner's patent application avoided the term "coal" to avoid violating James Young's 1850 patent on distilling coal for illuminants and oils. On Young see "Improvement in the Treatment of Certain Bituminous Mineral Substances, and in Obtaining Products Therefrom," *Mining Magazine* 5, no. 5 (November 1855): 412–15. For estimates of daily coal-oil production in the United States, see "Coal Oil Manufacture," *Hunt's Merchants' Magazine and Commercial Review* 42, no. 2 (February 1860): 245–46.

3. Abraham Gesner, *A Practical Treatise on Coal, Petroleum, and Other Distilled Oils*, 2nd ed., ed. George Weltden Gesner (1865), 9; and Paul Lucier, *Scientists & Swindlers: Consulting on Coal and Oil in America, 1820–1890* (2008), 143–61. On state-by-state and total U.S. crude production in barrels from 1859 to 1915, see Roy Cross, *A Handbook of Petroleum, Asphalt and Natural Gas*, Kansas City Testing Laboratory Bulletin 15 (1919): 14–15. In 1871 the United States exported 132,178,843 gallons of kerosene, which went up to 730,368,626 gallons by 1894. In the same period, kerosene (illuminating oil) exports in gallons averaged 36.53 percent of total U.S. crude oil production. See U.S. Dept. of the Interior, U.S. Geological Survey, *The Production of Petroleum in 1894* by Joseph D. Weeks (1895), from the *Sixteenth Annual Report of the Director, 1894–95*, part 4: *Mineral Resources of the United States, Calendar Year 1894*, 321.

4. See Harold F. Williamson and Arnold R. Daum, *The American Petroleum Industry: The Age of Illumination 1859–1899* (1959), 295–97 (hereafter Williamson and Daum, *API Illumination*); and S. F. Peckham, "Concerning Kerosene," *Popular Science Monthly* 27 (June 1885): 252, noted that of the "vast quantity" of U.S. crude oil production, the "largest proportion is consumed as illuminating oil, or kerosene." There were, he observed, oil refineries in cities such as Baltimore or Philadelphia that "convert . . . crude oil into various products, principally illuminating oil, for the home market and an export trade of vast proportions."

5. "An Act further to provide for the Safety of the Lives of Passengers on Board of Vessels propelled in Whole or in Part by Steam, to regulate the Salaries of Steamboat Inspectors, and for other Purposes," 14 Stat. 234 (1866), sec. 5, 227, 228 (quotations 1–2).

6. "An Act to provide for the better Security of Life on board of Vessels propelled in Whole or in Part by Steam, and for other Purposes," 16 Stat. 100 (1871), sec. 4–5, 441, 442 (quotations 1–2). See *Patterson* v. *Kentucky,* 97 U.S. 501, 504 (October term 1878), in 1 *A.R.C.* (American Ruling Cases) (1912), 1378–81, wherein the U.S. Supreme Court ruled that Kentucky did not violate a federal internal-revenue act of March 1867, when the state prohibited by statute in 1874 the sale of illuminating oils that had an ignition point less than 130 degrees Fahrenheit.

7. For periodization of illumination sources, see Mimi Sherman, "A Look at Nineteenth-Century Lighting: Lighting Devices from the Merchant's House Museum," *APT Bulletin* 31, no. 1 (2000): 40, cols. 2–3. On the chemistry and manufacturing of gas from the light products of petroleum see *Johnson's (Revised) Universal Cyclopædia: A Scientific and Popular Treasury of Useful Knowledge,* ed. Frederick A. P. Barnard and Arnold Guyot (1886), vol. 6: *Onion-Salt,* "Petroleum and Naphtha Gas," 230–31.

8. On camphene, a white, waxy solid, part of the terpene family, that occurs naturally in the essential oils of trees, plants, or plant parts, see Eduard Gildemeister and Frederick Hoffmann, *The Volatile Oils,* 2nd ed. (1913), vol. 1: 81–83, 301–5; and George Ripley and Charles A. Dana, eds., *The American Cyclopædia: A Popular Dictionary of General Knowledge* (1873), vol. 3: 667. See also Lucier, *Scientists & Swindlers,* 153; and a patent abstract describing Solomon Andrews's "spirit lamp, filled with alcohol or other highly inflammable fluid" that commences the generation of gas in his newly "invented gas lamp or machine" (*JFI* 8, no. 3 [September 1831]: 189–90).

9. For burning fluids, see E. N. Horsford, "On the Explosions of Burning Fluids," *Chemical Gazette* 8, no. 175 (February 1, 1850): 48-49; also James C. Booth and James Morfit, *On Recent Improvements in the Chemical Arts,* vol. 6, "Illumination," in *Smithsonian Miscellaneous Collections* 2 (1862), which describes burning fluids as illuminants composed of "solutions of camphine (purified spirit of terpentine) in alcohol." The generally careless public should not, the article explained, use these explosive fluids (170).

10. City of Boston, *The Charter and Ordinances of the City of Boston, Together with the Acts of the Legislature Relating to the City* (1856), "Camphene," 85–86; "Fire," sec. 22: 173. "An Act to Provide for the Inspection of Petroleum Oils and Fluids, and to Regulate the Sale Thereof," chap. 262: 246–47; and "An Act Concerning the Manufacture, Storage and Sale of Petroleum and Its Products," chap. 285: 268–69, in State of Massachusetts, *Acts and Resolves Passed by the General Court of Massachusetts, in the Year 1866* (1866); "An Act Regulating the Sale of Coal and Petroleum Oils," in State of Massachusetts, *Acts and Resolves Passed by the General Court of Massachusetts in the Year 1867* (1867), chap. 286: 684–85; City Document No. 95, City of Boston, *Rules and Regulations in Relation to Petroleum and Its Products, 1866,* 1–5, in *Documents of the City of Boston, for the Year 1866,* vol. 2: *Containing Documents from No. 67 to No. 138, Inclusive* (1867); "Petroleum, and Burning Fluids," 511–19, and Streets, "Rules and Regulations," sec. 4: 644, in *Ordinances and Rules and Orders of the City of Boston, Together with the General and Special Statutes of the Massachusetts Legislature Relating to the City* (1869); and "Petroleum and Burning Fluids," 177–80, Fire Department, Ordinances, "Petroleum and Its Products," 181–83, in *Documents of the City of Boston, For the Year 1872,* vol. 1: *Containing Documents from No. 1 to*

No. 40, Inclusive (1873); City Document No. 35, *Thirty-Fourth Annual Report of the Boston Fire Department, 1871* (1872).

11. Edwin T. Freedley surveyed the "Alcohol, Burning Fluid, and Camphene" business in Philadelphia on the eve of the Civil War and observed that burning fluid was inexpensive to make, but "its principal demerit" was "liability to explosion." See Edwin T. Freedley, *Philadelphia and Its Manufactures: A Hand-Book Exhibiting the Development, Variety, and Statistics of the Manufacturing Industry of Philadelphia in 1857* (1859), 146, 147 (quotation 1), 148–49. On the Philadelphia fire and poorly enforced regulations, see "The Great Fire in Philadelphia: Two Thousand Barrels of Coal Oil Burned," *Cleveland Daily Leader,* February 13, 1865, 2, col. 4; "Coal Oil: Law Defied and Life Imperilled," *Philadelphia Daily Evening Telegraph,* September 7, 1869, 5, col. 1 (quotation), pp. 2–3. Frederick C. Brightly, *Annual Digest of the Laws of Pennsylvania for the Years 1862 to 1868* (1868), "Petroleum," 1405–6, Act, March 2, 1865, P.L. 262; Act, March 24, 1865, P.L. 750; "Petroleum," 1523–24, Act, February 14, 1868, P.L. 147. Frederick C. Brightly, *Annual Digest of the Laws of Pennsylvania for the Years 1873 to 1878* (1878), "Petroleum," 1813–15, Act, April 4, 1873, P.L. 60; "Petroleum," 1935–38, Act, May 15, 1874, P.L. 189.

12. William H. Wahl, "The Utilization of the Light Petroleum Oils," *JFI* 63, no. 6 (June 1872): 425 (quotation), 426–32. On the institute's support of science for "material advance" in the United States, see Bruce Sinclair, *Philadelphia's Philosopher Mechanics: A History of the Franklin Institute 1824–1865* (1974), 1–2, 3 (quotation 1), 4.

13. J. B. Burleigh, "Report of the Committee of the Franklin Institute on the Causes of Conflagrations and the Methods of Their Prevention," *JFI* 65, no. 4 (April 1873): 261 (quotation), 262–66.

14. William H. Wahl, "The Light Petroleum Oils; Considered as to Their Safety or Danger, in Various Domestic Uses," *JFI* 65, no. 4 (April 1873): 267–70, 271 (quotation 1), 272–87.

15. Wahl, "Light Petroleum Oils," 272–74, 275 (quotation 1), 277–79, 282 (quotation 2), 284 (quotation 3), 285–287. For the distribution of the report, see "Storing Dangerous Petroleum," 95–96, *Proceedings of the Seventh Annual Meeting of the National Board of Fire Underwriters . . . April 23d and 24th, 1873* (1873); also "Minutes of a Meeting of the Executive Committee of the National Board of Fire Underwriters . . . May 14th, 1873," 142; and "Minutes of a Meeting of the Executive Committee of the National Board of Fire Underwriters . . . June 19th, 1873," 154, in *Proceedings of the Executive Committee of the National Board of Fire Underwriters, at Meetings Held in New York, April 23d and 24th, May 14th and 15th, June 19th, and September 17th and 18th, in Philadelphia, November 12th and 13th, and in New York, December 11th and 12th, 1873* (1873), wherein see 136; and National Board of Fire Underwriters, "Report of the General Agent," by Thomas H. Montgomery, *Proceedings of the Eighth Annual Meeting of the National Board of Fire Underwriters . . . April 22d and 23d, 1874,* (1874), 43.

16. See *Annual Digest of the Laws of Pennsylvania for the Years 1862 to 1868,* 1405–6; *A Digest of the Laws of Pennsylvania, from the Year One Thousand Seven Hundred to the Sixth Day of June, One Thousand Eight Hundred and Eighty-Three,* comp. John Purdon, 11th ed., vol. 2, rev. Frank C. Brightly (1885), 1326–32; Frank C. Brightly, *Laws and Ordinances of the City of Philadelphia, From the Year One Thousand Seven Hundred and One to the Twenty-First Day of June, One Thousand Eight Hundred and Eighty-Seven* (1887), "Petroleum," Article 3, Refining and Storage, 885 (quotations 1–2); and *A Digest of the Statute Law of the State of Pennsylvania From the Year 1700 to 1894,* 12th ed., vol. 2, rev. Frank F. Brightly (1895), 1674–77.

17. See William J. Fryer Jr., ed., *Laws Relating to Buildings, in the City of New York*, part 3: *Laws Relating to the Extinction and Prevention of Fire, and Explosives and Combustible Materials in Buildings, in the City of New York* (1887), New York City Consolidation Act, Chapter 410, Laws of 1882, 91–105; Mark Ash, *The New York City Consolidation Act, as in Force in 1891* (1891), Chapter 11, Title 3, § 457, pp. 211–12; Charles F. Birdseye, *The Revised Statutes, Codes and General Laws of the State of New York*, 2nd ed. (1896), vol. 1: *Domestic Commerce Law*, Article 2, Regulations of Trade and Business, §§ 22–26.

18. On turpentine, see "An Act to reduce several Laws, relating particularly to the City of New York, into one Act," passed April 9, 1813; and "An Act to amend the acts heretofore passed for the prevention of fires in the city of New York," passed April 20, 1830, in C. T. McClenachan, comp., *A Compilation of the Laws of the State of New York; Also, of the Ordinances, Resolutions, and Orders . . . Relating to the Fire Department of the City of New York, From 1812 to 1860* (1859), chap. 86, 14, § 72, pp. 21–22; and § 37, p. 63.

19. "Fire Marshal's Semi-Annual Report: To the Mayor, Common Council, and Police Justices of the City and County of New York," *USIG&M* 9, no. 52 (August 1859): 190–91, 205, 216; "Fire Marshal's Semi-Annual Report," *USIG&M* 14, no. 84 (April 1862): 289–91, 292 (quotation 1), 293–94, 308, 310; and *USIG&M* 23, no. 137 (September 1866): 210–12.

20. For consideration of a city ordinance to regulate oil, see "Petroleum and Crude Earth Oils," *NYDTB*, November 7, 1862, 3, col. 5. For state statutes, see "An Act to regulate the storage and keeping of crude petroleum, earth or rock oil, or any of its products, within the corporate limits of any city in the State of New York," chap. 773: 594a (quotations 1–2), 594b–d; passed June 15, 1865, as amended by Chapter 872, "An Act to amend an act entitled 'An Act to regulate the storage and keeping of crude petroleum,'" 877–78, passed May 4, 1866, in John W. Edmonds, ed., *Statutes at Large of the State of New York*, vol. 6 (1868). For calls for strict regulation of flammable illuminants, see "The Kerosene Murders—A Request to Our Common Council," *NYT*, April 22, 1868; "The Quality of Kerosene Oils," *NYT*, April 23, 1869; "Kerosene Oil: Labors of the Fire Commissioners in the Matter," *NYT*, August 22, 2; and "The Bureau of Combustibles: The Sale of Adulterated Kerosene," *NYT*, April 7, 1872. See also City of New York, Metropolitan Board of Health, *Manual of the Metropolitan Board of Health, and the Metropolitan Board of Excise, June, 1869* (1869), Code of Sanitary Ordinances, and Rules and Sanitary Regulations, sec. 170 (adopted January 25, 1869), 82.

21. For a contemporary classification of bitumens, see S. F. Peckham, "Notes of the Origin of Bitumens, Together with Experiments upon the Formation of Asphaltum," *Proceedings of the American Philosophical Society* 10, no. 80 (July 1868): 445–62.

22. On separating the "lighter oils" from kerosene in distillation, see "Coal Oils and Petroleum," *SciAm* 14, no. 21 (May 19, 1866): 345; on the dangers of kerosene, see "Dangerous Kerosene," *Chemical News and Journal of Physical Science*, American Reprint, vol. 5 (July 1869): 55–56; August, 113–114; October, 259; November, 318–19; December, 393; and a mandatory caution for retail illuminating oils, as noted in "Be it ordained by the Board of Health of the City of Chicago," January 4, 1870, in City of Chicago, *Report of the Board of Health of the City of Chicago, For 1867, 1868 and 1869; and a Sanitary History of Chicago; From 1833 to 1870* (1871), 329.

23. See Williamson and Daum, *API Illumination*, 236–37, 238 (quotation 1). For massive naphtha fires, see "A Big Blaze in Williamsburg: Pratt's Oil Works Destroyed," *NYHRD*, December 22, 1884, 5, cols. 5–6; and "Huge Gas Tanks in Danger," *NYDTRB*, June 20, 1894, 4, col. 6.

On an early effort to form a mutual fire-insurance company for gas producers, see *Proceedings of the American Gas-Light Association at Meetings Held on . . . 1873 . . . and . . . 1874* (1875), 7–9, 49–58; on gas-producers and insurance problems, see Thomas Hudson, "Municipal Fire Insurance," *Bankers' Insurance Managers' & Agents' Magazine* (London), vol. 64, no. 641 (August 1897): 294; "Answers at Large," *Proceedings of the American Gas Institute, Third Annual Meeting October 21, 22, 23, 1908 New York, N.Y.* (1909), 481–84; and F. W. Fisher, "Gas Making Not Extra Hazardous," *Gas Age-Record* 50, no. 11 (September 9, 1922), Insurance Rate Reduction, 331.

24. For Chandler's investigations, see State of New York, Board of Health, *Report on the Methods and Apparatus for Testing Inflammable Oils: Based Upon Investigations, Ordered by the State Board of Health, for Establishing a Standard Safety-Test of Illuminating Oils,* by Arthur H. Elliott, rpt. from *Second Annual Report State Board of Health of New York* (1882), 40–42.

25. For the Board of Health, see "An Act to create a metropolitan sanitary district and board of health," passed February 26, 1866, in Edmonds, ed., *Statutes at Large of the State of New York,* vol. 6 (1868), chap 74: 674–96. See also "Review of the Proceedings of the Board, November 1, 1868, to October 31, 1869, Explosive Burning Fluids," 10–12; and "Report on Dangerous Kerosene," 437–57 (Ordinance 443), submitted January 11, 1869, and a Supplement, January 21, by C. F. Chandler, in State of New York, Board of Health, Metropolitan Sanitary District, *Fourth Annual Report of the Metropolitan Board of Health of the State of New York, 1869* (1870); and New York City, Health Department, Board of Health, *First Annual Report of the Board of Health of the Health Department of the City of New York, April 11, 1870, to April 10, 1871* (1871), Appendix L, "Report on Petroleum Oil, with Special Reference to the Prevention of the Traffic in Dangerous Kerosene and Naphtha," by C. F. Chandler, chemist to the Department of Health of the City of New York, March 18, 1871, 561 (quotation).

26. Two instruments often stipulated in state laws were the Tagliabue cup and the Foster apparatus. See C. F. Chandler, "Report on Petroleum Oil, Its Advantages and Disadvantages," *American Chemist* 3, no. 2 (August 1872): 41, for engravings of the Tagliabue "Open Tester" and "Closed Tester" or pyrometer. On the Foster device, see Sir Boverton Redwood, *Petroleum: A Treatise on the Geographical Distribution and Geological Occurrence of Petroleum and Natural Gas* (1906), vol. 2, "Determination of Flashing-Point and Fire-Test," 581–82; on testing, see H. O. LaCount, "Memorandum on Tests of Petroleum Oil; Apparatus and Methods of Use," John Ripley Freeman Papers, MC 51, MIT Institute Archives and Special Collections, box 30, Association of Factory Mutual Fire Insurance Companies, folder "Petroleum Report, H. O. LaCount, 1898"; on legal setbacks to Board of Health regulations, see *The Metropolitan Board of Health* v. *M. Schmades* in Charles P. Daly, ed., *Reports of Cases Argued and Determined in the Court of Common Pleas of the City and County of New York* (1872), vol. 3: 282–85; and a discussion of the case in W. P. Prentice, *Police Powers Arising Under The Law of Overruling Necessity* (1894), 153–54.

27. New York City, Health Department, Board of Health, *First Annual Report of the Board of Health . . . City of New York,* 513, 517–18, 521, 523–25, 545, 616 (quotation 1).

28. Ibid., "Laws Regulating the Sale of Inflammable Oils," 562–66; and State of New York, Board of Health, *Report on the Methods and Apparatus for Testing Inflammable Oils,* "Petroleum tests required in various States and foreign countries," 37; and *Petroleum and Products,* International Library of Technology (1902), table 5: State Requirements, § 53, p. 34. On Michigan see "Facts and Opinions," *Weekly Underwriter* 48, no. 7 (February 18, 1893): 115 (quotation 1).

29. See successively City of Mobile, *The Charter and Code of Ordinances of the City of Mobile,* comp. Reuben A. Lewis (1866), chap. 27, § 250, p. 162; City of Galveston, *Charter and Revised Ordinances of the City of Galveston* (1875), chap. 26, "Petroleum Oil and Inflammable Commodities," 204–6; and *Jewell's Digest of the City Ordinances . . . of the City of New Orleans,* comp. Edwin L. Jewell (1882), title 16.—Police Regulations, chap. 6, sec. 1—Petroleum, 330–32 (hereafter *Jewell's Digest Ordinances New Orleans* plus date).

30. For a list of hazards and rates dating to 1819 for a New York City insurance organization, see Henry A. Oakley, "Early Records of Fire Underwriting in New York. No. 1," *InsTm* 1, no. 10 (October 1869): 693; "Early Records of Fire Underwriting in New York. No. 2," *InsTm* 1, no. 11 (November): 803.

31. HSP, Am 35812, Philadelphia Board of Fire Underwriters, "Minutes, 1853–1881," 120, October 23, 1861 (quotation 1); 121, November 20 and December 31; and 122, February 19, 1862 (quotations 2–3); (hereafter HSP, PBFU). Williams included with his report a newspaper clipping of a statement by Charles T. Jackson, state assayer, Massachusetts, in the *Boston Courier* regarding the safety of kerosene. See Charles T. Jackson, MD, state assayer, "Kerosene Explosion, *To the Editor of the Boston Courier,*" *DET,* April 8, 1861, 1, col. 6; and "Explosive Kerosene," *SDR,* April 9, 1861, 2, col. 2. See also a long article on the products of coal distillation and their attendant dangers, "Some Items about Coal Oil and Its Products," *Phil Inq,* October 11, 1861, 2, cols. 1–3.

32. HSP, FBFU, "Minutes, 1853–1881," 122, 134–35, May 12, 1863; 138, December 31, 1863, and January 5, 1864.

33. See Paul F. Caranci, *North Providence: A History and the People Who Shaped It* (2012), 42–44; [Zachariah] T. Allen, "Explosibility of Coal Oils" (1862), 330 (quotations 1–2); and [Hovey T. Freeman and Norris G. Abbott Jr.,] *The Factory Mutuals 1835–1935* (1935), 64–65, which cites Manufacturers Mutual records to the effect that Allen's task was to determine the dangers of kerosene. On Allen's fire-prevention efforts, see Richard Eric Greenwood, "'Scientific Engineering and Useful Improvements': The Manufacturing Career of Zachariah Allen, 1822–1872," 1996, 283–90.

34. Allen, "Explosibility of Coal Oils," 330 (quotation 1), 337 (quotation 2), 340 (quotation 3).

35. Ibid., 342 (quotation 1).

36. "Special Notices: At a Meeting of the New York Board of Fire Insurance Companies," *New York Commercial Advertiser,* December 24, 1861, 3, col. 3 (quotation). Also printed as "Board of Fire Underwriters of New York: Their Action in Relation to Petroleum, Rock Oil, Benzole, and Naphtha in Public Warehouses," *USIG&M* 14, no. 82 (February 1862): 182–84. For the process of making coal gas, see Derek Matthews, "The Technical Transformation of the Late Nineteenth-Century Gas Industry," *Journal of Economic History* 47, no. 4 (December 1987): 970.

37. "Board of Fire Underwriters of New York: Their Action," *USIG&M* 14: 183 (quotations 1–2), 184 (quotation 3).

38. Ibid.

39. See Brearley, *History NBFU,* 3–11, 14–15, 25–27, who noted that James M. McLean, president of the NYBFIC, was the first president of the NBFU, and that NBYFIC secretary Frank W. Ballard became the first secretary of the NBFU (14).

40. Brearley, *History NBFU,* 18; NBFU, *Journal of the Proceedings of the First Annual Meeting of the National Board of Fire Underwriters, Held in New York, Feb. 20th, and 21st, 1867* (1867), 7 (quotation 1), 14 (quotation 2), 39 (quotation 3); and "National Board of Fire Underwriters, Meeting of the Executive Committee," *NYEP,* February 19, 1867, 3, cols. 9–10. According to "Instruction to Local Boards of Underwriters . . . Miscellaneous: Coal Oil and Its Products," *InsTm* 2, no. 3 (March 1869), the NBFU executive committee asked the Committee on Local Boards, December 12–13, 1866, to secure action on "storage of coal oil and its products in places where . . . other property would be endangered" (209, col. 1).

41. NBFU, *Journal of the Proceedings of the First Annual Meeting of the National Board of Fire Underwriters,* 50 (quotations 1–3), 51 (quotations 4–5), 52.

42. "New York Insurance Law, Chapter 846: *An Act to Incorporate the New York Board of Fire Underwriters,* Passed May 9, 1867," *USIG&M* 25, no. 148 (August 1867): 165–70; "Report of the Board of Underwriters: Passage of Bills at Albany—Withholding of the Governor's Signature," no. 147 (July): 101–11; "Fire Insurance: New-York Board of Fire Underwriters vs. Gov. Fenton," *NYDTB,* July 24, 1867, 3, col. 3 (quotation 1); and H. R. Hayden, ed., *Annual Cyclopedia of Insurance in the United States, 1891* (1892), "New York Board of Fire Underwriters."

43. On the December 1861 NYBFIC committee, see above.

44. See NBFU, *Journal of the Proceedings of the First Annual Meeting of the National Board of Fire Underwriters,* Report of Committee on Gases, 14; "Board of Fire Underwriters of New York: Their Action in Relation to Petroleum," *USIG&M* 14: 182–84; NBFU, *Report of Proceedings of Session of Executive Committee . . . Sept. 11 and 12, 1867* (1867), 23, 26, 30 (adoption); "Gasoline and other Dangerous Substances," *CDG,* November 1, 1867, 4, col. 5; and "National Board of Fire Underwriters: Third Annual Meeting," *InsTm* 2, no. 4 (April 1869): 291; "Kerosene," InsTm 2, no. 4 (April 1869): 301, col. 2.

45. "Gasoline and other Dangerous Substances," *CDG,* November 1, 1867, 4, col. 5 (quotations 1–3); and NBFU, *Report of Proceedings of Session of Executive Committee . . . Sept. 11 and 12, 1867,* 23–24.

46. NBFU, *Proceedings of the Executive Committee of the National Board of Fire Underwriters . . . February 24th and 25th, 1869* (1869), 31 (quotations 1–2), 33 (quotation 3). For kerosene-related accidents in the New York City area, see "Fatal Kerosene Explosion—Prolific Cause of Death and Great Destruction of Property. A Step in the Right Direction—The Manufacturers and Vendor of the Oil Held to Await the Action of the Grand Jury," *USIG&M* 28, no. 164 (December 1868): 85–86.

47. "Annual Meeting of the National Board of Fire Underwriters of the United States," *USIG&M* 29, no. 169 (May 1869): 10–11; and NBFU, *Journal of the Proceedings of the Third Annual Meeting of the National Board of Fire Underwriters, Held in New York, April 21st, 22d and 23d, 1869, Together with the Proceedings of the Executive Committee, April 20th and 23d, 1869* (1869), "Report," 62–74; "Rules for Loading Vessels, &C., at Petroleum Yards and Depots," 70–72.

48. NBFU, *Journal of the Proceedings of the Third Annual Meeting of the National Board of Fire Underwriters,* 69 (quotations 1–2); for the carburetor manufacturers, see "National Board of Fire Underwriters: Third Annual Meeting," *InsTm* 2, no. 4 (April 1869): 301, col. 2.

49. "National Board of Fire Underwriters: Third Annual Meeting," *InsTm* 2, no. 4 (April 1869): 301, 302, col. 1 (quotations 1–2), col. 3 (quotation 3), 303.

50. New York Board of Fire Underwriters, Special Committee on Gas Machines, Carburetters and Matters Connected Therewith, *Report of the Committee on Gas Machines, Carburetters, and the Handling and Using of the Lighter Products of Petroleum, Made to the New York Board of Fire Underwriters, and Adopted October, 29, 1869* (1869), 5–6, 8 (quotation 1), 9, 10 (quotations 2–3), 13. I thank the Hagley Museum and Library, Wilmington, Del., for providing a digital copy, for which the institution retains the copyright. Several leading New York City fire-insurance companies distributed a circular in January 1870 based on the above-noted, gas-machine committee report. See "Gasoline: Portable Gas Machines," *InsTm* 3, no. 1 (January 1870): 41. For a contemporary view of the chemical properties and dangers of hydrocarbon-illuminants, see Symposiarch, "Gasoline Gas," *InsTm* 2, no. 6 (June 1869): 425–26; and NYBFU concern over gas machines in "Meeting of the Board of Underwriters. Efforts to Remove the Wanton Cause of Fires," *USIG&M* 34, no. 200 (December 1871): 59.

51. For NBFU adoption, see NBFU, *Proceedings of the Executive Committee of the National Board of Fire Underwriters . . . December 8th and 9th, 1869*, 18–24. In early 1873 the NBFU suggested rates for hazards in various states under the proviso that all local boards refuse to grant permits for the use of any petroleum fractions lighter than kerosene, and prohibit all kerosene sales or use by the insured "unless the filling and drawing be done only by daylight, and without the use of artificial light." See NBFU, *Proceedings of the Executive Committee of the National Board of Fire Underwriters At the Meeting of December 18th, 1872, and February 12th, 1873, Held at Underwriters' Hall, 158 Broadway, New York, And at that of March 27th, 1873, Held at the Rooms of the Philadelphia Association of Fire Underwriters 308 Walnut Street, Philadelphia* (1873), 34.

52. See "The National Board of Fire Underwriters: The Seventh Annual Meeting," *NYHRD*, April 24, 1873, 11, col. 6; "National Board of Fire Underwriters: Action of the Board in Reference to Local Boards, State Boards, the Building Laws, Fires and Incendiarism," *NYHRD*, April 25, 10, col. 3; "Chicago Fire Insurance: Indignation at the Action of the National Board of Underwriters," *NYTB*, September 25, 1874, 1, col. 3; "The National Board of Fire Underwriters: Their Action in Regard to Chicago," *NYEP*, October 3, 1874, 10, col. 2; and "Protection Against Fire: The Board of Trade of Concord, N.H., and the Superintending Agent of the National Board of Fire Underwriters in Conference," *BDA*, December 17, 1874, 1, col. 6.

53. On wind, see "Fire Underwriters Anxious about Chicago," *NYTB*, December 13, 1873, 1, col. 3; "Chicago's Risk From Fire: Insurance Companies Warned," *NYDTB*, July 28, 1874, 2, cols. 4–5; "Fire Risks in Chicago: A Withdrawal of Insurance Business Imminent," *NYDTB*, September 25, 2, col. 1; "The Grounds of Action: Statements of Insurance Officers," *NYDTB*, September 25, 2, cols. 1–2; "Insurance: No Particular Change in the Situation Yesterday," *DIO*, October 3, 1874, 12, cols. 1–3; "No Insurance for Chicago: Importance of the Recent Action of Underwriters," *New York Daily Graphic*, September 25, 1874, 615, cols. 1–2; "Chicago Fire Insurance: Views of Leading Underwriters," *New York Daily Graphic*, September 28, 633, cols. 1–2; and "The Chicago Water Supply Tested by the Fire Department," *NYDTB*, October 7, 1874, 7, col. 5.

54. Donald W. Linebaugh, "Illuminating a Country Retreat: Gas Lighting at President Lincoln's Cottage," *APT Bulletin* 43, nos. 2–3 (2012): 60–63.

55. On retail dangers, see "Modern Murder," *InsTm* 3, no. 9 (September 1870): 666; "Oils and Oily Rascals," *InsTm* 4, no. 2 (February 1871): 115, col. 3; "Board of Underwriters: How to Prevent Fires from the Reckless Use of Matches and Inflammable Oils—Annual Election of

Officers," *NYHRD,* April 24, 1875, 8, col. 3; "Fire Insurance: The National Board of Underwriters," *NYT,* April 24, 1875, 12; "National Board of Fire Underwriters: The Annual Meeting in New York." *BJ,* April 25, 1878, 2, col. 7; "Fuel for the Flames," *CDG,* August 25, 1881, 6, col. 1; and "Facts and Opinions," *Weekly Underwriter* 48, no. 7 (February 18, 1893): 115.

56. For local-board questions see NBFU, *Minutes of a Meeting of the Executive Committee of the National Board of Fire Underwriters . . . March 26th and 27th, 1874* , 336 in *Proceedings of the National Board of Fire Underwriters, and of the Executive Committee, From April, 1873, to March, 1874* (1874). See also NBFU, "Minutes of a Meeting of the Executive Committee of the National Board of Fire Underwriters . . . February 24th and 25th, 1876" (1876), 442–444. On NBFU concerns about kerosene and coal-oil storage, benzine, and naphtha, see "Causes of Fires," NBFU, *Proceedings of the Seventh Annual Meeting of the National Board of Fire Underwriters . . . 1873,* 25, for the observation that "forty classes of manufacturing risks" are using "products of petroleum"; as well as NBFU, *Proceedings of the Executive Committee of the National Board of Fire Underwriters, at Meetings, Held in New York, December 19th, 1873, January 29th & 30th, 1874, in Philadelphia, February 26th & 27th, 1874; and in New York, March 26th & 27th, 1874* (1874), 249; and NBFU, *Bulletin of the Executive Committee of the National Board of Underwriters,* no. 2 (December, 1873): 15, 22–23; no. 3 (January 1874): 40.

57. "Dangerous Kerosene," *Chemical News and Journal of Physical Science,* American Supplement 5, 55–56, 113–14, 259, 318–19, 359–60, 393; and C. C. Hine, *Fire Insurance: A Book of Instructions for the Use of the Agents of the United States* (1870), 23 (quotations 1–2), 24 (quotation 3), 25. For biographical information on Hine, see I. W. Holman, "Report of Committee upon the Death of Mr. Chas. C. Hine," in Fire Underwriters' Association of the Northwest, *Proceedings of the Twenty-Eighth Annual Meeting of the Fire Underwriters' Association of the Northwest, Chicago, Ill., September 29–30, 1897* (1897), 154–60.

58. On the Chicago fire and company losses, see *The Chicago Fire and the Fire Insurance Companies: An Exhibit of the Capital, Assets, and Losses of the Companies* (1871); for contemporary accounts of the Boston fire and insurance company losses and failures, see F. E. Frothingham, *The Boston Fire, November 9th and 10th, 1872: Its History, Together with the Losses in Detail of Both Real and Personal Estate* (1873); and *The Investor's Guardian and Insurance Almanac 1873* (1873): 71–75.

59. For early adoption of uniform fire-insurance policies that linked graded hazards, such as oil and burning fluid, to a rate schedule and other restrictions, see HSP, PBFU, "Minutes, 1853–1881," 42–43, September 28, 1855; 64–66, January 15, 1857; and 74, May 20, 1857.

60. On Oakley's speech, see "President's Address," in NBFU, *Proceedings of the Seventh Annual Meeting of the National Board of Fire Underwriters . . . 1873,* 11–24, 25 (quotations 1–4); "Storing Dangerous Petroleum," *Proceedings of the Seventh Annual Meeting of the National Board of Fire Underwriters . . . 1873,* 95–96. On the seventh meeting, see also "The National Board of Fire Underwriters: The Seventh Annual Meeting," *NYHRD,* April 24, 1873, 11, col. 6; and "National Board of Fire Underwriters Action of the Board in Reference to Local Boards, State Boards, the Building Laws, Fires and Incendiarism," *NYHRD,* April 25, 10, col. 3. The NBFU Executive Committee shared Oakley's apprehensions, voting in the fall of 1873 to allow mercantile storage of five barrels of "standard quality" kerosene and coal oil at no extra charges and a month later imposing additional charges for keeping quantities above this amount; see "Minutes of a Meeting of the Executive Committee of the National Board of Fire Underwriters Called to Consider the Matter of Legislation and Taxation . . . Decem-

ber 19th, 1873," in NBFU, *Proceedings of the Executive Committee of the National Board of Fire Underwriters at Meetings . . . December 19th, 1873*, 249 (quotation 1 above); and "Report of Thomas H. Montgomery, General Agent," in NBFU, *Reports of Thomas H. Montgomery, General Agent National Board of Fire Underwriters, September, 1872—April, 1877* (1877), 9–10, 23, report dated April 22, 1874. See also NBFU, *Bulletin of the Executive Committee of the National Board of [Fire] Underwriters*, no. 2: 14–15, 22 (quotation 2 above), 23; and no. 3: 40, for a note that the amended rule appeared in circular no. 111. See also note 51 above for NBFU concerns in 1873 about petroleum and its light products.

61. NBFU, "Second Day's Proceedings," *Proceedings of the Eighth Annual Meeting of the National Board of Fire Underwriters . . . April 22d and 23d, 1874*, 107, 113–21. For Oakley's remarks, see *Address of Henry A. Oakley, President of the National Board of Fire Underwriters of the United States, at the Eighth Annual Meeting, Held in the City of New York April 22d, 1874* (1874), 24–26. On Walton's request, see "National Board of Fire Underwriters: Approval of President Grant's Veto—Reports from Committees and Adoption of the Constitution," *NYEP*, April 23, 1874, 3, col. 9.

62. On the Executive Committee, see *Bulletin of the Executive Committee of the National Board of Underwriters*, no. 2: 22; and no. 3: 40. On the fire, see "New England: Massachusetts. Disastrous Fire at Haverhill," *BJ*, November 17, 1873, 4, col. 8; "Great Fire in Haverhill—The Business Centre in Ruins," *Lowell* (Mass.) *Daily Citizen & News*, 2, col. 2; "Disastrous Fire in Haverhill," *DET*, November 17, 1873, 1, col. 6; and "Great Fire in Haverhill, Mass.," *New Hampshire Patriot and State Gazette* (Concord), November 19, 3, col. 2. On the contents of the barrel, see "The Fire at Haverhill. Losses and Insurance," *BJ*, November 17, 1873, 2, col. 3.

63. On the Canajoharie fire, see "Disaster at Canajoharie. Arkell & Smith's Sack Factory Destroyed by an Explosion of Gasoline," *Albany* (N.Y.) *Evening Journal*, December 2, 1873, 3, col. 1; on the litigation, see "Mohawk Valley Notes," *The Argus* (Albany, N.Y.), April 16, 1875, 4, col. 2.

64. On the explosion, see "Appalling Calamity at Bennington," *St. Johnsbury* (Vt.) *Caledonian*, January 23, 1874, 2, cols. 3–4; "Awful Calamity at Bennington, Vermont," *North Star* (Danville, Vt.), 3, col. 4; "Terrible Explosion and Fire in Bennington," *Daily Messenger* (St. Albans, Vt.), 4, col. 6; and "The Bennington Calamity," *Argus and Patriot* (Montpelier, Vt.), January 29, 3, col. 6. For the theory on the origins of this disaster, see "The Late Bennington Disaster," *Daily Messenger*, January 28, 1874, 2, col. 4. On the technical features of an Excelsior gas machine that exploded in Warren, Massachusetts, in 1874, see C. O. Thompson, "Gas from Gasoline," *Journal of Gas Lighting, Water Supply, and Sanitary Improvement* 26, no. 648 (October 12, 1875): 552. For the history of the implementation of standard fire-insurance policies, see *Hayden's Annual Cyclopedia of Insurance in the United States 1902–1903* (1903), "Policy and Laws"; for the Massachusetts state policy (1887), see *The Annual Cyclopedia of Insurance in the United States, 1890*, "Fire Insurance Policy, Standard," Massachusetts, 57, 58 (quotation).

65. "Minutes of a Meeting of the Executive Committee of the National Board of Fire Underwriters Held at the Rooms of the Philadelphia Association of Fire Underwriters, No. 308 Walnut Street, Philadelphia, Feb. 26 and 27, 1874," in NBFU, *Reports of Thomas H. Montgomery, General Agent National Board of Fire Underwriters*, 273. On the September adoption, see "Gasoline," *Baltimore Sun*, October 25, 1867, 2, col. 1; and note 44 above.

66. "Minutes of a Meeting of the Executive Committee of the National Board of Fire Underwriters Held at the Rooms of the Philadelphia Association of Fire Underwriters . . . 1874,"

in NBFU, *Reports of Thomas H. Montgomery, General Agent National Board of Fire Underwriters,* 274. On Henry Woodward's carburetor, see the description on the U.S. Patent Office website, U.S. Patent Office, H. Woodward, of London, "Improvement in Apparatus for Carbureting," specification forming part of Letters Patent No. 81,238, dated August 18, 1868; and *Annual of Scientific Discovery; or, Year-Book of Facts in Science and Art For 1865,* ed. David A. Wells (1865), "Carbonization of Illuminating Gas," 64–65;

67. NBFU, *Report of the Ninth Annual Meeting of the National Board of Fire Underwriters, Held in Underwriters' Hall, 156 and 158 Broadway* (1875); "Report of Committee on Origin of Fires, Patents, and Gas Machines," 157 (quotation 1), 158–64, in NBFU, *Proceedings of the National Board of Fire Underwriters, and of the Executive Committee, from April 22, 1875, to April 25, 1876* (1876); and "Fire Insurance: The National Board of Fire Underwriters," *NYT,* April 24, 1875, 12. On technological artifacts, see Pablo Schyfter, "The Bootstrapped Artefact: A Collectivist Account of Technological Ontology, Functions, and Normativity," *Studies in History and Philosophy of Science* 40, no. 1 (March 2009): 101–2; Pieter E. Vermaas, "The Physical Connection: Engineering Function Ascriptions to Technical Artefacts and Their Components," *Studies in History and Philosophy of Science* 37, no. 1 (March 2006): 62–63; and Vermaas et al., *A Philosophy of Technology From Technical Artefacts to Sociotechnical Systems,* 5–13. I thank Sara Hart, LILAB, for digital copies of above NBFU materials.

68. NBFU, *Report of the Ninth Annual Meeting of the National Board of Fire Underwriters,* "Report of Committee on Origin of Fires, Patents, and Gas Machines," 158–59, 162–63; and "Fire Insurance: The National Board of Fire Underwriters," *NYT,* April 24, 1875, 12 (quotation); for the gas-machine resolution, see "Insurance Matters: Concluding Day's Session of the Annual Convention of the National Board of Fire Underwriters," *DIO,* April 24, 1875, 12, col. 3.

69. Henry A. Oakley and D. A. Heald, *Report to the Executive Committee of the National Board of Fire Underwriters on Gas Machines,* dated "New York, Jan. 27th, 1876," 4 (quotation 1), 5 (quotation 2). I thank the staff of the Rutherford B. Hayes Presidential Center, Spiegel Grove, Fremont, Ohio, for providing a copy of this report.

70. Donald W. Linebaugh made this identification in *The Springfield Gas Machine: Illuminating Industry and Leisure, 1860s–1920s* (2011), 44.

71. Oakley and Heald, *Report to the Executive Committee . . . on Gas Machines,* 6, 8 (quotation 2) , 9, 10 (quotations 1 and 3) and Cut B (illustration).

72. On late-twentieth-century reliance upon risk assessment and management, see J. H. Saleh and K. Marais, "Highlights from the Early (and pre-) History of Reliability Engineering," *RE&SS* 91, no. 2 (February 2006): 249–52; and Aven, "Risk Concept Applicable for Both Probabilistic and Non-Probabilistic Perspectives," 1081–82.

73. Oakley and Heald, *Report to the Executive Committee . . . on Gas Machines,* 3 (quotations 1–3), 9. See also Renn, "Role of Risk Perception for Risk Management," 51–52. "Technological artifacts" is my phrase and is not in the report. Nonetheless, it expresses my understanding of the report's engagement with the technical features of patented or well-designed gas machines and their capacity to mitigate risk.

74. Oakley and Heald, *Report to the Executive Committee . . . on Gas Machines,* 11.

75. For the NYBFU meeting, see "Gas Saving Machines," *NYHRD,* May 10, 1876, 10, col. 3 (quotation 1). The NBFU's reluctance to soften its stance on naphtha and gas production is in retrospect understandable. See "Brief Telegrams," *Manufacturers and Farmers Journal*

(Providence, R.I.), December 16, 1869, 1, col. 8; "Explosion of Naphtha Gas," *NYHRD,* August 4, 1878, 7, col. 4; "After the Fire. Sad Scenes at the Morgue and Undertakers' Shops," *NYHRD,* January 6, 1881, 5, cols. 4–6; "A Hero in Private Life: Mr. Daniel Schatt Almost Roasted Alive in Rescuing an Employee," *Phil Inq,* October 1, 1893, 7, col. 2; "Six Men Badly Burned: Explosion of Naphtha in Gas Works," *NYTB,* part 2, May 16, 1896, 13, col. 3.

76. See *The Insurance Cyclopaedia,* "Gas Machines," by Cornelius Walford (1878), vol. 5: 204; "Gasoline," 205; and Hayden, ed., *The Annual Cyclopedia of Insurance in the United States, 1891,* "National Board of Fire Underwriters," Proceedings of the Twenty-Fifth Annual Meeting, 208, on the "sale and use of oils and combustibles." Note that the public was in certain circumstances uneasy about gas machines in densely settled, urban areas, as in the case of the machine installed in the Continental Hotel supplied with gasoline from tanks on the premises and from new tanks erected at a stable at the corner of Raspberry Alley and Craig's Court, to be piped to the hotel. See HSP, PBFU, "Minutes, 1853–1881," 188, May 21, 1879; "Substitute for City Gas," *Phil Inq,* January 20, 1879, 2, col. 3; "Cheap Gaslight," *Phil Inq,* February 6, 1879, 2, col. 5; "The Stir That Gasoline Can Cause," *Phil Inq,* April 19, 1879, 2, col. 3; and "The Continental Hotel's Gas," *The North American* (Philadelphia), April 19, 1879, 1, col. 8.

77. See "A Distinction and Also a Difference," *InsTm* 34, no. 1 (January 1901): 16; also E. Van Ness, "The Gasoline Hazard," *InsEng* 2, no. 2 (August 1901): 118–26; and his criticisms, "National Board's Gasoline Rules," *InsEng* 2, no. 6 (December 1901): 517–22; William H. Merrill Jr., "Comments by W. H. Merrill, Jr., Secretary of the Underwriters' Laboratories," *InsEng* 2, no. 6 (December 1901): 521–27; "Comments by a New York Underwriter," *InsEng* 2, no. 6 (December 1901): 527–35; "Remarks by Insurance Engineering," *InsEng* 2, no. 6 (December 1901): 535–36; and "Rules and Requirements of the National Board of Fire Underwriters, for the Construction, Installation and Use of Gasoline Vapor Gas Lighting Machines, Lamps and Systems, as Recommended by Its Committee of Consulting Engineers, Edition of 1901," *InsEng* 2, no. 6 (December 1901): 557, 558 (quotations 1–2), 561 (quotations 3–5), 566 (quotations 6–7).

78. On dealing with appliances and patented devices and a proposal to allow brass or copper carburetors to be installed in cellars, see "National Board of Fire Underwriters: Annual Meeting for 1875 . . . Second Day," *Chronicle* 15, no. 17 (April 29, 1875): 270, col. 1. For NBFU discussions on dealing with manufacturers of gas machines, see "Tenth Annual Meeting of the National Board of Fire Underwriters . . . Origin of Fires, &c.," *InsTm* 9, no. 5 (May 1876): 348; see also 349, col. 3, on allegedly safe "fixed" gases for fuel. On carburetor manufacturers seeking NBFU approval, see "Minutes of a Meeting of the Executive Committee of the National Board of Fire Underwriters . . . June 14, 1876," in NBFU, *Reports of Thomas H. Montgomery, General Agent National Board of Fire Underwriters,* 293. For NYBFU dealings with gas-machine "proprietors and inventors," see "Report of the Committee on Gas Machines, Carburetters, and the Handling and Using of the Lighter Products of Petroleum . . . Conclusion," *InsTm* 2, no. 11 (November 1869): 782, col. 3 (quotation); as well as Hayden, ed., *Annual Cyclopedia of Insurance in the United States, 1890,* "Fire Insurance Policy, Standard," for NYBFU policy restrictions on generating illuminating gas or vapor in an insured building or a building adjacent to it (60).

79. "Insurance Matters: Concluding Day's Session," *DIO,* April 24, 1875, 12, col. 3. On insurance fears of gasoline, see "It Vitiates Fire Insurance: Many Companies Refuse to Accept a Risk Wherein Gasoline Is a Factor," *Phil Inq,* August 9, 1889, 2, col. 4. On the enforcement of

the rule that no gas machine could be "located on, or adjoining," insured premises, see "The National Board," *Chronicle* 15, no. 21 (May 27, 1875): 325, col. 3.

80. "Insurance: Annual Meeting of the Northwestern Association of Fire Underwriters at the Palmer House . . . Afternoon Session . . . The Lighter Products of Petroleum," *DIO,* September 23, 1875, 2, col. 1; on vapor lamps, see "Comments by a New York Underwriter," *InsEng* 2, no. 6 (December 1901): 529, 534 (quotation 1), 535; and "Rules and Requirements of the National Board of Fire Underwriters, for the Construction, Installation and Use of Gasoline Vapor Gas Lighting Machines, Lamps and Systems," Class D: Gasoline Vapor Lamps, *InsEng* 2, no. 6 (December 1901): 563–65.

81. On twentieth-century views of the development and subjectivity of rates in the nineteenth and early twentieth centuries, see Robert Riegel and H. J. Loman, *Insurance Principles and Practices,* 2nd ed. (1922), 234–35, 237, 239; and Mark Tebeau, *Eating Smoke: Fire in Urban America, 1800–1950* (2003), 63–64, 71–73, 76–77, 172–73, 196–98.

82. See "Introduction: Statebuilding in the Progressive Era: A Continuing Dilemma in American Political Development," in *Statebuilding from the Margins Between Reconstruction and the New Deal,* ed. Carol Nackenoff and Julie Novkov (2014), 3, 7–9. On calls from the insurance interests for appointment of municipal fire wardens "with police jurisdiction" who would meticulously inspect their districts and identify hazards, giving careful attention to special hazards (petroleum and its products fell into this classification), see Cyrus K. Drew, "Fire Underwriters' Association of the Northwest . . . President's Address," *TC* 34, no. 11 (September 11, 1884): 151, col. 1, given at the fifteenth annual meeting of the FUAN.

83. For state use of risk information on hazardous illuminants, see "Restrictions Against Unsafe Illuminating Oils," *SDR,* January 10, 1871, 4, col. 2; on the insurance interests' risk knowledge of hazardous illuminants, see "Gasoline Gas," *InsTm* 2, no. 1 (January 1869): 32; "Gasoline," *Baltimore Sun* (1867), 2, col. 1; E. W. Crowell, "Storing Dangerous Petroleum," *Proceedings of the Seventh Annual Meeting of the National Board of Fire Underwriters . . . 1873,* 95–96. For scientific-technical sources of information available to the insurance interests, see John Attfield, "The Igniting Point of Petroleum," *American Journal of Mining* 3, no. 3 (April 13, 1867): 42, an abstract of his complete paper published in the *London Mechanical Magazine;* and Benjamin H. Paul, "The Mode of Testing Mineral Oils Used for Lamps," *Chemical News and Journal of Physical Science* (American Supplement), vol. 6, no. 3 (March 1870): 129–31.

84. Petroleum prospecting and production in California moved forward between 1850 and 1900. The best coverage of the industry in California, with details of very early discoveries of asphaltum and crude-oil seeps, is Kenny A. Franks and Paul F. Lambert, *Early California Oil: A Photographic History, 1865–1940* (1985). See also "Asphaltum and Oil Deposits in Los Angeles and in Santa Clara," *SFEB,* March 1, 1882, 3, col. 8; "The Oil Fields of Los Angeles," *SFEB,* May 28, 1883, 4, col. 2; "Beginning a Great Enterprise," *Press and Horticulturist* (Riverside, Calif.), March 22, 1884, 5, col. 3; "Petroleum: Progress of the Oil Industry in California," *SFEB,* February 5, 1887, 4, col. 1; and Lionel V. Redpath, comp., *Petroleum in California* (1900), 9–13.

85. "The Coal Power of the World," *Western Reserve Chronicle* (Warren, Ohio), April 18, 1860, 1, col. 4; "Oil News," *Western Reserve Chronicle,* April 18, 1860, 1, col. 7; and statistics in Williamson and Daum, *API Illumination,* 290–91, 373. For early suggestions that petroleum could supplant coal as an illuminant or as fuel, see "Uses of Petroleum," *Weekly Portage Sentinel*

(Ravenna, Ohio), December 26, 1860, 1, cols. 5–6; and "The Coming Light and Heat," *PCDR*, January 28, 1873, 2, col. 1.

86. The basic narratives for SOC are Ralph W. Hidy and Muriel E. Hidy, *Pioneering in Big Business 1882–1911: History of Standard Oil Company New Jersey* (1955); and Williamson and Daum, *API Illumination.* My understanding of fuel-oil development draws heavily on newspaper sources and technical-professional journals.

87. *The Derrick's Hand-Book of Petroleum: A Complete Chronological and Statistical Review of Petroleum Developments During 1898 and 1899*, vol. 2 (1900), 289, February 4, 1879; 294, April 17, 1879; 306, September 8, 1879; 363, September 23, 1881. For Chicago, see "Petroleum in the Production of Steam," *PCDR*, February 20, 1871, 2, col. 2; for St. Louis, see "Petroleum Gas as Fuel," *PCDR*, October 30, 1871, 2, col. 3.

88. The basic examination of Ohio oil and gas fields is Edward Orton, *Geological Survey of Ohio: Preliminary Report Upon Petroleum and Inflammable Gas*, rpt. with supplement (1887); and "Ohio Wonderland, Natural Gas and Oil at Lima and Findlay," *CCG*, November 8, 1886, 1, cols. 1–7; 6, cols. 5–7; on the development of Ohio oil resources, see "Lima (O.) Oil Field in 1886," *STWPR* 15, no. 2 (February 11, 1887): 17, col. 1 (published as part of *AM&IW* 40, no. 6, February 11, 1887); "Notes from Lima," *AM&IW* 40, no. 12 (March 25, 1887): 17, col. 3; "Ohio Oil Fields: The Standard Company's Grip on the Lima Producers," *CCG*, April 2, 1887, 1, cols. 3–4; "Lima Oil for Fuel," *STWPR* 15, no. 51 (January 13, 1888): 17, col. 3 (published as part of *AM&IW* 42, no. 2, January 13, 1888); "Will Yield Millions: Immense Profit in Ohio Crude Oil," *Chicago Herald*, October 13, 1890, 8, col. 1 (quotation 1); on SOC controlling Ohio oil production, see "Ohio," *E&MJ* 47, no. 9 (March 2, 1889), 220, col. 1; and no. 16 (April 20, 1889): 376, col. 2. On problems using high-sulphur Lima oil to make water gas, see "Lima Oil and Water Gas," *AM&IW* 42, no. 16 (April 20, 1888): 17, cols. 1–2.

89. "Petroleum as Fuel," *CPD*, May 26, 1887, 4, col. 2; "Petroleum as Fuel—Experiments in Cleveland," *STWPR* 15, no. 22 (July 1, 1887): 21, col. 3 (published as part of *AM&IW* 40, no. 26, July 1, 1887); "Oil Fuel in Heating for Wire Rods," *STWPR* 52 (January 20, 1888): 18, cols. 3–4 (published as part of *AM&IW* 42, no. 3, January 20, 1888); "The Chicago Oil Pipe Line," *AM&IW* 42, no. 10 (March 9, 1888): 18, col. 2; "Manufacturing: Iron and Steel," *TIA* 51 (February 2, 1893): 255, col. 2, para. 4; "Best, Fox & Co.," *TIA* 51 (April 13, 1893): 848, col. 1; and "Large Contracts for Fuel Gas Plants," *TIA* 48, no. 11 (November 5, 1891): 778.

90. "The Chicago Oil Pipe Line," *AM&IW* 42, no. 10 (March 9, 1888): 18, col. 2; "A New Pipe Line: To Be Constructed from Lima, O, to Chicago to Transport Crude Oil for Fuel," *Boston Evening Journal*, April 16, 1888, 2, col. 3; and "The Lima-Chicago Oil Line," *AM&IW* 42, no. 17 (April 27): 11, col. 1.

91. "Lima Oil at Chicago," *AM&IW* 43, no. 7 (August 17, 1888): 7, col. 4; "Petroleum as Fuel," *BH*, August 20, 1888, 4, cols. 3–4; and "General Mining News," *E&MJ* 47, no. 4 (January 26, 1889): 94, col. 3; on municipal approval, see "Around the Offices . . . Aldermen Will Visit South Chicago to Look at Oil Pipes," *DIO*, December 14, 1888, 5, col. 1; "The Council Committee on Streets and Alleys," *Paint, Oil and Drug Review* 7, no. 12 (March 20, 1889): 10, col. 2; "No Oil Pipe Line: Mayor Cregier Vetoes the Ordinance Granted to the Hanford Oil Company," *DIO*, July 16, 1889, 3, cols. 1–3; "The Chicago City Council," *The Vindicator* (Atlanta–New Orleans), vol. 7, no. 2 (August 1, 1889): 45, col. 2; and "The Pipe Line Ordinance," *DIO*, August 2, 1889, 9, col. 2.

92. On the fuel-oil department, led by Civil War veteran and experienced oil man C. W. Owston, see "A More Hopeful Outlook for Lima Oil," *AM&IW* 40, no. 22 (June 3, 1887): 22, col. 1; and "The Liquid Fuel Brigade," *STWPR* 15, no. 19 (June 10, 1887): 17, col. 2, published as part of *AM&IW, vol. 40 noted above.* See also "Lima Oil For Fuel," *STWPR* 15, no. 51 (January 13, 1888): 17, col. 3 (published as part of *AM&IW* 42, no. 2, January 13, 1888); and "Lima Oil for Fuel," *AM&IW* 42, no. 14 (April 6, 1888): 22, col. 2 (quotation); on Midwest markets, see "The Lima-Chicago Oil Line," *AM&IW* 42, no. 17 (April 27, 1888), 11, col. 1.

93. Fred De Land, "The Power Plant (Concluded)," *Electrical Engineering: An Illustrated Monthly Magazine* 1 (January–June 1893): 64–68. Clayton O. Billow, "Fuel Oil: Comparative Results in Steam Generation," *Electrical Engineering* 5, no. 4 (April 1895): 202–9; no. 5, 235–39. "Insurance Circles . . . Defects Being Remedied," *DIO*, April 2, 1893, 9, col. 1; "World's Columbian Exposition.—VII. Boiler Plant," *Engineering News and American Railway Journal* 29, no. 15 (April 13, 1893): 342–44; and "Liquid Fuel: Interesting Govt. Report on Its Efficiency," *BDA*, May 31, 1894, 8, col. 1.

94. "Ohio Oil as Fuel," *AM&IW* 42, no. 20 (May 18, 1888): 9, col. 2; and "Crude Petroleum as Fuel," *AM&IW* 43, no. 1 (July 6, 1888): 9, col. 1 (quotations); on heavy industries in Chicago using fuel oil, see "The Progress of Oil Fuel," *AM&IW* 42, no. 14 (April 6): 7, col. 4; "Petroleum Fuel," *AM&IW* 43, no. 21 (November 23): 17, col. 1; "Use of Crude Petroleum for Raising Steam at the South Chicago Works, *AM&IW* 44, no. 20 (May 17, 1889): 17, col. 4; and "Electricity in the Works of Fraser & Chalmers," *TIA* 51 (March 1, 1893): 479, col. 3. For the adoption of the rules, see H. S. Tiffany, *Tiffany's Book of Policy Forms for Fire Insurance Agents*, 4th rev. ed. (1890), "Form of Crude Petroleum Permit for Fuel," 234a.

95. "Crude Petroleum as Fuel," *AM&IW* 43, no. 1 (July 6, 1888): 9 (quotation 1).

96. Ibid. (quotations 1–3).

97. "Chicago's Oil Fuel and Insurance," *AM&IW* 43, no. 7 (August 17, 1888): 3, col. 1 (quotation 1).

98. "Oil Fuel and Insurance," *AM&IW* 43, no. 21 (November 23, 1888): 10, col. 2. For how the "insurance industry governs" (applied herein to nineteenth-century circumstances), see Richard V. Ericson, Aaron Doyle, and Dean Barry, *Insurance as Governance* (2003), 5 (quotation), 6; and Richard V. Ericson and Aaron Doyle, "Risk and Morality," in *Risk and Morality*, ed. Richard V. Ericson and Aaron Doyle (2003), 4–5.

99. For California, see "Mines and Mining: The Oil District of Los Angeles County," *SFC*, January 8, 1884, 6, col. 7; "Petroleum for Fuel: Successful Experiments on Steamers," *SFC*, September 17, 1885, 5, col. 4; "The Economy of Liquid Fuel," *SFEB*, November 30, 1885, 2, col. 2, 4th ed.; "Industrial Topics . . . Petroleum as Fuel," *SFC*, July 8, 1886, 3, col. 7; "Petroleum in Southern California," *SFEB*, September 20, 1886, 2, col. 2, 4th ed.; and "Oil as Fuel: The Necessity of a Cheap and Harmless Fuel in Fresno," *Fresno Republican*, June 8, 1888, 5, cols. 6–7.

100. "Insurance Circles: Rules Governing the Storage of Crude Petroleum," *DIO*, January 5, 1892, 3, col. 1; "Insurance Circles: Rules for the Safe Burning of Crude Petroleum," *DIO*, March 6, 1892, 7, cols. 1–2; and "The Storage of Fuel Oil," *TIA* 49 (January 21, 1892): 109, cols. 2–3. In 1885 the old Chicago Board of Underwriters (1872) merged with the Chicago Insurance Underwriters' Exchange (1880) to form the Chicago Fire Underwriters' Association; see A. T. Andreas, *History of Chicago From the Earliest Period to the Present Time*, vol. 3: *From the Fire of 1871 Until 1885* (1886), 461, 466.

101. "Insurance Circles: Rules Governing the Storage of Crude Petroleum," *DIO*, January 5, 1892, 3, col. 1; "Insurance Circles: Rules for the Safe Burning of Crude Petroleum," *DIO*, March 6, 1892, 7, cols. 1–2; "The Storage of Fuel Oil," *TIA* 49 (January 21, 1892): 109; on the fuel to be used at the exposition, see "World's Columbian Exposition.—VII. Boiler Plant . . . Oil Fuel Supply," *Engineering News and American Railway Journal* 29, no. 15 (April 13, 1893): 342.

102. "Insurance Circles: Rules Governing the Storage of Crude Petroleum," *DIO*, January 5, 1892, 3, col. 1; and "Insurance Circles: Rules for the Safe Burning of Crude Petroleum," *DIO*, March 6, 1892, 7, col. 1; and "The Storage of Fuel Oil," *TIA* 49 (January 21, 1892): 109, col. 2.

103. "Insurance Circles: Rules Governing the Storage of Crude Petroleum," *DIO*, January 5, 1892, 3, col. 1; "Insurance Circles: Rules for the Safe Burning of Crude Petroleum," *DIO*, March 6, 1892, 7, cols. 1–2; and "The Storage of Fuel Oil," *TIA* 49 (January 21, 1892): 109.

104. "Gas at Cheaper Rates: Edward Atkinson Warmly Indorses the Plan of Mr. Whitney," *BDA*, March 11, 1896, 7, col. 3; also "Fuel Gas vs. Coal Gas for Heat and Power," *AM&IW* 38, no. 22 (May 28, 1886): 12, cols. 1–2; "Manufacturing. Comparative Cost of Some Different Fuels," *BDA*, November 6, 1888, p. 8, col. 1; "Manufacturing: Amount and Cost of Fuel in New England Mills," *BDA* (morning ed.), September 7, 1889, 8, col. 2. On SOC, New England markets and mills, and the use of fuel oil, see "Standard Oil: Its Encroachment in New England," *BDA*, September 10, 1888, 8, col. 2; "Interesting Gossip of the Day," *New York Sun*, June 12, 1889, 4, col. 7; and "Lowell Experiment. The Oil-Burning Question Still of Great Interest," *BDA*, February 7, 1891, 2, col. 1.

105. Julius W. Adams, "Report to the Trustees of the Petroleum Light Company, by Col. Julius W. Adams, Engineer, of experiments made at the Morgan Iron Works, New York, Oct. 11, 1865," *Scientific American* 14, no. 6 (February 3, 1866): 82–83; and "American Petroleum Furnace," *TMM*, n.s., vol. 16 (July 20, 1866): 41–42.

106. U.S. Department of the Navy, *Report of the Secretary of the Navy, with An Appendix, Containing Bureau Reports, Etc.*, December 1867 (1867), Appendix, Bureau of Steam Engineering, 24, 173–81. "Petroleum as a Fuel for Steam Engine Furnaces," *Engineering Facts and Figures for 1866: An Annual Register of Progress in Mechanical Engineering and Construction* (1867), 134–36, 137 (quotation 1), 138. "Richardson's Petroleum Furnace," *TMM* n.s., vol. 16 (July 13, 1866): 19–20; *TMM* n.s., vol. 16 (September 14, 1866): 164–65; "Liquid Steam Fuel," *TMM* n.s., vol. 16 (November 2, 1866): 271; and "Naval, Military, and Gunnery Items," *TMM* n.s., vol. 16 (December 14, 1866): 376, col. 1, para. 5.

107. "Petroleum in the Production of Steam," *PCDR*, February 20, 1871, 2, col. 2; and "Successful Experiments with Petroleum as Fuel," *PCDR*, December 18, 1872, 2, col. 3.

108. "Successful Puddling with Petroleum," *AM&IW* 32, no. 11 (March 16, 1883): 9, col. 3 ("after years of experimenting"); "Petroleum as Fuel," *AM&IW* 38, no. 11 (March 12, 1886): 13, cols. 1–2; "Vaporized Petroleum as Fuel. A Fully Developed Invention of the Utmost Interest to Manufacturers and Householders," *SR*, July 8, 1887, 6[5], col. 1; "Holly Steam by Petroleum: The Aerated Fuel Company's Apparatus in Operation at the New Boiler-House," *SR*, November 30, 6, cols. 1–2; "Experimenting with Aerated Fuel," *SDR*, July 23, 1888, 6, col. 3; "Petroleum Vapors for Engines: Description of the Mode of Generating It, and How It Has Been Successfully Used in Boston," *Phil Inq*, December 18, 1888, 4, col. 6; "Crude Petroleum as a Fuel," *Science* 13, no. 322 (April 5, 1889): 249–50; "Crude Petroleum for Manufacturing Purposes,"

AM&IW 44, no. 22 (May 31, 1889): 5, col. 4; "Improved System for Burning Oil," *AM&IW* 45, no. 6 (August 9, 1889): 3, cols. 1–4; "Manufacturing: Petroleum Oil as a Fuel Under Boilers," *BDA*, September 11, 1889, 8, col. 2; "Manufacturing News . . . Connecticut," *BDA*, October 12, 8, col. 3; "Aerated Fuel Adopted by G. W. Prentiss & Co.," *SDR*, November 19, 1890, 6, col. 4; and "Fuel Gas in New England," *AM&IW* 48, no. 3 (January 16, 1891): 60, col. 2.

109. George H. Barrus, "A Paper on Boiler Tests, Illustrated by the Results of 140 Evaporative Tests Conducted by the Author," in NECMA, *Proceedings of the Semi-Annual Meeting, Held at Boston, October 31, 1888*, no. 45 (1888): 34–230; and "Oil Against Coal: A New Kind of Mill Fuel Very Probable, *BDA*, December 29, 1890, 8, col. 3. On the potential of petroleum fuel for energy, see "Cheaper Fuel Wanted," *BDA*, June 3, 1889, 4, col. 3; "Manufacturing: Amount and Cost of Fuel in New England Mills," *BDA*, September 7, 1889, 8, col. 2; and "The Burning of Crude Petroleum Oil as a Substitute for Coal," *BDA*, January 20, 1891, 4, col. 3 (quotations 1–2).

110. A. G. Cumnock, "The Distribution of Oil in the Mills," NECMA, *Proceedings of the Fourth Semi-Annual Meeting, Held at Boston, July 15, and, by Adjournment, October 21, 1868*, no. 5 (1876): 22–24, with discussion, 24–27; and "Statistics on the Distribution of Oil in the Mills," NECMA, *Proceedings of the Fourth Annual Meeting, Held at Boston, April 21, 1869*, no. 6 (1872): 43–47; Zachary Allen, "The Transmission of Power From the Motor to the Machine," in NECMA, *Proceedings of the Sixth Annual Meeting, Held at Boston, April 19, 1871*, no. 10 (1871): 27; Daniel Hussey, "Gearing, Shafting, Belting, and Pulleys," NECMA, *Proceedings of the Sixth Annual Meeting, Held at Boston, April 19, 1871*, no. 10 (1871): 37–46; and discussions on lubrication, 6–12; also William A. Burke, "The Transmission of Power," in NECMA, *Proceedings of the Semi-Annual Meeting, Held at Boston, October 18, 1871*, no. 11 (1871): 26-31, with discussions on lubrication, 6–7, 10, 13–14.

111. Dane Yorke, *Able Men of Boston: The Remarkable Story of the First 100 Years of the Boston Manufacturers Mutual Fire Insurance Company . . . 1850–1950* (1950), 154, 156 (quotations 1–3). On the context and importance of fire-prevention engineering in the late nineteenth century, see Edward V. French, *1860–Fifty Years–1910 Arkwright Mutual Fire Insurance Company: The Development of the Prevention Idea* (1912), 33–34. For a late-twentieth-century appraisal of fire-protection engineering and science, see Harold E. Nelson, "The Potential of Scientifically Based Fire Protection Engineering?" (1985), 3–7.

112. [Freeman and Abbott,] *Factory Mutuals*, 86 (quotations 1–2), 91 (quotations 3–4), 99–101, 117.

113. John M. Ordway (MIT), "Professor Ordway," [comments on his oil investigation], NECMA, *Proceedings of the Semi-Annual Meeting, Held at Boston, October 30, 1878*, no. 25 (1879), 17–35, interspersed with and supplemented by comments and questions from Edward Atkinson, 29, 34–36, 41. Ordway did not cover the "lubricating power" of the oil samples solicited by Edward Atkinson from various mills but only "certain chemical qualities" and their "facility of oxidation" (17). See also Edward Atkinson, "Appendix to Oil Report," NECMA, *Proceedings of the Semi-Annual Meeting, Held at Boston, October 30, 1878*, 59–79, dated January 30, 1879, with data assembled by Ordway.

114. The contract, dated May 25, 1878, is in Edward Atkinson, "The Methods and Appliances for Preventing and Extinguishing Fires," NECMA, *Proceedings of the Thirteenth Annual Meeting, Held at Boston April 24, 1878*, no. 24 (1878): 14–16, 17–19 (contract). Atkinson observed that "spontaneous combustion and friction brings up the whole subject of oil in use for

the lubrication of machinery" (17). See also Edward Atkinson, "Oil," BMMFIC Special Report (1880), 5–9 (identified/dated by Peckham, 203); and his "Preliminary Report Upon the Lubrication of Heavy Bearings," in BMMFIC, Special Report no. 14, *Report Upon Coverings for Steam Pipes By Prof. John M. Ordway, of the Massachusetts Institute of Technology, Also a Preliminary Report on the Lubrication of Heavy Bearings* (1883), 25–28; and "Reminiscences," *TNECMA*, Semi-Annual Meeting . . . Lenox, Mass., September 29—October 1, 1903 (1903), 137–38. The special reports are bound in *Boston Manufacturers Mutual Fire Insurance Co. Reports*, vol. 19 (HG8018 .I59, Cornell University Library).

115. Zachary Allen, "The Transmission of Power From the Motor to the Machine," NECMA, *Proceedings of the Sixth Annual Meeting . . . April 19, 1871*, 13–35; Daniel Hussey, "Gearing, Shafting, Belting, and Pulleys," NECMA, *Proceedings of the Sixth Annual Meeting . . . April 19, 1871*, 37–46; William A. Burke, "The Transmission of Power," NECMA, *Proceedings of the Semi-Annual Meeting . . . October 18, 1871*, 27–29; and W. S. Kelley, "Application of Electro-Motive Power to Mill Work," NECMA, *Proceedings of the Twenty-Fourth Annual Meeting, Held at Boston, April 24, 1889*, no. 46 (1889): 16–49; "Cost of Long-Distance Electrical Power Transmission," *Proceedings of the Semi-Annual Meeting, Held at Boston, October 30, 1889*, no. 47 (1889): 85–108; also D. M. Thompson, "Opening Address," *TNECMA, Annual Meeting (Providence Meeting,) Held at Savoy Hall, Providence, R.I. April 24–25, 1895*, no. 58 (1895): 77–90; George I. Rockwood, "The Ideal Cotton Mill Steam Plant: Its Design and Management," *TNECMA, Annual Meeting (Providence Meeting,) Held at Savoy Hall, Providence, R.I. April 24–25, 1895*, no. 58 (1895): 95–110; A.C. Rice, "Water Power Development and Its Cost," *TNECMA, Annual Meeting (Providence Meeting,) Held at Savoy Hall, Providence, R.I. April 24–25, 1895*, no. 58 (1895): 127–32; Eckley B. Coxe, "Some Thoughts upon the Economical Production of Steam, with Special Reference to the Use of Cheap Fuel, by a Miner of Coal," *TNECMA, Annual Meeting (Providence Meeting,) Held at Savoy Hall, Providence, R.I. April 24–25, 1895*, no. 58 (1895): 133–210; and Sidney B. Paine, "Electrical Driving of Textile Establishments," *TNECMA, Annual Meeting (Providence Meeting,) Held at Savoy Hall, Providence, R.I. April 24–25, 1895*, no. 58 (1895): 216–39.

116. AFMFIC, "Burning Petroleum Under Boilers," [*Circular*] no. 22, sec. 4, 5, dated "February, 1888," in bound vol., BMMFIC pubs., LILAB. See also "Petroleum Fuel," *AM&IW* 42, no. 16 (April 20, 1888): 8, col. 2, for fuel-oil rules in San Francisco issued by the Insurance Union. J. W. Lord, an engineer for the BMMFIC, recalled that AFMFIC-insured factories first used fuel oil in furnaces in 1888, advising their members the next year about adequate safeguards; see "The Fire Hazard of Fuel Oil," *American Concrete Institute: Proceedings of the Fifteenth Annual Convention Held at Atlantic City, N.J., June 27 and 28, 1919* (1919), vol. 15: 179. ADDED TO BIBLIOG

117. AFMFIC, Special Report, "Petroleum for Fuel," 150 (quotation 1), in bound vol. BMMFIC pubs., LILAB, and dated "August, 1888." A text nearly identical to this Special Report, attributed to the AFMFIC, is in "How to Safely Use Petroleum for Fuel," *AM&IW* 44, no. 11 (March 15, 1889): 6, col. 2.

118. "Petroleum as Fuel: And How the Insurance Companies Regard It," *BDA*, July 25, 1888, 8, col. 2; "Petroleum Fuel: How a Conservative Man Regards It," *BDA*, July 31, 1888, 8, col. 1; "Petroleum Fuel Systems," *TIJ* 18, no. 5 (June 1889): 227–29. On the history of the exchange, see "Ex-President Crosby Relates the History of the Exchange," *Weekly Underwriter* 48, no. 2 (January 14, 1893): 21–23; and Hayden, *Annual Cyclopedia of Insurance in the United States 1915*, "New England Insurance Exchange."

119. "Petroleum Fuel Systems," *TIJ* 18, no. 5 (June 1889): 227, 229.

120. Ibid., 227, 229, 230 (quotations 1–2). Surveys, certificates, and inspections were fundamental ways to gather risk knowledge in the fire insurance business; see Ace Insurance Co. (Philadelphia), Archives, Records of the Insurance Co. of North America (INA), "Directors Minutes Feb 1803–Feb. 1812," book 2: 163–64, December 1, 1807, for a report of a "committee appointed to consider the limits within which it would be proper for the Company to confine its insurance against Fire," wherein it was stipulated that the company should appoint reliable men "whose duty it will be to survey and certify the situation of all bldgs & property on which insurance is required at expense of applicant. No insurance can be issued with[out] return of the survey & certificates to the office & the President."

121. AFMFIC, "Fuel Oil," BMMFIC Publications, vol. 3, 18v–19r, dated April 29, 1889, LILAB; "Insurance: Petroleum as Fuel and Insurance," *BDA,* May 22, 1889, 5, col. 2; "Petroleum as Fuel: Mr. Edward Atkinson's Controversy with the Standard Oil Company," *BH,* June 14, 1889, 4, col. 8; and "Insurance News," *DIO,* June 16, 1889, pt. 2, 12, col. 2.

122. On Atkinson persuading SOC to produce high-quality lubricants suitable for safe mill use, see his "Reminiscences," *TNECMA,* Semi-Annual Meeting, 1903, 137, in *Boston Manufacturers Mutual Fire Insurance Co. Reports,* vol. 19 (HG8018 .I59, Cornell University Library). On the nature of his relationship to SOC, see "There Was No "Combat": Mr. Edward Atkinson's Conference with the Standard Oil People," *BH,* June 13, 1889, 8, col. 4 (quotations 1–2); and "Petroleum as Fuel. Mr. Edward Atkinson's Controversy," *BH,* June 14, 1889, 8.

123. On traditional hazard mitigation, see David R. Godschalk, "Urban Hazard Mitigation: Creating Resilient Cities," *Natural Hazards Review* 4, no. 3 (August 1, 2003): 139–40. See also "Copy of Survey of Mill furnished Protection Fire Ins. Co. Hartford Brick Mill 1842," in the Hagley Museum and Library, Insurance Policies, 1842–63, George H. Gilbert & Co. Collection, Accession no. 2178; which features printed, standard questions, and a broad column for handwritten responses—entries describing the facilities, machinery and fire-protection apparatuses and the location and orientation of buildings.

124. On the duties, such as identification of hazards through inspection and investigations of those who supervised fire-insurance agencies and company activities, that contributed to risk knowledge, see "The Underwriters: Second Day's Session of the Northwestern Association," *DIO,* September 24, 1875, 8, cols. 1–2, "Agency Supervision" and "Supervising Agents." On rules issued by Hartford Insurance Company, see "Insurance and Oil Fuel," *AM&IW* 44, no. 8 (February 22, 1889): 5, col. 4; on fuel-oil rules from the Philadelphia Fire Underwriters' Association, see "Insurance: Insurance and the Use of Crude Petroleum as Fuel," *BDA,* June 14, 1889, 2, col. 2; for similar requirements from the Underwriters' Association of the Middle Department (founded 1881), see "Earth Oil for Fuel," *TIA* 45 (February 20, 1890): 293, cols. 2–3; Hayden, *Annual Cyclopedia of Insurance in the United States 1915,* "Underwriters' Association of the Middle Department"; and *The Insurance Year Book, 1892–93: [Fire and Marine*], 20th annual issue (1892), "Permit to Use Earth Oil for Fuel Without Gravity Pressure," 86–87; identical to "Earth Oil for Fuel," *TIA* 45: 293 above.

125. "National Board of Fire Underwriters: Meeting of the Executive Committee," *NYEP,* February 19, 1867, 3, col. 10 (quotation 1); "Annual Meeting of the National Board of Fire Underwriters of the United States: Report of Executive Committee," *USIG&M* 29, no. 169 (May 1869): 10; and Oakley and Heald, *Report to the Executive Committee . . . on Gas Machines,* " 3 (quotation 2).

126. "National Board of Fire Underwriters: Extracts from the Address of President Heald at the Twenty-Third Annual Meeting," *InsTm* 23, no. 6 (June 1889): 297 (quotation 1); "Traps to Catch Firebugs," *NYHRD,* May 17, 1889, 4, col. 2; and "Report of Committee on Lighting, Heating and Patents," in *Proceedings of the Twenty-Third Annual Meeting of the National Board of Fire Underwriters* (1889), "Petroleum as Fuel," 136.

127. On rationalization, see Torsten Björkman, "The Rationalisation Movement in Perspective and Some Ergonomic Implications," *Applied Ergonomics* 27, no. 2 (1996): 111–17; R. H. Westgaard and J. Winkel, "Occupational Musculoskeletal and Mental Health: Significance of Rationalization and Opportunities to Create Sustainable Production Systems—A Systematic Review," *Applied Ergonomics* 42, no. 2 (January 2011): 262, 265–66; Michele Chwastiak, "Rationality, Performance Measures and Representations of Reality: Planning, Programming and Budgeting and the Vietnam War," *Critical Perspectives on Accounting* 17, no. 1 (January 2006): 29–55; Blandine Laperche, Gilliane Lefebvre, and Denis Langlet, "Innovation Strategies of Industrial Groups in the Global Crisis: Rationalization and New Paths," *Technological Forecasting & Social Change* 78, no. 8 (October 2011): 1319, 1324–25, 1329; R. J. Ormerod, "Critical Rationalism in Practice: Strategies to Manage Subjectivity in OR Investigations," *European Journal of Operational Research* 235, no. 3 (June 16, 2014): 784–97; and Tomas Gorner and Michal Simon, "Using the Theory of Technical Systems to Describe the Interaction between Human and Technical Systems with the Ergonomic System," *Procedia Engineering* 100 (2015): 593.

128. "Preface to Volume Twenty-Third, New Series," *USIG&M* 23 (1866): vi; "Fire Insurance Advance-Rate Movement: New Tariff of Rates of Premiums for 1866," *USIG&M* 23, no. 133 (May): 21–24; "Report of the Committee of Providence Under-Writers, on Woolen Mill Risks," *USIG&M* 23, no. 134 (June): 56–58; "Atlanta Board of Underwriters' Address to the Citizens," *USIG&M* 24, no. 144 (April 1867): 282; "National Board of Fire Underwriters: The Meeting of the Executive Committee at Philadelphia," *USIG&M* 25, no. 146 (June): 54, 56; "Editorial Notes: National Board of Fire Underwriters—Meeting of the Executive Committee at the Board of Trade Rooms, Boston, September 11, 1867," *USIG&M* 25, no. 149 (September): 244–45; "Report of the Committee on Local Boards, Rates, and Commissions," *USIG&M* 25, no. 150 (October): 269–72; Henry A. Oakley, "President's Address," *USIG&M* 35, no. 205 (May 1872): 3–6; and "Report of Thomas H. Montgomery, General Agent," *USIG&M* 37, no. 218 (June 1873): 54–55; see also James M. DeCamp's presidential address in "Fire Underwriters' Association of the Northwest: Fourteenth Annual Meeting," *TIJ* 11, no. 8 (September 1883): 245–46. On managerial supervision of agents and risks, see Ace Insurance Company, Records of the Insurance Company of North America, "Directors Minutes Jan 1877–Sept 1887," vol. 7, November 4, 1884, 163, "Committee on Fire Agency Expenses," for the observation that "Each risk as taken by the Agent is reported to the Manager who examines the nature of the hazard . . . position of the property insured, both as regards to its hazardous surroundings and its proximity to other risks . . . assumed by the Company"; and "Directors Minutes Jan 1895–Apr 1904," vol. 9: 224–25, "Statement of Expenses Relating to the Business of Fire Insurance" for the responsibility of regional managers to analyze agent reports on the "nature of the hazard" and to "decline such risks as may be undesirable."

129. See "National Board of Fire Underwriters: Reception of the Annual Report—Election of Officers and Committees," *NYHRD,* April 21, 1870, 4, col. 2, operations report by E. W. Crowell, chairman of the Executive Committee (quotation); "Insurance Problems: Session of the

National Board of Fire Underwriters—An Increase in the Rates of Premiums Recommended," *NYTRB,* November 23, 1872, 1, col. 5; "National Board of Fire Underwriters Action of the Board in Reference to Local Boards, State Boards, the Building Laws, Fires and Incendiarism," *NYTRB,* November 23, 1872, 10, col. 3; Brearley, *History NBFU,* 47–69; for state supervision, see "A Remedy Wanted for the Abuses of State Supervision," *USIG&M* 35, no. 209 (September 1872): 233–36; and H. Roger Grant, *Insurance Reform: Consumer Action in the Progressive Era* (1979), 71–77, 155–59.

130. "Meeting of the Board of Underwriters: Efforts to Remove the Wanton Cause of Fires," *USIG&M* 34, no. 200 (December 1871): 59–60; and "Giving Up Schedule Rates," *NYDTRB,* March 26, 1877, 5, col. 3 (quotations 1–3).

131. See "Board of Underwriters: How to Prevent Fires," *NYHRD,* April 24, 1875, 8, col. 3 (quotation 1); and "Address of U. C. Crosby," in *Tenth Anniversary of the New England Insurance Exchange Hotel Vendome, Boston, Mass., January 6, 1893* (1893), 9 (quotation 3), 10 (quotation 2), 18–19.

132. See "Tariff and Classifications," Minutes of First Annual Meeting of the National Board of Fire Underwriters, NBFU, *Journal of the Proceedings of the First Annual Meeting of the National Board of Fire Underwriters,* 35–37; William Frazier Ross, *The Estimation of Exposures, or External Hazards, in Fire Insurance* (1874), 4 ("experience of fire insurance companies"), 7-10, 13–14, 19, 37–39, 50–51; and "Fire Insurance: Meeting of New England Companies—Endeavor to Increase Rates—The Schedules Prepared," *NYHRD,* February 15, 1879, 3, col. 4. The distinctions between "tariffs" and "schedules" are sometimes ambiguous, since they were often used interchangeably in nineteenth-century insurance literature. Both are used in *Tariffs and Schedules, for Rating Risks,* comp. H. S. Tiffany (1891). Schedules are tables or lists with detailed descriptions of conditions, processes or features of a large number of hazards ("risks" in insurance narratives). As a system for establishing plausible rates based on uniform descriptions of the distinctive features of various risks, schedules broadcast minimum rates and the factors that justify added charges for major classes of risks such as A-, B-, or C-type buildings (60, 216–17). Tariffs are charges added to rates for "standard" buildings or factories devoted to a variety of processes (142–43, 298–99, 366, 410–11), as well for a huge number of hazards made especially severe by specific circumstances.

133. On the duty of local agents in the absence of adjusters to gather risk information, see Jeremiah Griswold, *Hand-Book of Adjustments of Loss or Damage by Fire, for the Use of Fire Underwriters,* 2nd ed. (1877), 46–51. On the importance of inspections for fire underwriters and their contributions to risk knowledge, see "A Desirable Reform," *BH,* May 27, 1887, 4, cols. 2–3. On the hazards of petroleum and its products, see Tiffany, *Tariffs and Schedules, for Rating Risks,* 7, 9, 69–70, 133–34, 241, 276–77, 309, 336, 347. On the detailed knowledge required of special agents, see *Tiffany's Special Agents' and Adjusters' Assistant,* comp. H. S. Tiffany, 10th rev. ed. (1891), in particular petroleum and its products, gas machines, and machinery in manufacturing (301–2).

134. Griswold, *Fire Underwriters' Text-Book,* 225–30, 235–39; Thomas H. Montgomery, "General Agent's Report," April 21, 1875, Rules of the National Board, 66, in *Reports of Thomas H. Montgomery, General Agent National Board of Fire Underwriters;* and F. O. Affeld, "Report of a special committee on classification of fire records by local boards," *AmEx&Rev* 48, no. 6 (June 1894): 274. See also S. S. Heubner, "Fire-Insurance . . . (2) Schedule Rating," in *The Encyclopedia*

Americana (1919), vol. 11: 239, col. 1, for what was necessary to make "scientific determination of rates" which included "good and bad features" of location, construction, fire appliances, occupancy, and "management of the risk." For the West Coast, see Grayson Dutton, "Inspection," *Proceedings of the Twenty-Seventh Annual Meeting of the Fire Underwriters' Association of the Pacific San Francisco, Cal., February 17–18, 1903* (1903), 63–67.

135. For the text of the rules the NYBFU issued in October 1881, see "Electrical Inspection and the National Electric Code," *Quarterly of the National Fire Protection Association* 7, no. 1 (July 1913): 53–54, based on a paper by J. C. Forsyth, electrical inspector for the NYBFU; for BMMFIC circulars in 1882, see Thomas H. Day, "Concentric Wiring," *Electrical Review and Western Electrician* 68, no. 9 (February 26, 1916): 383; and Ralph Sweetland, "Electrical Fire Hazards . . . Lecture 2: Early Recognition of Hazards of Electricity," in *Lectures on Fire Insurance* (1912), 10–12. For the NBFU rules, based on NYBFU standards, issued May 1882, see "Electrical Hazards: Annual Meeting of the Electrical Committee of the Underwriters' National Electric Association—Some History," *TIP* 5, no. 119 (December 8, 1897): 2; and Henry Morton and William A. Anderson, *Electric Lighting, and the Underwriters' Standard Requirements in Reference Thereto* (1882), wherein is the text (63–65) of the electrical requirements which the NYBFU had adopted January 12, 1882, and subsequently endorsed, May 25, by the NBFU.

136. For the NELA, NEIEx, and C. M. Goddard's role in codification, see "National Electric Code History: Exhaustive Review Made by F. Elliot Cabot of Boston Board Before Electrical Engineers in Providence, R.I.," *TS* 91, no. 18 (October 28, 1922): 682, col. 2; "Electric Hazards: Last Week's Meeting of the Electric Committee of the Underwriters' National Electric Association—Full Report of the Proceedings," *TIP* 1, no. 15 (December 18, 1895): 1–3; and Thomas Henry Day, *Practical Instruction on "Electrical Fire Hazards"* (1918), 23, on Goddard and the NEIEx. For the NELA, see William Brophy, "Report of the N.E.L.A. Committee on Safe Wiring," *The Electrical Engineer* 23, no. 476 (June 16, 1897): 672–73; and C. J. H. Woodbury, who claimed that the rules of fire underwriters gave the "various electrical interests . . . . a standard" so they could "meet on a common ground." See *The Influence of the Underwriters' Rules on Electrical Development: A paper Read before the New York Electrical Society, March 27, 1907* [1907], 4, summarized in "Electrical Development and Underwriters' Rules," *InsEng* 13, no. 4 (April 1907): 347–48; and in *Western Electrician* 40, no. 26 (June 29, 1907): 574, col. 3; as well as his printed speech, *History of the National Electrical Code: An Address Before The National Electrical Contractors' Association At Its Sixth Annual Convention . . . 1906* [1906], which provides details on the role of Cleveland inventor Charles F. Brush and others in that city in devising electrical rules. I thank the staff of the Boston Athenaeum for supplying copies of the above speeches by Woodbury.

137. On insurability, see C. J. H. Woodury, "The Relation of Electric Lighting to Insurance," NELA, *Proceedings of the National Electric Light Association . . . At Its Semi-Annual Convention, August 1885*, vol. 1 (1886): 187–96 (hereafter *NELA*, convention number, place, date, vol. number). For a defense of the insurability of central power stations, see "Electric Stations as Insurance Risks," *Western Electrician* 11, no. 20 (November 12, 1892): 249; George P. Low, "Generating Stations and Their Physical Hazards," in FUAP, *Proceedings of the Sixteenth Annual Meeting of the Fire Underwriters' Association of the Pacific, San Francisco, February 16th and 17th, 1892* (1892), 277–85.

138. On reconciling insurance and electrical interests, see P. H. Alexander, "Report of the Committee on Harmonizing Electrical and Insurance Interests," *NELA*, 10th Convention,

Semi-Annual Meeting, Niagara Falls, N.Y., August 1889, vol. 7 (1890): 4 (committee members), 73–86, and discussion, 94–96; "Report of the Committee on Harmonizing Insurance and Electric Lighting Interests," *NELA,* 11th Convention, Annual Meeting, Kansas City, February 1890, vol. 8 (1890): 36–37, discussion, 55, 58; and C. J. H. Woodbury, "Construction of Central Stations," *NELA,* 11th Convention, Annual Meeting, Kansas City, February 1890, vol. 8 (1890): 78–81.

139. S. S. Leonard, "Petroleum Fuel," *NELA,* 8th Convention, Semi-Annual Meeting, New York City, August 1888, vol. 4 (1888): 54–60; and "Fuel Oil," *NELA,* 9th Convention, Annual Meeting, Chicago, February 1889, vol. 5 (1889): 200–205; also M. J. Francisco, "Liquid Fuel," *NELA,* 9th Convention, Annual Meeting, Chicago, February 1889, vol. 5 (1889): 206–11; and C. M. Ransom, "Advantages of Oil as Fuel," *NELA,* 9th Convention, Annual Meeting, Chicago, February 1889, vol. 5 (1889): 212–18, as well as Leonard's comment that, "to comply with the requirements of the insurance companies, would make the use of oil almost prohibitory" (219).

140. S. E. Barton, "The Relation of Electric Lighting to Fire Insurance," *NELA,* 4th Annual Convention, Pittsburgh, February 1888, vol. 3 (1888): 448–49; "Electric Light Stations as Fire Risks," *NELA,* 9th Convention, February 1889, vol. 5 (1889): 155–59, 157 in particular; and George P. Low, "Electric Stations as Insurance Risks," FUAW, *Proceedings of the Seventeenth Annual Meeting of the Fire Underwriters' Association of the Pacific . . . 1893* (1893), 62–64, 69, 73–74; and "Concerning Electric Station Risks," FUAW, *Proceedings of the Twenty-Fourth Annual Meeting of the Fire Underwriters' Association of the Pacific San Francisco . . . 1900* (1900), 68–71.

141. See also Thomas Henry Day, "Popularizing the National Electrical Code: Electrical Inspector Day of the Hartford Board Gives an Illustrated Talk on Class A of the Code," *TS* 61, no. 16 (October 19, 1907): 979, 981; and "National Electric Code History: Exhaustive Review Made by F. Elliot Cabot," *TS* 91 (1922): 682–84. For the NBFU's ongoing role in promoting the NEC, see W. H. Merrill Jr., "State Control of the Electrical Hazard," *InsEng* 4, no. 6 (December 1902): 531–33; "The Electrical Bureau," NBFU, *Proceedings of the Thirty-Fifth Annual Meeting of the National Board of Fire Underwriters,* 104–5; and "State and Municipal Control of Electric Wiring," Appendix, letter from W. H. Merrill Jr., Electrician (and Secretary of the Electrical Bureau, NBFU), to F. C. Moore, Chairman, Committee on Lighting and Heating, NBFU, *Proceedings of the Thirty-Fifth Annual Meeting of the National Board of Fire Underwriters,* 111–12, with discussion of electrical hazards, 114–24.

142. See NBFU, "Report on Lighting, Heating and Patents," *AmEx&Rev* 52, no. 6 (June 1898): 266, wherein a subcommittee investigating acetylene sought to "formulate rules and regulations based upon recognized technical and scientific knowledge of the subject matter" (col. 1); and "Address of President U.C. Crosby," NFPA, *National Fire Protection Association: A Partial Record of the Transactions at Third Annual Meeting Held in Boston . . . 1899* (1899), 17, for a declaration that the NFPA existed to reduce "fire waste . . . eliminate the causes of fire" and "provide for its control and extinguishment," to be accomplished by bringing together the "experts and practical men who are especially interested in these questions." On field experiences and judgments about petroleum risks, see "Petroleum Fuel Systems," *TIJ* 18, no. 5 (June 1889): 229–30. For examples of important insurance figures who had extensive backgrounds in the fire insurance business and experience as surveyors, engineers and mediators between field risks and fire underwriters, see *Hayden's Annual Cyclopedia of Insurance in the United States 1910–1911 (1911);* for Samuel P. Cochran (126), partner in the important insurance agency Trezevant & Cochran (Dallas, 1888), who surveyed many Ohio cities, including

Cincinnati, in 1873, for the NBFU, served in various insurance agencies, and was named in 1881 special agent for the Phoenix of Hartford for its Texas underwriting. Cochran acquired extensive experience in and knowledge of fire risks and the hazard mitigation stance of many major fire insurance companies. His Dallas associate, John T. Trezevant (*Hayden's Annual Cyclopedia 1910–1911,* 610–11), a civil engineer who found success in insurance, participated in many fire-insurance organizations in Texas. As knowledgeable leaders who understood the details of surveying, inspection, and risk management, they circulated risk knowledge and awareness throughout the fire insurance fraternity and professional organizations such as the NBFU.

143. For the NFPA's early activities and objectives, see "National Fire Protection Association: Recognition of the Demand for the Standardization of Materials, Devices and Methods," *InsEng* 3, no. 4 (April 1902): 323 (quotation 1), 324–25 (quotations 2–3). According to a contemporary, the NFPA grew out of a New York City meeting to discuss rules for installing automatic sprinklers; see Charles A. Hexamer, "Fire Insurance—Rates and Schedule Rating," *Annals of the American Academy of Political and Social Science* 26 (September 1905): 222.

144. E. C. Irvin, "President's Address," NBFU, *Proceedings of the Thirty-Third Annual Meeting of the National Board of Fire Underwriters, Held at the Rooms of the New York Board . . . 1899* (1899), 24, 25 (quotation 2), 26 (quotation 3), 27, 28 ("legal and scientific advice"),"Report of the Executive Committee," 68 (Circular no. 542, issued March 25, 1899, by NBFU, Executive Committee). And "Report of the Committee on Lighting, Heating, and Patents," NBFU, *Proceedings of the Thirty-Third Annual Meeting of the National Board of Fire Underwriters, Held at the Rooms of the New York Board . . . 1899,* 86, 89 (quotation 1).

145. NBFU, *Proceedings of the Thirty-Third Annual Meeting of the National Board of Fire Underwriters,* 28, 65 (list of special committee members), 68; and NBFU, *Proceedings of the Thirty-Third Annual Meeting of the National Board of Fire Underwriters,* "Report of Committee on the President's Recommendations," 135–36.

146. "Board of Consulting Engineers," NBFU, *Proceedings of the Thirty-Fourth Annual Meeting of the National Board of Fire Underwriters . . . May 10, 1900* (1900), 21 (quotations 1–4); the Executive Committee appointed George E. Bruen, an engineer, to represent the Committee on Lighting and Heating on the Board of Consulting Engineers (72). See also NBFU, *Proceedings of the Thirty-Fourth Annual Meeting of the National Board of Fire Underwriters . . . May 10, 1900,* "Report of the Committee on Lighting and Heating," dated May 10, 1900, with a list of appointments to the Board of Consulting Engineers (81–82).

147. NBFU, *Proceedings of the Thirty-Fourth Annual Meeting of the National Board of Fire Underwriters,* 73–74; and NBFU, *Proceedings of the Thirty-Fourth Annual Meeting of the National Board of Fire Underwriters,* "Report of the Committee on Lighting and Heating," 82 (quotation 1), 83; also "Report of the Executive Committee," NFPA, *National Fire Protection Association, Proceedings of the Fourth Annual Meeting Held in New York . . . 1900* (1900), 23–25; and a letter dated May 15, 1900, New York, from H. K. Miller, general agent, NBFU, to Everett U. Crosby, Secretary, NFPA, in NBFU, *Proceedings of the Thirty-Fourth Annual Meeting of the National Board of Fire Underwriters,* "National Board of Fire Underwriters," 74.

148. See NFPA, *Proceedings of Fifth Annual Meeting Held in Chicago . . . 1901* (1901), "Report of the Executive Committee," 22 (quotation 2); and Edward Atkinson, *The Prevention of Loss by Fire: Fifty Years' Record of Factory Mutual Insurance* (1900), 15 (quotation 3).

149. See "Underwriters' Requirements," *InsEng* 1, no. 1 (April 1901): 88–93 (fire alarm wiring); no. 2 (May): 183–89 (extinguishers); no. 3 (June): 282–85 (mill-fire hose); vol. 2, no. 1 (July 1901): 71–94 (fire doors and shutters); no. 2 (August): 168–90 (automatic sprinklers); no. 4 (October): 374–76 (watchmen's time recording); no. 5 (November): 466–76 (wired and prism glass). For the NFPA committee work behind the published requirements, see "Carbonic Acid Gas Extinguishers," 40–46; "Hose, Hydrant and Cast-Iron Pipe," 47–50; "Automatic Sprinklers," 76–120; "Report of Committee on Wired and Prism Glass," 123–30; "Report of Committee on Fire Doors and Shutters," 130–42; "Report of Committee on Electric Fire Pumps," 171–77; "Automatic Sprinklers in Grain Elevators," 177–97; and "Report of the Committee on Fire Pumps," 198–205, all in NFPA, *Proceedings of Fourth Annual Meeting . . . 1900;* as well as "National Fire Protection Association: Recognition of the Demand for the Standardization of Materials, Devices and Methods," *InsEng* 3, no. 4 (April 1902): 326–40; and NFPA, *Proceedings of Fifth Annual Meeting . . . 1901*, 25.

150. "Suggestions for the Organization and Equipment of Mill Fire-Brigades," *InsEng* 2, no. 3 (September 1901): 277 (quotation 1), 278–83; "Construction, Installation and Use of Gasoline Vapor Gas Lighting Machines," vol. 2, no. 6 (December): 557–66. George P. Sheldon, "President's Address," *Proceedings of the Thirty-Fourth Annual Meeting of the National Board of Fire Underwriters,* "Clauses and Forms," 20; and "Board of Consulting Engineers," 21; "Report of the Committee on Lighting and Heating," 82; and Letter, Goddard to Miller, 89.

151. NBFU, "Report on Lighting, Heating and Patents," *AmEx&Rev* 52, no. 6 (June 1898): 266; "Report of the Committee on Lighting, Heating, and Patents," NBFU, *Proceedings of the Thirty-Third Annual Meeting of the National Board of Fire Underwriters,* 86–87, 88 (quotation 2), 89 (quotation 1); "Appendix to Report of the Committee on Lighting and Heating. Report of the Committee of Consulting Engineers," NBFU, *Proceedings of the Thirty-Fifth Annual Meeting of the National Board of Fire Underwriters,* "Acetylene Gas," 108 (quotations 3–4), 109.

152. On inquiries see AFMFIC, "Burning Petroleum Under Boilers," [Circular] no. 22, sec. 4, 5, dated "February, 1888," in bound vol. of BMMFIC pubs., LILAB; and information on oil hazards, BMMFIC, Special Circular, No. 29, "Storage of Oil," 3rd ed., April 1893, in vol. 3: 38, bound volume of BMMFIC pubs, LILAB (digital copies courtesy of the LILAB); on trails and experiments, see "An Apparatus for the Manufacture of Vapor Fuel," *AM&IW* 34, no. 25 (June 20, 1884): 5, col. 4; "Petroleum as Fuel," *AM&IW* 38, no. 11 (March 12, 1886): 13, cols. 1–2; "Report of Petroleum Oil Burning at Salem Pumping Station at North Beverly, Mass., Oct., 1886," *AM&IW* 42, no. 14 (April 6, 1888): 19, cols. 1–2; "Oil as a Metallurgical Fuel," *AM&IW* 44, no. 10 (March 8, 1889): 17, cols. 2–3; and "Oil Fuel in the Puddling Furnace," *AM&IW* 45, no. 7 (August 16, 1889): 17, cols. 1–2. For experiments with fuel oil for marine propulsion, see "Oil Fuel in Ocean Steamers," *TIA* 53 (March 22, 1894): 543, col. 1; and "Liquid Fuel on Pacific Steamers," *STWPR* 15, no. 5 (March 4, 1887): 21, col. 3, as published in *AM&IW* 40, no. 9 (March 4, 1887).

153. See "Insurance and Oil Fuel," *AM&IW* 44, no. 8 (February 22, 1889): 5, col. 4; "Petroleum as Fuel: And How the Insurance Companies Regard It," *AM&IW* 44, no. 8 (February 22, 1889): 8, col. 2 (quotation 1), and "Insurance Circles . . . Defects Being Remedied," *AM&IW* 44, no. 8 (February 22, 1889): 9, col. 1.

154. Beatrice D'Ippolito, Marcela Miozzo, and Davide Consoli, "Knowledge Systematisation, Reconfiguration and the Organisation of Firms and Industry: The Case of Design,"

Research Policy 43, no. 8 (October 2014): 1334–35; Danièle Bénézech et al., "Completion of Knowledge Codification: An Illustration through the ISO 9000 Standards Implementation Process," *Research Policy* 30, no. 9 (December 2001): 1396; and Patrick Cohendet and Frieder Meyer-Krahmer, "The Theoretical and Policy Implications of Knowledge Codification," *Research Policy* 30, no. 9 (December 2001): 1564–65.

155. Hayden, ed., *Annual Cyclopedia of Insurance in the United States, 1890,* "Fire Underwriters, National Board of," 69, 74 (quotations 1–3). See also "Louisville Has a $100,000 Fire," *Lexington* (Ky.) *Leader,* February 18, 1902, 1, col. 2, regarding the destruction of the plant of the Louisville Bolt and Iron Company, which was initiated by an exploding pipe that dumped raw oil on "molten metal" in the furnaces.

156. It was noted that in terms of the BCE's contribution to petroleum-risk management the NBFU decision to "formulate rules and requirements governing the storage and use of fuel oil" was an "important step toward standardizing the practice and reducing its dangers." See A. M. Schoen, "Texas Fuel Oil," *InsEng* 4, no. 6 (December 1902): 534. On the UL, see "How Underwriters' Tests are Made," *InsEng* 4, no. 6 (December 1902): 99; and "Growth of the Underwriters' Laboratories," *InsEng* 4, no. 6 (December 1902): 194.

157. On Schoen's SETA appointment, see "Personals. Agency Appointments," *The Weekly Underwriter,* 48, no. 2 (January 14, 1893): 18, col. 2.

158. "The Study of Fire Hazards: Work of Boston Fire Underwriters—Oil Hazards," *TIP* 13, no. 324 (November 13, 1901): 6, col. 1 (quotations 1–2); NBFU, *Proceedings of the Thirty-Seventh Annual Meeting of the National Board of Fire Underwriters, Held in the Assembly Room of the New York Board of Fire Underwriters . . . 1903* (1903), Appendix: Consulting Engineers of the National Board of Fire Underwriters, Annual Report, dated April 20, 1903, "Fuel Oil," 72 (quotations 3–4). See also South-Eastern Tariff Association, *Quarter-Centennial History of The South-Eastern Tariff Association: 1882-1907* (1907), 186.

159. "Schoen's Fuel Oil Investigation Is Completed," *TS* 51, no. 3 (July 18, 1902): 57, col. 1; "The Use of Fuel Oil: Underwriters Agree Upon Changes in Insurance Policies," *NODP,* October 28, 1902, 6, col. 3; NBFU, *Proceedings of the Thirty-Seventh Annual Meeting of the National Board of Fire Underwriters,* 79; and "Rules and Requirements of the National Board of Fire Underwriters for Regulating the Hazard of the Storage and Use of Fuel Oil, and for the Construction and Installation of Oil Burning Equipments, as Recommended by Its Committee of Consulting Engineer, Edition of 1902," *InsEng* 4, no. 6 (December 1902): 596–603.

160. On fires, see "Lucas Oil Burns," *GDN,* February 6, 1901, 4, col. 1; "A Great Oil Fire," *GDN,* March 4, 1901, 1, cols. 6–7; "Spindle Top Fire," *GDN,* September 20, 1901, 6, col. 1; and "Safety Committee," *GDN,* September 29, 1901, 4, col. 1; also "Texas Fuel Oil," *InsEng* 4, no. 6 (December 1902): 534–35.

161. "Texas Fuel Oil," *InsEng* 4, no. 6 (December 1902): 535–37.

162. Ibid., 541; and William H. Booth, *Liquid Fuel and Its Combustion* (1903), appendix 7, "Extracts from Report by Mr. Schoen on Installations of Fuel Oil," 351.

163. "Texas Fuel Oil," *InsEng* 4, no. 6 (December 1902): 538–39, 540 (quotation 1); and Booth, *Liquid Fuel,* 351. On technical artifacts, see note 67, above.

164. "Texas Fuel Oil," *InsEng* 4, no. 6 (December 1902): 534 (quotation 3), 540–41; Booth, *Liquid Fuel,* 350 (quotations 1–2); and "Engineer Schoen on the Fuel Oil Hazard," *The Vindicator* 30, no. 2 (July 15, 1902): 30 (quotations 4–5).

165. "National Board Rules Regarding Use of Fuel Oil," *TS* 51, no. 24 (December 12, 1902): 539, col. 1 (quotations 1–3).

166. NBFU, "Rules and Requirements Fuel Oil," *InsEng* 4, no. 6 (December 1902): 598.

167. Ibid., 599 (quotation 1), 600 (quotation 2), 601.

168. Nancy Leveson, "A Systems Approach to Risk Management through Leading Safety Indicators," *RE&SS* 136 (April 2015): 19, col. 1 (quotations 1–3).

169. For Pacific Insurance Union rules that governed the installation of a fuel-oil power plant at the California Midwinter International Exposition, in San Francisco in 1894, see A. M. Hunt, "California Crude Oil and Its Use as Fuel," *"Industry": A Magazine Devoted to Science . . . on the Pacific Coast,* no. 77 (December 1894): 731–44. On debates about an ordinance to regulate burning petroleum to raise steam, see "Fire Committee: Underwriters and Oil Men Present," *SFC,* November 23, 1887, 6, col. 5.

170. On complexity and efficiency in energy transitions, see Joseph A. Tainter, "Energy, Complexity, and Sustainability: A Historical Perspective," *Environmental Innovation and Societal Transitions* 1 (2011): 89–90. On the role of technology and "institutional/organizational" changes in the establishment and diffusion of petroleum as an end-use energy carrier that pushed forward "historical energy transitions," see Arnulf Grubler, "Energy Transitions Research: Insights and Cautionary Tales," *Energy Policy* 50 (November 2012): 10; and Charlie Wilson, "Up-Scaling, Formative Phases, and Learning in the Historical Diffusion of Energy Technologies," *Energy Policy* 50 (November 2012): 81–94. On the historical transitions in energy systems, see Rutter and Keirstead, "A Brief History and the Possible Future of Urban Energy Systems," 72–80; and the complex and "drawn-out" processes that explain energy transitions in Fouquet and Pearson, "Past and Prospective Energy Transitions," 1–7.

CHAPTER TWO

1. "President's Address," FUAN, *Proceedings of the Twelfth Regular Meeting Fire Underwriters' Association of the Northwest . . . 1881* (1881), 12–13; and U. C. Crosby, "Necessity for Concerted Action in Improvement and Inspection of Risk," *Proceedings of the Twenty-Second Annual Meeting of the Fire Underwriters' Association of the Northwest . . . 1891* (1891), 49–50; and "Report of the Executive Committee," *National Fire Protection Association: Proceedings of the Fourth Annual Meeting . . . 1900*, 23. On fire waste, see "President's Address," NBFU, *Proceedings of the Twenty-Ninth Annual Meeting of the National Board of Fire Underwriters . . . May 9, 1895* (1895), 54–55; and "Edward Atkinson on the Fire Waste and Its Consequences," *The Weekly Underwriter* 50, no. 2 (January 13, 1894): 22–23. On the duty to protect property through fire insurance, see Henry A. Oakley, *Addresses as President of the National Board of Fire Underwriters of the United States on Several Occasions . . . 1871–1876* (1876), "Address at the special meeting of the National Board in New York, November 21, 1872," [3]. On statistics, see Oakley, *Addresses as President,* "National Board of Fire Underwriters: Address of the President at the Seventh Annual Meeting of the Board April 23rd, 1873," Statistics of Business, 27–29; A. D. Smith, "The Lack of Proper Statistics," FUAP, *Proceedings of the Twelfth Annual Meeting of the Fire Underwriters' Association of the Pacific, San Francisco, February 20th and 21st, 1888* (1888), 96–99; and F. O. Affeld, "Reliable Statistics, the Basis of Efficient and Economical Fire

Protection and Good Government," National Association of Fire Engineers, *Proceedings of the Twenty-Second Annual Convention* (1894), 91–104.

2. On the Mobile insurance interests employing an "expert," see Letter, G. B. Thames, Charles R. Garnett, John Gaillard, James K. Glennon, to Walter Wash Walsh, Chairman Committee No. 4, Mobile, Ala., dated December 13, 1901, printed in see "The Editor's Letter-Box: Storage of Oil," *MDR*, February 25, 1902, 5, col. 3; and "Hotel Arrivals," *MDR*, Sunday, February 16, 1902, 14, col. 7, Hotel Bienville, A. M. Schoen, Atlanta.

3. See Jane H. Pease and William H. Pease, "The Blood-Thirsty Tiger: Charleston and the Psychology of Fire," *South Carolina Historical Magazine* 79, no. 4 (October 1978): 284–89.

4. For fire statistics, see David D. Dana, *The Fireman: Fire Departments of the United States with a Full Account of All Large Fires, Statistics of Losses and Expenses* (1858), 122–37, 162–64, 165 (quotation). For city fires, see Clarence H. Forrest, *Official History of the Fire Department of the City of Baltimore Together with Biographies and Portraits of Eminent Citizens* (1898), 204–13; Pease and Pease, "The Blood-Thirsty Tiger," 281–83; Thomas Frederick Davis, *History of Jacksonville, Florida* (1925), 99–100, 104, 152, 159, 168, 173, 179, 190–95, 219–21, 225–27; and Alan D. Watson, *Wilmington, North Carolina, to 1861* (2003), 109.

5. The fire on March 31, 1840, started in W. H. Kelly's furniture store on St. Francis above Water Street, wrapped the building in a "complete mass of flame," and destroyed the entire block. See "Destructive Fire," *Mobile Daily Commercial Register & Patriot*, April 1, 1840, 2, col. 1; and Dana, *The Fireman*, 155 (quotation 1), 156 (quotations 2–4), 157 (quotations 5–7). On September 10, 1854, a fire spread from a "small frame house on Conception street" to "Bloodgood's factory," moved on to several more homes, and threatened the railroad depot. Wind and a lack of water allowed the unimpeded fire to destroyed "thirty to forty" houses at a "total loss from $50,000 to $60,000." The next day the fire burned several uninsured houses, and in the afternoon another fire damaged the Orphan Asylum on Dauphin Way. See "Fires in Mobile," *Montgomery Advertiser & Gazette*, September 16, 1854, 2, col. 2 (quotations in note); and Dana, *The Fireman*, 158.

6. A prominent New York insurance periodical distinguished a conflagration from an "ordinary fire by the extent of territory burned over, by the aggregate of the money lost" (*The Spectator* 68, no. 20 [May 15, 1902]: 275, left col.). S. H. Lockett, well-known Chicago fire investigator, asserted that a conflagration was an "impending and always existing danger where there are present combustible buildings or objects." See "The Conflagration Hazard," FUAN, *Proceedings of the Thirty-Fifth Annual Meeting of the Fire Underwriters' Association of the Northwest, Chicago, Ill., September 28–29, 1904* (1904), 91–97. On Mobile, see David E. Alsobrook, "Alabama's Port City: Mobile during the Progressive Era, 1896–1917," PhD diss., Auburn University, 1983, iv, 86. Regarding negotiations with the Bienville Water Supply Company to furnish water to the city, see "Good, Sound Law," *MDR*, October 5, 1886, 4, col. 2; "The Police Board Meeting," *MDR*, October 23, 1886, 4, col. 3; as well as "Tired Out," *MDR*, October 27, 1886, 4, col. 4, for a letter, dated October 26, 1886, from the company which specifically tied the anticipated water to fire protection. One solution was to dig a cistern to store water on one's property in case of fire. See "Municipal Matters: Standing Committees," *MDR*, July 21, 1876, 4, col. 2; "Papers From the Common Council," *MDR*, July 26, 1876, 4, col. 2; and "Disastrous Fire," *MDR*, July 2, 1876, 1, col. 4, which blamed a destructive fire on inadequate water supplies, an allegation underscored by a newly dug but empty cistern.

7. On kerosene accidents, see "Fires Caused by Kerosene," *The Vindicator* 6, no. 3 (November 1, 1888): 847–48; and "Deaths resulting from the practice of starting fires with kerosene," *The Vindicator* 28, no. 10 (November 15, 1901): 151. There were approximately eighty-one references in the *NYT* from 1867 to 1870 to kerosene-related accidents (illuminating oil). On state laws regulating kerosene flash points to mitigate such hazards, see "Preventing Oil Explosions," *NYT,* June 24, 1882, 8, col. 4; and C. F. Chandler, *Report on Petroleum as an Illuminator* (1871), 38–39, 43, 47, 53–57; on the definition of flash point, see U.S. Department of the Interior, Bureau of Mines Technical Paper 49, *The Flash Point of Oils: Methods and Apparatus for Its Determination,* by Irving C. Allen and A. S. Crossfield (1914), 5; and J. H. Thomson and Sir Boverton Redwood, *Handbook on Petroleum for Inspectors under the Petroleum Acts,* 3rd. ed. (1913), 68–71. On dangerous elements in fuel oil, see H. Tweddle, "Crude Petroleum and Its Products as Fuel," *E&MJ* 68, no. 36 (October 14, 1899): 459; and Albert W. Gunnison, "Petroleum and Its Relation to Fire Underwriting," FUAP, *Proceedings of the Twenty-Eighth Annual Meeting of the Fire Underwriters' Association of the Pacific* (1904), 34–37.

8. On the warehouse fires, see "Destruction of Oil Warehouses," *NYTB,* February 19, 1878, 2, col. 3 (quotation); and "A Big Blaze," *MDR,* February 19, 1878, 4, col. 2. For subsequent fires, see "Fire Record: At Mobile-Loss $85,000," *NYTB,* April 5, 1878, 8, col. 4; "A Large Fire in Mobile," *NYTB,* June 20, 1879, 5, col. 5; and "Great Damage Done in Mobile," *NYTB,* January 9, 1886, 1, col. 4.

9. On One Mile Creek, see map by D. M. N. Ross, city engineer, dated February 21, 1888, in the case file of "McDonald vs Otis," Mobile County Chancery Court Case, #4485, Mobile County Circuit Court Records, University of South Alabama, McCall Rare Book & Manuscript Library, Mobile. According to *George Matzenger's Mobile Directory for 1888* (1888), 412, and *Maloney's Mobile City Directory 1903* (1903), 70, the northern boundary of the city during the last two decades of the nineteenth century ran along One Mile Creek and Bayou Marmotte. The 1901 Pillans-Ross Map (see note 10, below) shows that the Standard Oil works was 770 yards to the west-northwest of the shingle mill. For fire accounts, see "Fire in Mobile," *NODP,* October 27, 1890, 1, cols. 2 (quotation 2), 3; "Fire Echoes," *MDR,* October 28, 1890, 1, cols. 2–5; "River News," *MDR,* October 28, 1890, 4, col. 2; "Day of Gloom," *MDR,* October 28, 1890, 5, cols. 1 (quotation 1), 2–5, 6 (quotations 3–4), 7–8; "What the Reporters Saw," *MDR,* October 28, 1890, 5, cols. 4–6; "The Steamer Jewel," *MDR,* October 29, 1890, 4, col. 2; and "Mobile Swept by Destructive Flames," *NYTB,* October 27, 1890, 1, col. 5.

10. "The Oil Monopoly," *MDR,* October 19, 1886, 1, col. 7 (quotations 1-2); "Kentucky," *ManRecord* 10 (October 23, 1886): 370, col. 3; "Dissolution of the Firm of Chess, Carley & Co.," *NODP,* October 13, 1886, 1, col. 5; and "F. D. Carley Dies in Brokers' Office," *NYT,* April 5, 1922, 1, col. 2. For locations, see MMA, *Map of the City of Mobile, Compiled from an Original Map made by P. J. Pillans City Engineer, and Harry Pillans, Ass't Eng'r, With Additions to date, by D. M. N. Ross, City Engineer, 1901* (1901). The SOC works on block 376 was bound by Sumpter, Marion, Conception and St. Joseph streets. As Harriet E. Amos noted in *Cotton City: Urban Development in Antebellum Mobile* (1985), 99 (quotation 3), the land immediately north of the SOC plant between One Mile and Three Mile creeks was a "swamp" not suited "for development" (99).

11. SOC acquired People's Oil in 1898. See Williamson and Daum, *API Illumination,* 693–94. For New York City, see "An Act in relation to storage and the keeping of combustible material in the city of New York," chap. 742 of *Laws of New York for 1871,* rpt. in Chandler, *Report on*

Petroleum as an Illuminator, 107–10; and *McKinney's Consolidated Laws of New York Annotated,* book 19: *General Business Law* (1917), secs. 302–7. For Chicago see City of Chicago, *The Municipal Code of Chicago,* codified and rev. Egbert Jamieson and Francis Adams (1881), chap. 15, "Miscellaneous Ordinances," Article 9, Section 1068, 267. See also "The Gomez Building: Its Erection Ratified by the General Council," *MDR,* October 6, 1894, 7, col. 4; and an untitled, handwritten petition, dated October 5, 1894, MMA, RG 5, ser. 2, box 4, folder 9, 1894, document #5493 (quotation).

12. Maupin, president of R. L. Maupin Packing Co., was one of eight councilmen, each of whom represented a city ward. See *George Matzenger's Mobile Directory for 1895,* 264; also a letter, dated November 3, 1894, to Judge R. L. Maupin, MMA, RG 5, ser. 2, box 4, folder 9, 1894, document #5515 (quotations).

13. On Huger, see Don H. Doyle, *New Men, New Cities, New South: Atlanta, Nashville, Charleston, Mobile, 1860–1910* (1990), 129–31. See also "Report of Committee No. (4) on Petition of Peoples Oil Co. to erect a storage tank," November 5, 1894, MMA, RG 5, ser. 4, box 13.1, folder 8, document #5494 (quotations); and "The General Council," *MDR,* November 10, 1, col. 4.

14. On jobbers as regional intermediaries or "commission agents" who buy quantities of oil from refineries and then resell it to specific industries or retail outlets, see Williamson and Daum, *API Illumination,* 535. See also "Petition from People's Oil Company," May 16, 1895, MMA, RG 5, ser. 2, box 4, folder 10, 1895, document #5495 (quotations); "The New Guardhouse: Petitions," *MDR,* June 6, 1895, 4, col. 3; "Report of Committee No. (4) on Favorably on petition of the Peoples Oil Co. to erect a tank," July 5, 1895, MMA, RG 5, ser. 4, box 13.2, folder 10, document #5496; and cover of document #5495. On the Chicago Underwriters, see Tiffany, *Tiffany's Book of Policy Forms for Fire Insurance Agents,* "Form of Crude Petroleum Permit for Fuel," adopted November 14, 1889, 234a.

15. For Gomez, see the advertisement "Tinware Factory," *MDR,* September 1, 1885, 6, col. 3, and the paragraph "F. Gomez," 8, col. 4. On Thomas, see "Further Agitation: Mr. Thomas Gives Mobile a Further Lift as a Cotton Export City," *MDR,* October 13, 1894, 5, col. 3; "Well-Known Mobilian Causes Sensation," *MDR,* February 1, 1902, 6, cols. 1–2; and "George B. Thomas Sends Ambassador," *MDR,* February 18, 1902, 5, col. 1. On Ross's approval, see "Building Permits," *MDR,* September 2, 1894, 5, col. 3; and "The War into Africa: Thomas Wants an Iron Shell Building on Magnolia St.," *MDR,* October 3, 1894, 4, col. 3. See also the first two letters printed in "Mr. Geo. Thomas' Kick," *MDR,* September 26, 1894, 4, col. 3; "The Gomez Building: More Correspondence Concerning Its Construction," *MDR,* September 28, 1894, 3, col. 3; and "The Gomez Building: Its Erection Ratified," *MDR,* October 6, 1894, 7, cols. 2–3.

16. See the letters, McIntosh & Rich to Hon. C. L. Lavretta, dated September 20, 1894; McIntosh & Rich to C. L. Lavretta, Mayor, September 22, 1894; C. L. Lavretta to Joseph C. Rich, September [25], 1894; McIntosh & Rich to Mr. F. Gomez, September 24, 1894; McIntosh & Rich to D. M. N. Ross, September 24, 1894; D. M. N. Ross to C. L. Lavretta, September 25, 1894, in "Mr. Geo. Thomas' Kick," *MDR,* September 26, 1894, 4, cols. 3–4; as well as McIntosh & Rich to Hon. C. L. Lavretta, dated September 25, 1894, in "The Gomez Building: More Correspondence," *MDR,* September 26, 1894, 3, col. 3; see also "Ignoring the Fire Limits Law," *MDR,* September 5, 1894, 2, col. 2.

17. See the letter, P. J. Hamilton to Hon. C. L. Lavretta, dated September 21, 1894, in "Mr. Geo. Thomas' Kick," *MDR,* September 26, 1894, 4, col. 3. The newspaper argued that the

defective city charter imposed contradictory duties upon the mayor, who as both executive and judge in permit matters, deferred to the city engineer's determinations of violations of the fire limits and general-council committee's power to approve permits. See "The Mayor's Responsibility," *MDR* September 28, 1894, 2, cols. 1–2; and "The Law Needs Amendment," October 6, 1894, 4, col. 1. The court case was *Montgomery* v. *Louisville and Nashville Railroad Co.*, 84 Alabama 127.

18. See the letter, McIntosh & Rich to C. L. Lavretta, dated September 22, 1894, in "Mr. Geo. Thomas' Kick," *MDR*, September 26, 1894, 4, col. 3 (quotations 3–4); and City of Mobile, *Charter and Code of Ordinances* (1889), Charter, sec. 21, pp. 18 (quotation 1), 19 (quotation 2); and chap. 10, sec. 75: "no buildings shall be erected or constructed of wood" (16). For the charter statute "An Act To amend an act entitled an act to incorporate the port of Mobile, and provide for the government thereof. Approved February 11th, 1879," see *Acts of the General Assembly of Alabama Passed at the Session of 1886–87* (1887), 223–53. William J. Novak pointed out that in the 1890s some state courts required municipalities to have "explicit statutory delegation of power" from the state, if they passed ordinances establishing fire limits or destroyed wooden structures if in prohibited areas. See *The People's Welfare: Law & Regulation in Nineteenth-Century America* (1996), 279n102.

19. "The General Council," *MDR*, November 10, 1894, 1, col. 6; p. 7, col. 5; "Amendments to the City Charter," 4, col. 1 (quotation); and "The Law Needs Amendment," 4, col. 1.

20. "Proposed Amendments," *MDR*, November 11, 1894, 5, col. 1 (quotation); "Mobile's City Charter: City Charter of Mobile," *MDR*, December 8, 1894, 1, col. 1; and *Acts of the General Assembly of Alabama Passed at the Session of 1894–5*, no. 230, "An Act To amend sections six, eleven, fifteen, twenty-one, twenty-five, thirty-four, forty of an act entitled an act to amend an act entitled an act to incorporate the port of Mobile, and to provide for the government thereof, approved December 10th, 1886," sec. 4, pp. 384–87; approved February 6, 1895 (389).

21. *The Charter and the Code of Ordinances of The City of Mobile, With an Appendix*, comp. Peter Joseph Hamilton (1897), chap. 11, article 1, sec. 337.

22. For the statute, see *Acts of the General Assembly of Alabama Passed at the Session of 1894–5*, no. 230, sec. 4, p. 385 (quotation 1); for codes, see *The Code of Ordinances of the City of Mobile, With the Charter, and an Appendix* (1859), sec. 123, p. 118 (quotation 4); *Code of Ordinances of the City of Mobile*, comp. Reuben A. Lewis (1866), sec. 95, p. 113 (wardens); sec. 96, p. 113 (danger of fire); sec. 250, p. 162 (quotations 2–3); City of Mobile, *Charter and Code of Ordinances* (1889), chap. 10, sec. 80, p. 66 (quotation 4); chap. 25, sec. 242, p. 115; and *The Charter and the Code of Ordinances of The City of Mobile*, (1897), chap. 11, sec. 342 (quotation 4); chap. 27, secs. 435–36.

23. "The Storage of Fuel Oil," *American Gas Light Journal* 56, no. 5 (February 1, 1892): 150.

24. Department of the Interior, USGS, *Mineral Resources of the United States 1902* (1904), 536–37, 544–45, 568–69; *Mineral Resources of the United States 1900* (1901), 580; and "Significance of the Texas Oil Field," *ManRecord* 39, no. 12 (April 11, 1901): 212, col. 4. On calorific estimates, see USGS, *Mineral Resources of the United States 1901* (1902), 573–74; and Henry H. Humphrey, "On the Use of Beaumont Oil as Fuel," *Journal of the Association of Engineering Societies* 30, no. 3 (March 1903): 101. On the physical and chemical qualities of Beaumont oil, see USGS, *Oil Fields of the Texas-Louisiana Gulf Coast Plain*, Bulletin 282, by N. M. Fenneman (1906), 125–27, 131–36.

25. On Mobile businesses, see "As From a Tower," *MDR*, September 1, 1899, 1, cols. 1–4; "Shipping in Port," *MDR*, July 10, 1900, 4, col. 3; "As to Mobile," *MDR*, August 3, 1900, 4, col. 3; "Electrical Betterments," *MDR*, September 1, 1900, 3, cols. 1–3; "Thirteen Steamer Lines," *MDR*, September 1, 1900, 4, cols. 4–5; and "The Saw Mill Interests," *MDR*, September 1, 1900, 7, cols. 7–8. For Mobile companies, see "Fuel of the Future: Test of Texas Crude Oil Is Made Here by Steam Tug," *MDR*, May 22, 1901, 5, col. 1; "The Advantages of Petroleum Fuel," *MDR*, May 23, 1901, 5, col. 1; "Another Test Given Crude Oil," *MDR*, May 24, 1901, 1, col. 7; and "Test of Crude Oil," *MDR*, May 25, 1901, 8, col. 4. For coal consumption see United States, see Bureau of Mines, *Annual Reports of the Department of the Interior for the Fiscal Year Ended June 30, 1900, Twenty-First Annual Report of the United States Geological Survey, Part 6* (1901), 416–18. On the Fuel Pipe Line company, see MMA, RG 5, ser. 16, "General Council Minutes, 1900–1903," vol. 2 (May 15, 1901): 309–10; "The General Council," *MDR*, May 16, 1901, 5, col. 2; and "Council," *Mobile Daily Item*, May 16, 5, cols. 1–2.

26. "A New Enterprise," *MDR*, May 5, 1901, 5, col. 3 (quotation 1); Mobile County, Probate Judge Office, Incorporation Book 3, "Declaration of Incorporation," dated May 4, 1901, p. 565—not received for recording by the county until May 27 (see upper left marginalia of Declaration)—the date used for incorporation in the online file of Alabama, Secretary of State, "Corporate Details," arc-sos.state.al.us/CGI/SOSCRP10.mbr.

27. On Spotswood's family in Mobile, see. T. C. De Leon, *Belles Beaux and Brains of the 60's* (1907), 251–54; and Thomas McAdory Owen, *History of Alabama and Dictionary of Alabama Biography*, 4 vols. (1921), "Spotswood, William Augustine Washington," vol. 4: 1607–8.

28. To obtain transportation for the Spanish-American War, the Quartermaster's Department chartered Spotswood's tugboat *Captain Sam* in June 1898. See U.S. House of Representatives, *Annual Reports of the War Department for the Fiscal Year Ended June 30, 1898*, "Report of the Secretary of War," *Miscellaneous Reports*, 55th Congress, 3d sess., 1898, doc. no. 2, 435. Spotswood also won government contracts for shore fortifications, waterway obstruction clearance, shoreline protection, and seawall construction. See U.S., *Annual Reports of the War Department for the Fiscal Year Ended June 30, 1898*, part 1, appendix 4-P, "Defenses of Mobile and Mississippi Sound" (1898), 737–38; *Annual Reports of the War Department for the Fiscal Year Ended June 30, 1899*, "Report of the Chief of Engineers," part 1, appendix 4-P, "Defenses of Mobile and Mississippi Sound," (1899), 924–26, 929-30, and 911, 913 on Spotswood's bid proposals for fortifications, Tampa, Florida; 56th Congress, 2nd sess., 1900, doc. no. 2, "Shore Protection," 948–49; part 2, appendix S-18, "Removing Sunken Vessels or Craft Obstructing or Endangering Navigation," 1783–84; and *Annual Reports of the War Department for the Fiscal Year Ended June 30, 1901*, "Report of the Chief of Engineers," part 1, appendix 3, "Fortifications" (1901), 841.

29. *George Matzenger's Mobile Directory for 1901* (1901), 458, 490; Spotswood was also manager of the Mobile & Gulf Telegraph Company (341); see *The Mobile City Directory 1902* (1902), 245, 581.

30. "Crude Oil for Cuba," *MDR*, July 11, 1901, 8, col. 4.

31. "Fuel of the Future," *MDR*, July 11, 1901, 5, col. 1; "The Advantages of Petroleum Fuel," 5, cols. 1–2.

32. "Petition from W. C. Spotswood . . . To erect reservoirs & Tanks on property at or near the foot of Palmetto street," May 15, 1901, MMA, RG 5, ser. 2, box 5, "Petitions," folder 10, 1901,

document #8396 (quotation 1); *The Charter and the Code of Ordinances of 1897 of The City of Mobile* (1897), chap. 6, "Legislative Department," article 2—"Franchises," sec. 99, 80–81 (quotation 2); and "Notice of Application to Mayor and General Council for Right to Lay and Operate a Fuel Oil Pipe Line on Water Street, Etc.," *MDR*, May 24, 1901, 5, col. 4 (quotations 3–4).

33. MMA, "General Council Minutes, 1900–1903," vol. 2 (June 17, 1901): 321, 324; vol. 2 (July 15, 1901): 336–37, (quotations); "Indiscreet Enforcement," *MDR*, June 18, 1901, 2, col. 7; and "Peter Burke Recorder Pro Tem: Market Ordinance Amended," *MDR*, July 16, 1901, 3, col. 1.

34. Although not extant in the city archives, the underwriters' letter, dated August 19, 1901, appeared six months later within a longer contemporary communication, dated February 24, 1902, from "X.Y.Z." See "The Editor's Letter-Box: Storage of Oil," *MDR*, February 25, 1902, 5, col. 3.

35. "Paving Bonds Authorized," *MDR*, August 22, 1901, 6, col. 4 (quotation 2); MMA, "General Council Minutes, 1900–1903," vol. 2 (August 21, 1901): 350, (quotation 1); *George Matzenger's Mobile Directory for 1899*, 50, 361.

36. Many English and eastern-based fire, marine, and casualty insurance companies operated in the United States in the nineteenth century and had an interest in oil-storage issues. Selected examples with date of commencement of business include: Caledonian Insurance Company, Edinburgh (1890); Commercial Union Assurance Company, London (1871); Commercial Union Fire Insurance Company, New York (1891); Germania Fire Insurance Company, New York (1859); Hamburg-Bremen Fire Insurance Company, Hamburg, Germany (1873); Hartford Fire Insurance Company, Hartford, Connecticut (1810); Home Insurance Company, New York (1853); Insurance Company of North America, Philadelphia (1792); Liverpool and London and Globe Insurance Company, Liverpool, England (1848); London and Lancashire Fire Insurance Company, Liverpool (1879); London Assurance Corporation, Fire Branch, London (1872). This does not include several active fire-insurance companies in New Orleans and Mobile, such as: Germania Insurance Company, New Orleans (1866, reorganized 1891); Hibernia Insurance Company, New Orleans (1871); Merchants Insurance Company, New Orleans (1830, reorganized 1893); Sun Insurance Company, New Orleans (1856); and Teutonia Insurance Company, New Orleans (1871). See *Best's Insurance Reports upon all American and Foreign Joint-Stock Companies, and American Mutual Companies and "Lloyds" Associations* (New York, 1901).

37. *George Matzenger's Mobile Directory for 1901*, 26, 167, 169, 178, 481, 589–90, 630. On James Glennon, see "Change of Firm," *MDR*, January 3, 1886, 5, col. 4; and "Notice," p. 1, col. 6, for an announcement of the dissolution of Williams & Glennon after the death of Col. Price Williams.

38. On the New Jersey fire, see "Stream of Flaming Oil in a Village," *NODP*, March 25, 1901, 2, col. 4 (quotation 1); "New Jersey Village in a Sea of Fire," *NYT*, March 25, 1901, 1, col. 5; "Torrents of Flame Descends on Town, *NYHRD*, March 25, 1901, 4, col. 5; "Town Almost Wiped Out," *NYTB*, March 25, 1901, 1, col. 2. For an announcement of rate increases "on all stocks of merchandise without exception" from the tariff association, see "Insurance Rates Go Up," *MDR*, February 22, 1902, 5, col. 1; on the SETA disclosure that fire underwriters had moved up an anticipated rate increase by several months to satisfy Jacksonville insurance people, see *Proceedings of the Twenty-First Annual Meeting of the South-Eastern Tariff Association . . . May 21–22, 1902* (1902), 35 (quotations 2–3); hereafter *PSETA 1902*.

39. On the Mobile fire, see "Oil House Burned," *MDR*, June 22, 1901, 8, col. 5 (quotations 1–3). On the Point Breeze conflagration, see "Destructive Oil Fire: Gets Beyond the Control of the Firemen," *MDR*, August 21, 1901, 1, col. 7; p. 2, col. 1.

40. For barge movement, see "Marine News," *MDR*, August 10, 1901, 4, col. 1; p. 7, col. 6; and "The Fuel Oil Barge," August 23, 1901, 8, col. 5 (quotation). For the 1901 refitting of the *Helen Buck* at the East Side Docking Company from lumber to oil barge, see "The Advantages of Petroleum Fuel," *MDR*, August 23, 1901, 5, col. 2; the *Helen Buck* carried 8,797–9,000 barrels of oil according to the *OIJ* 1, no. 10 (October 1, 1902): 10; and "Helen Buck Found," *MDR*, December 19, 8, col. 4.

41. State of Texas, Secretary of State, Corporation Section, "Charter of the Mississippi-Texas Oil Co.," record #10224, filed May, 1901. On the company's oil-field drilling, see "Refinery at Port Arthur," *NODP*, July 12, 1901, 1, col. 3; "Gusher Still Running Wild," *NODP*, August 28, 1901, 1, cols. 1–2; "The Oil Country," *NOTD*, July 12, 1901, 2, col. 1; and "The Mississippi-Texas Oil Company," *NOTD*, July 21, 1901, 8, col. 6.

42. MMA, "General Council Minutes, 1900–1903," vol. 2 (October 15, 1901): 385, 412–13; "Cheap Light and a Union Depot," *MDR*, December 17, 1901, 2, col. 3. C. F. Woods, Alex McRaven, and H. T. Gorch of the "Mississippi and Texas Oil Company," who controlled a "gusher in the Beaumont field," were in Mobile. Reportedly, the company planned to build a wharf and a "mammoth storage tank on the east side of the Mobile river for loading oil into vessels for export to foreign countries." See "Will Build a Wharf and Immense Storage Tank," *NOTD*, October 11, 1901, 2, col. 1.

43. Letter, G. B. Thames, Charles R. Garnett, John Gaillard, James K. Glennon, to Walter Wash [Walsh], Chairman Committee No. 4, Mobile, Ala., dated December 13, 1901, printed in "The Editor's Letter-Box: Storage of Oil," *MDR*, February 25, 1902, 5, col. 3.

44. I thank Dr. Michael Thomason, University of South Alabama (retired), Mobile, for this information.

45. "There Was No Quorum," *MDR*, December 31, 1901, 8, col. 4.

46. MMA, "General Council Minutes, 1900–1903," vol. 2 (December 31, 1901): 437. The Mississippi-Texas Oil Company did not disappear. It was a defendant in the suit filed by the heirs of Pelham Humphries against the Hogg-Swayne syndicate. See "Latest Suit for Title to Spindletop Property," *OIJ*, 1, no. 4 (July 5, 1902): 3, col. 1. According to Joseph Cullinan, the company had contracts with the Texas Company in April 1902 to deliver oil from Spindletop, transporting it from the field through the Sun Company. See letter, Joseph Cullinan to McKie & Autry, Corsicana, Texas, April 19, 1902, University of Houston Libraries, Special Collections, Joseph Cullinan Papers (hereafter UH-SC-JCP), box 1, vol. 1, p. 213.

47. John O. King, *Joseph Stephen Cullinan: A Study of Leadership in the Texas Petroleum Industry, 1897–1937* (1970), 159–60. On Lucas remaining a stockholder in the J. M. Guffey Petroleum Co., see "$15,000,000 Oil Company," *MDR*, May 17, 1901, 3, col. 3.

48. Hartwell was secretary-treasurer of the Mobile Towboat Association, and by 1910 vice-president and secretary-treasurer of the Mobile Towing & Wrecking Company. See advertisement for "Mobile Towing & Wrecking Company," *MDR*, September 1, 1910, sec. 2, p. 3, cols. 1–3. He and lumber-exporter Horace Turner in 1909 built the Turner-Hartwell Dock in a parcel of swamp between Water and Virginia streets near Choctaw Point. On Turner, see advertisement, "Horace Turner & Co.: Export Lumber," *MDR*, September 1, 1909, sec. 2, p. 1,

cols. 5–7; also "New Facilities for our Commerce," *MDR*, September 1, 1909, sec. 2, p. 6; and "Turner-Hartwell Company's New Docks," *MDR*, September 1, 1910, sec. 1, Trade edition, p. 7.

49. For Guffey's aggressive business plans, see "Beaumont Oil Field," *MDR*, May 24, 1901, 2, col. 7; and "Lake Charles: Lands Leased by Oil Prospectors," *NODP*, March 2, 1901, 9, col. 2. For Riter-Conley steel fabrication, see "The Riter-Conley Mfg. Company," *TIA* 69 (March 20, 1902): 39, col. 2; vol. 70 (October 9, 1902): 24, col. 2; and "Riter-Conley Contracts," vol. 86, no. 6 (August 11, 1910): 326. For a picture of an enormous gas-storage tank that the company constructed, accompanied by names of ten major cities with Riter-Conley-branch offices, see the advertisement for "riter-conley co. pittsburg, pa." in *American Gas Engineering Journal* 112, no. 19 (May 8, 1920): 19.

50. "Guffey Company Places $5,000,000 of Bonds," *OIJ*, 1, no. 3 (June 21, 1902): 13. I thank Fiona Flett, assistant archivist, Shell Services International, Shell Archives, London Centre, London, for Shell information in a letter dated May 13, 1998.

51. MMA, "General Council Minutes, 1900–1903," vol. 2: 437–39 (quotations); and "Resolution to Grant the Petition of the J. M. Guffey Petroleum Company for Authority to Construct and Maintain a Fuel Oil Storage Tank upon Certain Lands within the City of Mobile," RG 5, ser. 6, "Resolution."

52. "The General Council: Grants Petitions for a Storage Oil Tank and Railroad Switch," *MDR*, January 1, 1902, 8, col. 5; "Only One Voted 'No,'" *MDR*, February 23, 6, col. 1 (quotations).

53. "An Ordinance to Regulate the Storage and Use of Fuel Oils in the City of Mobile," MMA, RG 5, ser. 7, Ordinances, document #8884.

54. On New Orleans see "The Underwriters of New Orleans," *The Vindicator* 28, no. 4 (August 15, 1901): 53, col. 2; "Ordinance Adopted for New Orleans," *The Vindicator* 28, no. 9 (November 1, 1901): 138–39; on Galveston, see "Pending Ordinances at Galveston," *The Vindicator* 28, no. 9 (November 1, 1901): 139–140; "Scheme of the Oil Men, *The Vindicator* 28, no. 9 (November 1, 1901): 140; "Radical Measures Necessary," *The Vindicator* 28, no. 10 (November 15, 1901): 152.

55. See *PSETA 1902*, 2, for "resident" members of the Executive Committee for 1902–3. The list included representatives of L&L&G Insurance Co., Sun Insurance Co., London Assurance, and the Insurance Company of North America. For telegrams sent to M. J. Sanders in New Orleans, see "Marine Companies on Fuel Oil," *MDR*, March 20, 1902, 3, col. 2. For the SETA and the founding of the NFPA, see Percy Bugbee, *Men against Fire: The Story of the National Fire Protection Association 1896–1971* (1971), 4.

56. For Schoen's role in "thoroughly" protecting the interests of SETA member-companies throughout the South regarding the "question of oil storage," particularly in Mobile, see *PSETA 1902*, 26–27. For his NBFU assignment and portions of his report, see "Texas Fuel Oil," *InsEng* 4: 534–44; NBFU, *Proceedings of the Thirty-Seventh Annual Meeting of the National Board of Fire Underwriters*, May 14, 1903, "Report of the Committee on Lighting, Heating and Patents," 67, 72, 74; and "Engineer Schoen on the Fuel Oil Hazard," *The Vindicator* 30: 30. Schoen was the elected second vice-president of the NFPA in 1924–25 and 1925–26, elected first vice-president for 1926–27 and president in 1928–29. I am grateful to Sue Marsh, National Fire Protection Association, Quincy, Mass., for information on Schoen's NFPA attainments. For the 1902 guidelines, see Booth, *Liquid Fuel and Its Combustion*, appendix 8, "Rules and Requirements of the National Board of Fire Underwriters for Regulating the Hazard of the

Storage and Use of Fuel Oil . . . As Recommended by Its Committee of Consulting Engineers, Edition of 1902," 353 (quotation 1), 354–358. On the risk industry Schoen represented, see Knowles, *Disaster Experts*, 8, 11–12, 32–37, 43–44, 47, 51, 61.

57. On Schoen communicating with major fire-insurance companies, see "The Editor's Letter-Box: Storage of Oil," *MDR*, February 25, 1902, 5, col. 3. For excerpts of insurance responses, see "Oil Storage, a Menace," *MDR*, February 20, 1902, 4, col. 2; and "As to Storage of Fuel Oil," *MDR*, February 21, 1902, 4, col. 2. For the claim that Mobile insurance agents received forty telegrams from their companies about the Guffey tank, see the letter to the editor signed "Thames & Batre," dated February 21 and printed in "The Editor's Letter-Box: Objections to the Location of the Tank," *MDR*, February 22, 1902, 3, col. 1. On state insurance regulation, see *The Code of Alabama 1896* 1 (1897): 771–72, chap. 63, article 3, sec. 2619. It provided that insurance companies in a tariff association had to pay beneficiaries "in the event of loss or damage" the amount of "actual loss" plus a 25 percent premium; see also "Provision Designed to Prevent Tariff Associations to Fix Rates of Insurance," 21, part of a legal pamphlet on insurance law, for which see ADAH, Insurance Department, State Publications, *Annual Report of the Insurance Commissioner of the State of Alabama . . . December 31, 1901* (1902).

58. "Oil Storage, a Menace," *MDR*, February 20, 1902, 4, col. 2.

59. See Dolores Greenberg, "Energy, Power, and Perceptions of Social Change in the Early Nineteenth Century," *American Historical Review* 95, no. 3 (June 1990): 694 (quotations 1–3), 695–97, 701, 706, 710.

60. "As to Storage of Fuel Oil," *MDR*, February 21, 1902, 4, col. 1 (quotations 4–10). With the approval of Mobile insurance people, Committee No. 4 recommended at a June meeting of the General Council that it grant permission for construction of a 3,000-barrel turpentine tank at Charleston and Palmetto. See "Petitions Considered," *MDR*, April 16, 1902, 5, col. 2.; "The Paving Scheme: Reports of Committees," *MDR*, June 17, 1902, 5, col. 5; and MMA, "General Council Minutes, 1900–1903," vol. 2 (June 16, 1902): 526–27. For Pillans as an admiralty lawyer, see *Blue Book of American Shipping: Marine and Naval Directory of the United States 1901* (1901), 280.

61. "As to Storage of Fuel Oil," *MDR*, February 21, 1902, 4, col. 1 (quotations); "Marine Companies on Fuel Oil, *MDR*, March 20, 1902, 3, col. 2; "Fuel Oil Questions under Discussion," *NODP*, February 19, 1902, 10, cols. 6–7; and "Fuel Oil and Insurance," *NODP*, February 26, 1902, 9, cols. 5–6. On M. J. Sanders, see "Insurance Statements," *NODP*, February 19, 1902, 6, col. 2.

62. "As to Storage of Fuel Oil," *MDR*, February 21, 1902, 4, col. 2 (quotations). For an informative advertisement from the Factors & Traders company, see *MDR*, June 20, 1901, Southern Railway Edition, 7, cols. 6–7. On Michael J. McDermott see *George Matzenger's Mobile Directory for 1901*, 302–3.

63. "The Editor's Letter-Box: Mr. Pillans Thinks Underwriters Are Misinformed," *MDR*, February 22, 1902, 3, col. 1.

64. For Titusville, Pa., see "Disastrous Oil Fire," *NYT*, June 12, 1880, 1, col. 6; "Disastrous Oil Fires," *NYT*, June 13, 1880, 7, col. 2; "Losses by Fire," *NYT*, June 14, 1880, 5, col. 1. And later, Oil City, Titusville, and Corry, "Scores of Lives Lost," *NYT*, June 6, 1892, 1, cols. 5–6; p. 5, col. 3; "Destruction in Corry," cols. 6–7; "Hours of Peril in Titusville," col. 7. "Two Cities of Mourning," *NYT*, June 7, 1892, 1, cols. 5–6; "Sad Scenes at Oil City," *NYT*, June 8, 1892, 5, col. 1; "Are Still Full of Pluck," *NYT*, June 9, 1892, 5, col. 1; "Still Terror-Stricken," *NYT*, June 10,

1892, 5, col. 1; "The Dead at Oil City" and "Aid for the Sufferers," *NYT*, June 12, 1892, 8, col. 4; "The Oil City Disaster," *NYT*, June 14, 1892, 5, col. 3.

65. "The Editor's Letter-Box: Objections to the Location of the Tank," *MDR*, February 22, 1902, 3, col. 1 (quotations). Letter writer "X.Y.Z." observed that A. M. Schoen "reported the situation to the various companies doing business in Mobile." In response they telegraphed their Mobile agents, which, the writer argued, showed that the opposition to the Guffey tank came from the insurance companies, not a coal dealer as Harry Pillans had intimated. See "The Editor's Letter-Box: Storage of Oil," *MDR*, February 25, 1902, 5, col. 3.

66. "The Editor's Letter-Box: Objections to the Location of the Tank," *MDR*, February 22, 1902, 3, cols. 1, 2 (quotations). For an editorial supporting Thames and Batre, see "The Fuel Oil Question," *MDR*, February 22, 1902, 4, col. 1.

67. "License Ordinance," *MDR*, March 15, 1902, 3, cols. 4–5; and *PSETA 1902*, 26–27.

68. See "The Editor's Letter-Box: Danger of Fire from Oil Tanks," *MDR*, March 15, 1902, 8, col. 4 (quotations 1–4). For the petition, see "The Fuel Oil Tank: Petitions Against Its Erection Presented to the Council Last Night," *MDR*, March 21, 1902, 4, cols. 4 (quotation 5), 5. For a meeting scheduled for March 21 to hear the underwriters and other foes of the Guffey decision, see MMA, "General Council Minutes, 1900–1903," vol. 2 (March 19, 1902): 484, Council Meeting.

69. "Fuel Oil Tank Must Move," *MDR*, March 22, 1902, 2, col. 1 (quotations 1–2). On Z. M. P. Inge (1856–1920), see ADAH, Private Records, Zebulon Montgomery Pike Inge Papers 1833–1918, box 103, folder 1, for information about his law degree from the University of Virginia (1876) and license to practice law from Hale County, Alabama, dated October 3, 1876.

70. MMA, "General Council Minutes, 1900–1903," vol. 2 (March 21, 1902): 488–89, (quotations 3–5); "Fuel Oil Tank Must Move," *MDR*, March 22, 1902, 2, col. 1 (quotations 1–2, 6).

71. "The Oil Tank Grant Rescinded," *MDR*, March 22, 1902, 4, col. 2 (quotations 1–3). On operating in Alabama, see ADAH, Secretary of State, Corporation Division, Corporation Records, SG 5469, Domestic & Foreign 1890–1920, folder "JM Guffey Company" (quotations 4–5).

72. See "Franchise," in *Judicial and Statutory Definitions of Words and Phrases*, 2nd ser., vol. 2: *Deposition—Lamp* (1914), 626–30. For a telephone franchise, see "Electrical Ordinances," *MDR*, June 25, 1901, 3, cols. 4–5; on Mobile Light, see "Peter Burke Recorder Pro Tem," *MDR*, June 25, 1901, 3, col. 2; "The Franchise Granted," *MDR*, September 13, 1901, 3, cols. 1–3; and "The Editor's Letter Box," *MDR*, November 15, 1901, 3, cols. 1–2. Mobile Gas-Light & Coke Co. renewed a contract with the city in April 1882. See "New Business," *MDR*, April 13, 1882, 1, col. 6.

73. "An Ordinance Granting a Franchise to Lay an Oil Pipe Line Within the City of Mobile," *MDR*, April 4, 1902, 4, col. 5; April 5, p. 5, col. 4; "The General Council," *MDR*, April 16, 1902, 5, col. 3; MMA, "General Council Minutes, 1900–1903," vol. 2 (April 15, 1902): 498.

74. For publication of the second franchise ordinance, see "An Ordinance Granting a Franchise to Lay Oil Pipe Lines Within the City of Mobile," *MDR*, April 25, 1902, 5, col. 4; April 26, p. 5, col. 4; and May 9, p. 7, col. 5. On council deliberations, see "The General Council," *MDR*, May 16, 1902, 8, col. 4; "War Declared on Mosquitoes," *MDR*, May 17, 1902, 5, col. 2; and MMA, "General Council Minutes, 1900–1903," vol. 2 (May 15, 1902): 510–11.

75. "Pipe Line Franchise: Granted Last Night to the Guffey Oil Company," *MDR*, May 21, 1902, 3, col. 1 (quotation); and "The Editor's Letter-Box: Objections to the Location of the Tank," *MDR*, February 22, 1902, 3, col. 1 (drainage into swamp). On council members who

asked for changes in the original ordinance see MMA, "General Council Minutes, 1900–1903," vol. 2 (May 20, 1902): 514.

76. "Pipe Line Franchise," *MDR*, May 21, 1902, 3, col. 1 (quotations 1–2); MMA, "General Council Minutes, 1900–1903," vol. 2 (May 20, 1902): 514–16. In 1907, Councilman Patrick Lyons pushed through an amendment to the ordinance "governing the storage of illuminating and lubricating oils within certain defined limits." The amendment allowed lubricants to be stored in tanks of "lower quality of steel" than what had been stipulated for "illuminating oils." But this did not involve changes in the "boundaries of the oil-storage limits." See "City Council Special Meeting," *MDR*, March 22, 1907, p. 7, col. 3; MMA "General Council Minutes," vol. 3 (March 21, 1907): 280.

77. "Texas Oil in Bayonne," *NYT*, March 29, 1902, 1, col. 5 (quotation 1); and "The Fuel Oil Tank," *MDR*, August 2, 1902, 8, col. 4. For Cullinan, see Joseph Cullinan to Arnold Schlaet, New York City, September 11, 1902, 413 (quotation 2), UH-SC-JCP, Private Letters, vol. 2 (June 20, 1902–October 10, 1902), whose trip is explained in Private Letters, vol. 2: 330–400, August 27–September 8, 1902.

78. "The Fuel Oil Tank," p. 8, cols. 4–5.

79. Ibid., col. 4.

80. "Guffey Oil Tank," *MDR*, October 31, 1902, 8, col. 5; "Mobile Budget: Guffey Oil Company Demolishing Its Storage and Will Withdraw From the Field," *NODP*, November 7, 1902, 6, col. 3.

81. "A Guffey Scoop," *NODP*, November 7, 1902, 7, col. 3; and "Oil Business Growing," part 1, *NODP*, November 23, 1902, 15, col. 2 (quotation). According to "River & Marine Intelligence," *MDR*, November 7, 1902, 7, col. 4, Guffey already owned part of the tug and purchased the remaining interest from Spotswood and Hartwell.

82. See "Mobile the Coal and Iron Chute of America," *MDR*, September 1, 1910, Trade Edition, sec. 2, p. 1. Receipts for Alabama coal grew between 1890 and 1894 from 39,320 to 135,212 tons. By 1904 coal receipts from all sources was 165,000 tons, of which 100,000 tons went for export and "foreign steamers" and 43,000 tons for "local steam and coastwise" traffic. Aside from bunkering, coal volume through the Port of Mobile grew very slowly. See Frederick E. Saward, *The Coal Trade: A Compendium of Valuable Information*, 22nd ed. (1895), 71; and *The Coal Trade* (1905), 106 (quotations in note).

83. Paul H. Giddens, *Standard Oil Company (Indiana): Oil Pioneer of the Middle West* (1955), 9; Williamson and Daum, *API Illumination*, 694 (quotation 1).

84. "Oil Depot Burned," *MDR*, January 3, 1904, p. 5, col. 2 (quotation 2).

85. "City Council Adopts Paving Proposition," *MDR*, January 7, 1904, 5, col. 4 (quotation 3); MMA, "General Council Minutes, 1904," vol. 3: 57; see MMA, "Sketch Showing Location of Proposed Track to Texas Oil Co. Crossing Emanuel St," attached to "Petition from The Texas Company, July 12th, 1910," July 15, 1910, RG 5, ser. 2, box 7, folder 8, document #4452.

86. "Guffey Sells Out," *NODP*, August 29, 1902, 6, col. 3; "Guffey Company Superintendent Resigns," *NODP*, October 29, 1902, 2, col. 4; "Guffey Oil Company," *NODP*, November 25, 1902, 2, col. 5; "Guffey Company Gossip," *OIJ* 1, no. 13 (November 15, 1902): 5; ADAH, Secretary of State, Corporation Division, Corporation File, SG 5469, "Articles of Incorporation of The Gulf Refining Company," as well as a letter to the Secretary of State of Alabama, dated August 4, 1905, accompanied by signed statements designating a legal agent in Alabama; King, *Joseph*

Stephen Cullinan, 160; and Craig Thompson, *Since Spindletop: A Human Story of Gulf's First Half-Century* (ca. 1951), 14, 22.

87. On Gulf Refining, see State of Louisiana, Secretary of State, Business Services, "Charter of Gulf Refining Company of Louisiana," Article 9, pp. 5–6; Kenny A. Franks and Paul F. Lambert, *Early Louisiana and Arkansas Oil: A Photographic History, 1901–1946* (1982), 37; *R. L. Polk & Co.'s Mobile City Directory 1907,* 804; and *Mobile City Directory 1910,* 338.

88. The General Council approved construction of a "frame shed" on Government and Hamilton streets, which was within the fire limits, but did not mention an iron exterior or roof to justify the decision. See "Report of Committee No. (4) on Annual Report & Recommendation of the Chief Engineer of Fire Department," May 8, 1899, MMA, RG 5, ser. 4, box 14.1, folder 2, document #7227.

89. "National Underwriters Make Report on Mobile," *MDR,* May 20, 1905, 1, cols. 1–3; p. 3, cols. 1–2.

90. For SETA's claim, see *Proceedings of the Twenty-Second Annual Meeting of the South-Eastern Tariff Association . . . 1903* (1903), "Fuel Report of Mr. Schoen," 23 (quotations).

91. Robert H. Wiebe, *The Search for Order, 1877–1920* (1967), xiii (quotation), 295; David S. Landes, *The Unbound Prometheus,* 2nd ed. (2003), 199.

CHAPTER THREE

1. In late April 1901 two twenty-thousand-gallons tanks arrived at the pumping plant of Cow Bayou Canal and Irrigation Company, having ridden to the site first on railroad flatcars and then on a barge from nearby Orange, Texas. The company stored fuel oil in them to fire its boilers. See "Orange News Notes," *HDP,* May 1, 1901, 5, col. 4; "Figuring on Petroleum as Fuel," *HDP,* March 2, 1901, 3, col. 4; "To Use Oil for Fuel," *HDP,* March 31, 1901, 7, col. 2; and "Mills of Houston," *HDP,* April 7, 1901, 11, col. 1. On Gulf Coast oil as fuel and illuminant, see F. H. Oliphant, "Petroleum," 510, 513.

2. "Bailing the Oil Wells," *HDP,* July 27, 1901, 3, col. 4. For rice farmers and sugar planters obtaining "machinery to burn oil" to power irrigation pumps and mills, see *Rice Belt Journal* (Welsh, La.), June 27, 1902, 4, col. 3; and an *HDP* editorial, August 22, 1902, 4, col. 1 (quotation).

3. On out-of-state visitors, see "Well Is Under Control," *HDP,* January 18, 1901, 5, col. 3, and "New Rush on Beaumont," March 29, 1901, 5, col. 4. Initially the Beaumont field accounted for roughly 16,000,000 of Texas's 1902 production of 18,083,658 barrels, but owing to a falloff in field pressure, it contributed only half as much oil in 1903. The entire state turned out 17,955,572 barrels, with Sour Lake and Saratoga making up the difference. See D. H. Hewland, ed., *The Mineral Industry During 1903,* vol. 12 (1904), "Petroleum," 284–85.

4. On the dominance of coal and the challenge of fuel oil, see David Stradling, *Smokestacks and Progressives: Environmentalists, Engineers, and Air Quality in America, 1881–1951* (1999), 6–11, who points out that fuel oil became a cost-competitive energy source early on in Texas and California owing to abundant local supplies. Regarding the difficulty of obtaining reasonably priced coal in Houston in the 1880s, see Harold L. Platt, "Energy and Urban Growth: A Comparison of Houston and Chicago," *Southwestern Historical Quarterly* 91, no. 1 (July 1987): 3. On lignite, see "Houston, Texas," *HDP,* June 8, 1901, 7, cols. 3–4; and wood consumption, "How Houston Burns Wood," *HDP,* November 10, 1901, 12, col. 2; see also Joseph A. Pratt,

"Letting the Grandchildren Do It: Environmental Planning During the Ascent of Oil as a Major Energy Source," *Public Historian* 2, no. 4 (Summer 1980): 29.

5. See *Hayden's Annual Cyclopedia of Insurance in the United States 1902–1903*, "Texas Fire Prevention Association"; for the TPFA's NFPA membership and appearance in its publications, see NFPA, *Proceedings of Sixth Annual Meeting Held in Philadelphia May 13th, 14th and 15th, 1902* (1902), 7; "Houston, Texas," *InsEng* 3, no. 1 (January 1902): 103; "Texas Fire Prevention Blank," 6, no. 4 (October 1903), 390. For the origins of the NFPA and the NBFU see Brearley, *History NBFU*, 88–89.

6. On rational interface, see Ji-Hyun Lee and Tian-Chiu Li, "Integrated Physically Based Manipulation and Decision-Making Tree for Navigation to Support Design Rationale," in *Human Interface and the Management of Information: Methods, Techniques and Tools in Information Design*, Symposium on Human Interface 2007, Beijing, "Proceedings," part 1, ed. Michael J. Smith and Gavriel Salvendy (2007), 481–83.

7. Koen Steemers, "Energy and the City: Density, Buildings and Transport," *Energy and Buildings* 35, no. 1 (January 2003): 3, 9–10; Sam C. M. Hui, "Low Energy Building Design in High Density Urban Cities," *Renewable Energy* 24, no. 3–4 (November 2001): 628, 630–31; William E. Rees, "Urban Ecosystems: The Human Dimension," *Urban Ecosystems* 1, no. 1 (1997): 72; N. E. McIntyre, K. Knowles-Yánez, and D. Hope, "Urban Ecology as an Interdisciplinary Field: Differences in the Use of 'Urban' between the Social and Natural Sciences," *Urban Ecosystems* 4, no. 1 (January 2000): 14; and Ibrahim Dincer, "The Role of Energy in Energy Policy Making," *Energy Policy* 30, no. 2 (January 2002): 137. See also James B. McSwain, "Fire Hazards and Protection of Property: Municipal Regulation of the Storage and Supply of Fuel Oil in Mobile, Alabama, 1894–1910," *Journal of Urban History* 28 (July 2002): 599–628.

8. For lumber, cotton, and petroleum activities, see David G. McComb, *Houston: The Bayou City* (1969), 110–17.

9. Claudio Soares De Magalhães, "Social Agents, the Provision of Buildings and Property Booms: The Case of São Paulo," *International Journal of Urban and Regional Research* 23, no. 3 (September 1999): 46 (quotation). The technologies included rotary drilling rigs, field-gathering networks, metal tanks, compressors, pipelines, and burners as well as railroad motive power, maritime vessels, boilers, rigging, steel barges, and pumps; all were the means to produce, store, and transport oil to markets.

10. Sidney A. Shapiro and Robert L. Glicksman, *Risk Regulation at Risk: Restoring a Pragmatic Approach* (2003), 51 (quotation). For urban infrastructure as "basic to all economic life," embracing "streets and public transportation; water supply and sewage removal; police and fire protection; judicial, educational, and health facilities; and parks and other recreational facilities," see Martin J. Beckmann, "An Economic Model of Urban Growth," in Jesse H. Ausubel and Robert Herman, eds., *Cities and Their Vital Systems: Infrastructure Past, Present, and Future* (1988), 98. For the natural environment as the "life-support systems of the earth: the atmosphere, oceans and water systems, soil and biological life," see Karl Georg Høyer and Petter Næss, "The Ecological Traces of Growth: Economic Growth, Liberalization, Increased Consumption—and Sustainable Urban Development?" *Journal of Environmental Policy and Planning* 3 (July–Sept 2001): 177. On the built environment as the "ensemble of our humanly created surroundings," see Michael Benedikt, "Complexity, Value, and the Psychological Postulates of Economics," *Critical Review* 10 (Fall 1996): 551.

11. Eric H. Monkkonen, *America Becomes Urban: The Development of U.S. Cities & Towns 1780–1980* (1990), 161–64.

12. Disasters take place at the intersection of "society, technology, and environment." They represent "the failure of a society to adapt successfully to certain features of its natural and socially constructed environment." See Anthony Oliver-Smith, "Anthropological Research on Hazards and Disasters," *Annual Review of Anthropology* 25 (1996): 303–5 (quotation on 303); and Mark Manion and William M. Evan, "Technological Catastrophes: Their Causes and Prevention," *Technology in Society* 24 (August 2002): 208, 216.

13. F. N. Gray, secretary of the Houston Manufacturers' Association, argued that expensive bituminous coal had impeded capital investment into local industrial projects. But now, he claimed, Houston would benefit from the Beaumont petroleum strike because "oil solves the fuel problem for manufacturing uses." See "Will Help Houston," *HDP*, January 15, 1901, 3, col. 3.

14. *Morrison and Fourmy's General Directory of the City of Houston, 1902–1903* (Galveston, n.d.), 500–501; "An Oil Company," *HDP*, January 20, 1901, 9, col. 2; "To Promote Oil Industry," *HDP*, January 24, 1901, 6, col. 4; "For Oil Development," *HDP*, January 25, 1901, 6, col. 1; "Echoes of the Oil Field," *HDP*, April 28, 1901, 2, col. 4; "The Local Oil Situation," *HDP*, July 4, 1901, 3, col. 2; "Harris County Oil Notes," *HDP*, July 5, 1901, 3, col. 2; "Oil Field News: Contract for New Well," *HC*, March 22, 1902, 4, col. 5; "Another Attempt to be Made," *HC*, September 10, 1901, 5, col. 3; "Oil Wegll [*sic*] Development," *HDP*, April 6, 1901, 3, col. 2; "More Big Oil Companies," *HDP*, April 26, 1901, 3, col. 6; "Eight More Oil Companies," *HDP*, April 28, 1901, 4, col. 5; "Another Oil Company," *HDP*, April 30, 1901, 3, col. 3; "Chicago-Texas Oil Co.," *HDP*, July 26, 1901, 3, col. 2; "A New Houston Oil Company," *HDP*, July 26, 1901, 3, col. 3; "They Own Two Wells," *HC*, February 8, 1902, 6, col. 4; "Oil Companies Chartered," *HDP*, April 7, 1901, 6, col. 1; "The Fever Is Cooling Off," *HDP*, May 1, 1901, 3, col. 1. According to "Oil Companies Chartered," *HDP*, April 28, 1901, 4, cols. 3–4, there were, to date, fifteen oil companies with home offices in Houston. For incorporation data, see the Texas Secretary of State (SOS) website, and stock advertisements in the principal newspapers, especially March–June 1901.

15. "Burning Fuel Oil," New Orleans *Sunday States*, March 24, 1901, 19, cols. 1–2; "Another Oil-Consuming Plant," *HDP*, April 9, 1901, 7, col. 3; "Stribling Pipe Line Completed," *HDP*, February 21, 1901, 3, col. 4; "The Use of Oil for Fuel," *HDP*, April 26, 1901, 6, col. 5; "Will Use Fuel Oil," *HDP*, September 7, 1901, 7, col. 4; "Oil Field News," *HC*, March 11, 1902, 7, col. 1; "Hotel Using Oil for Fuel," *HDP*, May 30, 1901, 3, cols. 2–3; H. P. McLaughlin and Co. advertisement, *HDP*, June 1, 1901, 1, col. 3; "Meeting of City Council," *HDP*, June 4, 1901, 6, col. 3 (quotation); "Pumping Station in Operation," *HC*, December 24, 1901, 7, col. 3; "Another Oil Consumer," *HDP*, August 1, 1901, 6, col. 4. See also Houston Office of the City Secretary, CCM, book 2L, June 3, 1901, 134.

16. D. R. Beatty, Texas oilman and a director of Beatty Oil Company, remarked during a visit to Houston in May 1901 that he felt "assured that if we can convince the great fuel consuming centers of the world that they can depend upon a practically unlimited supply of fuel oil," many industries and manufacturers "will adopt this oil" and buy every barrel produced. See "Extent of the Oil Field," *HDP*, May 23, 1901, 5, col. 5; see also "Letters from the People: Transportation of Fuel Oil," *HDP*, July 8, 1901, 4, col. 5. "The New Treasure Field of Texas: Its Discovery, Its Development, Its Present Status and Its Great Future," part 8: "Permanency of

the Field," *HC*, February 8, 1902, 8, col. 5, described Beaumont and Houston people as "men who are familiar with the actual conditions in the [oil] field."

17. See "Oil Burned at Corsicana," *HDP*, January 21, 1901, 3, col. 1; and "Loss at Corsicana," *HDP*, January 22, 1901, 2, col. 2, wherein a five-thousand-barrel tank "partially filled with partly refined oil" caught fire in a refinery explosion. Regarding fires and fire prevention in the Spindletop field, see "The Committee Is At Work," *HDP*, September 1, 1901, 6, cols. 3–4; "Oil Men's Meeting," *GDN*, September 1, 1901, 4, cols. 4–5; "Want of Pressure," *GDN*, September 6, 1901, 2, cols. 2–3; "The Beaumont Oil Field on Fire," *NODP*, September 12, 1902, 1, cols. 1–3; "The Beaumont Oil Field Fire At a Standstill," *NODP*, September 13, 1902, 1, cols. 5–6, and p. 2, cols. 3–6; "The Oil Fire at Beaumont," *NODP*, September 13, 1902, 4, col. 1; "News and Notables at New Orleans Hotels: Manager of the Keith-Ward Oil Company Describes Conditions on the Tract," *NODP*, September 13, 1902, 6, col. 1; "Local Interests Affected by the Beaumont Blaze," *NODP*, September 13, 1902, 7, cols. 1-2; "The Beaumont Oil Fire Will Soon Be Extinguished," *NODP*, September 14, 1902, 1, cols. 3–6, and p. 2, cols. 3–6; "The Fire in the Oil Field," *NODP*, September 14, 1902, 4, col. 3; and "The Beaumont Oil Fire Stopped All in a Minute," *NODP*, September 15, 1902, 1, cols. 5–6.

18. Joseph A. Pratt, "Growth or a Clean Environment? Responses to Petroleum-related Pollution in the Gulf Coast Refining Region," *Business History Review* 52 (Spring 1978): 6–7.

19. Harold L. Platt, *City Building in the New South: The Growth of Public Services in Houston, Texas, 1830–1910* (1983), 90–103, 155–80.

20. "Effect of Low Oil Rates," *HDP*, August 15, 1901, 4, col. 2. See also Joseph A. Pratt, "The Ascent of Oil: The Transition from Coal to Oil in Early Twentieth-Century America," 11, 13.

21. "Industrial Houston," *HDP*, September 1, 1901, 10, col. 3. A *Houston Chronicle* editorial cast the whole process in a framework of energy and market fundamentals. Fuel, the newspaper argued, was the most important factor in the decision to build or not to build manufacturing facilities and the "primary factor in factory operation." When all the ingredients—raw materials, labor, cheap fuel—were present, transportation to market and profits were assured under the "great law of commercial evolution." See "Expansion in the Oil Field," *HC*, January 20, 1902, 4, cols. 1–2.

22. CCM, book 2K, January 28, 1901, 485–87; "An Ordinance, granting right to Construct Mains & Pipes, for furnishing Oil, Gas etc.," City Ordinance Book, 1901, January 28, 578, 579 (quotations), 580; CCM, book 2K, February 4, 1901, 493–94; "An Ordinance granting right to construct and lay Mains and Pipes for furnishing Oil Gas etc," City Ordinance Book, 1901, February 4, 1901, 581–83; "Mayor Woolford's Gavel," *HDP*, March 12, 1901, 9, col. 2; "Meeting of City Council," *HDP*, March 19, 1901, 6, col. 2; "Some Startling Reforms," *HDP*, February 20, 1901, 6, col. 2; City Ordinance Book, 1901, March 18, 594–95; and CCM, book 2L, March 11, 1901, 38, and March 18, 48. For a trenchant analysis of these deals, see Dr. [C]. S. Helsig, "Pipe Line Franchises," *HDP*, February 21, 1901, 6, cols. 1–2.

23. See "[Pipeline] Proposition," *HDP*, January 27, 1901, 13, col. 1; "To Pipe Oil From Fields," *HC*, October 18, 1901, 6, col. 1.

24. "Houston's Pipe Line," *HC*, January 4, 1902, 6, col. 3. There is no clear evidence that the bill actually became law, though the legislature may have drawn up or even passed this bill. The applicable state law, "An Act to provide for the organization of corporations for the

purpose of the storage and transportation, and purchase and sale of oil, gas, salt, brine and other mineral solutions," sec. 4, stated "that such pipe line so laid shall not exceed eight inches in diameter." See "Incorporations—Oil, Gas, Salt, Etc., Companies," chap. 117, *General Laws of the State of Texas Passed at the* Regular Session of the *Twenty-Sixth Legislature* (1899), 202–3; the statute is also in "Corporation Laws of Texas," *HDP*, April 29, 1901, 4, col. 4.

25. For citizen petitions for fire protection and recommendations from city officials, see CCM, book 2L, March 18, 1901, 43; April 8, 1901, 65; May 2, 1901, 117; July 1, 1901, 160, 168; and Houston, Texas, Office of the City Secretary, Miscellaneous Papers, 1900–1903 (hereafter Houston Misc. Papers), "Petition of Citizens for Extension of Water Mains to Silver & Spring & Sabine & Weber," dated August 5, 1901.

26. For a proposed contract between the city of Houston and the Houston Water Company, in which maintenance of water pressure, water-hydrant placement, and emergency use of Buffalo Bayou water "in event of conflagration" were prominent topics, see "A New Water Contract," *HDP*, May 7, 1901, 6, cols. 4–5. See also the editorial, "For Additional Fire Protection," *HDP*, June 26, 1901, 4, col. 1; CCM, book 2L, June 25, 1901, 155–56; and Martin V. Melosi, "Sanitary Services and Decision Making in Houston, 1876–1945," *Journal of Urban History* 20 (May 1994): 370–71.

27. Lawrence O. Gostin, *Public Health Law: Power, Duty, Restraint* (2000), 47–50; Christopher G. Tiedeman, *A Treatise on the Limitations of Police Power in the United States* (1886; rpt. New York, 1971), 2 (quotation). Cities suppressed and controlled unpleasant activities that produced unwholesome odors or were bothersome, irksome, or irritating, including soap or bone-boiling, animal slaughtering, or disposing of human wastes, as well as operating a house of prostitution or staging immoral plays, exhibitions, or shows.

28. "An Act to consolidate in one act, and amend the several acts incorporating the city of Houston, in Harris county," chap. 60, sec. 27, *Special Laws of the Ninth Legislature of the State of Texas* (Austin, 1863), rpt. in *The Laws of Texas 1822–1897*, comp. and arranged by H. P. N. Gammel (1898), vol. 5: 5, 552–53 The pagination used in this book reflects the pagination of the Gammel volumes, not the originals.).

29. "An Act to consolidate in one Act and amend the several Acts incorporating the City of Houston, in Harris county," chap. 99, sec. 27, *Special Laws of the State of Texas Passed by the Eleventh Legislature* (Austin, 1866), rpt. in Gammel, comp., *Laws of Texas*, vol. 5: 1391–92; "An Act to consolidate in one Act and amend the several Acts incorporating the City of Houston, in Harris county," chap. 3, sec. 24, *Special Laws of the State of Texas Passed at the Session of the Fourteenth Legislature* . . . (Houston, 1874), rpt. in Gammel, comp., *Laws of Texas*, vol. 8: 277 (quotation 1). Defining and suppressing nuisances became "to define what shall be deemed nuisances in said city, and to abate them by summary proceedings" (277). For the observation that "fire insurance interests formed an extralegal body of decision makers" in the post-1870 period, who lobbied municipal officials for good public services such as "improved fire protection," see Jon C. Teaford, "New Life for an Old Subject: Investigating the Structure of Urban Rule," *American Quarterly* 37, no. 3 (1985): 346, 347 (quotation above), 348–56.

30. For the amending acts see *Special Laws of the State of Texas Passed at the Regular Session of the Seventeenth Legislature* (1881), chap. 2, rpt. in Gammel, comp., *Laws of Texas*, vol. 9: 20; chap. 16 (S.B. No. 339). See also "An Act to provide a charter for the City of Houston, Harris County, Texas," chap. 7 (S.B. No. 306), *Special Laws of the State of Texas Passed at the Regular*

Session of the Twenty-Fifth Legislature (1897), rpt. in Gammel, comp., *Laws of Texas,* vol. 10: 1407); and *Charter of the City of Houston, Harris County, Texas, As Amended in 1899* (n.d.), 34; and *Charters and Rev. Code of Ordinances of the City of Houston 1897–1905,* bound volume of pamphlets, Office of the City Attorney, Houston, kindly provided by Paul Bibler in 2003.

31. *Charter of the City of Houston, Harris County, Texas, As Amended in 1899,* 34, 41, 42 (quotation 1); and TSA, Records of case no. 3796, *Jacob Binz, Appellant,* v. *National Supply Company, Appellee*; Court of Civil Appeals, 4th Supreme Judicial District, San Antonio, Texas, 1907; from District Court, Harris County, 55th Judicial District, Case no. 34,485 (55th JDC), *National Supply Company* v. *Jacob Binz,* "The City of Houston 1st amended original answer to cross-action of Jacob Binz, filed May 17, 1906," Transcript, §17 (quotations 2–3).

32. "An Act to provide a charter for the City of Houston, Harris County, Texas" (S.B. No. 306) in Gammel, comp., *Laws of Texas,* vol. 10: 1402.

33. CCM, book 2L, July 15, 1901, 176–77; "Meeting of City Council," *HDP,* July 16, 1901, 6, col. 1.

34. "Petition of Citizens, protesting against Erection of Oil Tanks, at Wood & Vine Sts.," in envelope dated August 12, 1901, Houston Misc. Papers (quotation 1); CCM, book 2L, July 29, 1901, 191; August 12, 1901, 217–18; "The City Council," *HDP,* July 30, 1901, 6, col. 1; and "The City Council," *HDP,* August 13, 1901, 6, col. 1.

35. Richard W. Gamble was a principal in the company. Its activities included stock brokerage, petroleum sales, and work as contractor for fuel-oil equipment, supplies, and installations (*Morrison and Fourmy's General Directory of the City of Houston, 1902–1903,* 172). See also the advertisements "Fuel Petroleum" and "Fuel Oil Equipment" in *HDP,* June 17, 1901, 5, col. 6; and August 18, p. 4, col. 3. The two-page typed letter from R. W. Gamble and Co. to Hon. Jno. D. Woolford, Mayor of the City of Houston, July 30, 1901, is in an envelope dated August 12, 1901, in Houston Misc. Papers.

36. Gamble and Co. to Woolford, July 30, 1901, Houston Misc. Papers.

37. CCM, book 2L, August 12, 1901, 217–18; and notations from the fire committee, city engineer, city attorney, and Fifth Ward aldermen (with date of report adoption "Aug 12 1901") on "Petition of Citizens, protesting against Erection of oil Tanks, at Wood & Vine Sts.," envelope dated August 12, 1901, Houston Misc. Papers (quotations 1–2). The petition contains a handwritten notation under the initials "jaw" (alderman W. J. A. Wherlin?), indicating that the records showed Vine Street to be within the city limits; see *Harris County Deed Records,* vol. R: 543–44; and a map, W. E. Wood, ed., *City of Houston, Harris Co., Texas: published from actual surveys* (1869); also Gammel, comp., *Laws of Texas,* vol. 10: 1407 (quotation 3).

38. "The City Council in Session," *HDP,* August 20, 1901, 6, col. 2 (quotation 1); "Petition from RW Gamble to Erect Oil Tank," envelope dated August 19, 1901, Houston Misc. Papers (quotations 2–8).

39. "Petition from RW Gamble to Erect Oil Tank," Houston Misc. Papers (quotations). To assess the validity of R. W. Gamble's viewpoint, the council again called upon the Fifth Ward aldermen, as well as the committees on fire and on streets and bridges, and invested them "with power to act." See CCM, book 2L, August 19, 1901, 223.

40. "The City Council in Session," *HDP,* August 20, 1901, 6, cols. 1–2; and CCM, book 2L, August 19, 1901, 224. For Cochran's business interests, see *Morrison and Fourmy's General Directory of the City of Houston, 1902–1903,* 108.

41. Genn Saji, "Safety Goals in 'Risk-Informed, Performance-Based' Regulation," *RE&SS* 80 (May 2003): 163 (quotation 1), 165; "Burning of Oil as Fuel," *HDP*, August 21, 1901, 6, col. 1 (quotation 2).

42. "Burning of Oil as Fuel," *HDP*, August 21, 1901, 6, col. 1.

43. "Oil and Fire Insurance," *HDP*, February 3, 1901, 28, cols. 5–6. For forms such as Form No. 2 (1902), "Buyers' Transit Cotton Form," see South-Eastern Tariff Association, *Book of Policy Forms, January 1, 1903* (1903), 10. For the Chicago rules, see "Insurance Circles: Rules Governing the Storage of Crude Petroleum," *NODS*, August 7, 1901, 3, col. 1.

44. On experience and the recursive use of rational and reliable knowledge in engineering practice, see Edward W. Constant II, "Reliable Knowledge and Unreliable Stuff: On the Practical Role of Rational Beliefs," *Technology and Culture* 40, no. 2 (April 1999): 337.

45. CCM, book 2L, September 9, 1901, 241; and "A Deadlock in the City Council," *HDP*, September 10, 1901, 7, cols. 1–2. Maffit, Burnett, and Semmes streets and New Orleans Avenue enclosed the proposed TNORR tank site, while Texas and Prairie avenues and Hutchins and St. Emanuel streets marked off the proposed GCSFR tank.

46. CCM, book 2L, September 16, 1901, 259, 262, 265–66 (quotations); and "Council Again Deadlocked," *HDP*, September 17, 1901, 6, cols. 2–3.

47. Anthony Giddens, *The Consequences of Modernity* (1990), 27 (quotation).

48. CCM, book 2L, September 30, 1901, 294 (quotation); and "Meeting of the City Council," *HDP*, October 1, 1901, 6, col. 4.

49. The title of the detailed thirteen-section rendition is different from the ordinance that the city attorney put forward on September 16. Further, the text of this thirteen-section version is unsigned and has no endorsements recording an adoption date or dates of relevant committee reports. Since a fuel ordinance did not surface in council deliberations until October 7, this thirteen-section text may be a version whose handwritten changes and additions reflected the ordinance's evolution up to September 30, when Alderman Robinson offered his alternative. Typed on three unnumbered leaves with handwritten interlineations, the thirteen-section version of the fuel-oil ordinance, titled "An Ordinance to Regulate the Use and Storage of Crude Petroleum and Fuel Oil by Consumers in the City of Houston," ends with the typed notation "Adopted October 1901" and is extant in Houston, Office of the City Secretary, folder "Crude Oil Ordinance," hereafter cited as "Crude Oil Ordinance folder." I thank Mrs. Anna Russell, city secretary and custodian of records for the City of Houston, for granting me access to this material.

50. "An Ordinance to Regulate the Use and Storage of Crude Petroleum and Fuel Oil by Consumers in the City of Houston"; Ernest H. Peabody, "Oil Fuel," *Transactions of the International Engineering Congress, 1915* (San Francisco, 1916), vol. 10: 504. The figures speculating about the size of small supply tanks are somewhat arbitrary. No specific figures appeared in any ordinances for supply tanks.

51. W. J. Hardee to Charles E. Rice, president, Morris Building and Land Improvement Co., May 5, 1902, New Orleans Public Library, Louisiana Division, City Archives, KJ 510, City Engineer, Outgoing Correspondence, vol. 7: 289 (hereafter NOPL/LD/CA, CE, OC, plus volume and page number). Hardee told Rice that Hardee's department did not object to "extending the vent pipe from the oil tank" (permit to build granted April 17, 1902) "over to the wall of the

Morris Bldg." Hardee observed that running the vent pipe up to the top of the building would "insure immunity against offensive odors to the occupants of intervening buildings."

52. "An Ordinance to Regulate the Use and Storage of Crude Petroleum and Fuel Oil by Consumers in the City of Houston," Crude Oil Ordinance folder. Note that section 1 of this version required tanks to have a "vapor proof screw top manhole cover." Chicago engineer Clayton Billow included this same feature in his description of the "component parts of a fuel oil plant." See "The Use of Oil For Fuel," *HDP,* March 31, 1901, 28, col. 4.

53. "An Ordinance to Regulate the Use and Storage of Crude Petroleum and Fuel Oil by Consumers in the City of Houston," Crude Oil Ordinance folder. The NBFU "Rules and Requirements" (1902) allowed owners to put their oil tanks that supplied boilers inside a vault but cautioned that this increased the "danger of a mixture of gas with air in a confined space," increasing the risk of an explosion from lightning or other ignition sources. See Booth, *Liquid Fuel,* appendix, "Class B 'Individual Oil-Burning Equipments: For other than Household Purposes,'" 14, and "Location," paragraph a., p. 355.

54. "An Ordinance to Regulate the Use and Storage of Crude Petroleum and Fuel Oil by Consumers in the City of Houston," Crude Oil Ordinance folder; oil pumped from a tank passed through a relief valve so that excessive oil sent to burners could be returned directly to the supply chamber. Gases resident in the oil would bleed off through the relief valve, insuring "a steady oil supply to the burners" or any intermediate device such as a standpipe. For a useful description of supplying fuel oil to an industrial furnace, see Matthew H. Mawhinney, *Practical Industrial Furnace Design* (1928), 30 (quotation 1).

55. "An Ordinance to Regulate the Use and Storage of Crude Petroleum and Fuel Oil by Consumers in the City of Houston," Crude Oil Ordinance folder. On the importance (and difficulty) of maintaining internal oil and air pressure in atomizing burner systems, see Wilibald Trinks, *Industrial Furnaces,* 2nd ed., 2 vols. (1925), vol. 2: 62–90, 211–12.

56. "AN ORDINANCE TO REGULATE THE USE AND STORAGE OF CRUDE PETROLEUM AND FUEL OIL BY CONSUMERS (AND THE ERECTION & LOCATION OF OIL STORAGE TANKS) IN THE CITY OF HOUSTON," four typed pages, numbered in top left corner beginning on the second page, Crude Oil Ordinance folder; see also CCM, book 2L, October 7, 1901, 306 (quotation); "Important Council Meeting," *HDP,* October 8, 1901, 5, col. 4.

57. Buffalo Bayou, a watercourse that early in the nineteenth century was on the edge of Houston, was seasonally navigable to the upper Galveston Bay, and was subsequently the course of the Houston Ship Channel; see *The Port of Galveston and the State of Texas,* ed. Andrew Morrison (1890), 64. White Oak Bayou was at the foot of Taylor Street.

58. "An Ordinance to Regulate the Use and Storage of Crude Petroleum and Fuel Oil by Consumers in the City of Houston," Crude Oil Ordinance folder. Both versions of the ordinance had the ten-thousand-barrel limit for a bulk tank, amended from two thousand, in section 9. However, officials who worked with the second rendition originally had written two thousand barrels, changed that to five thousand, and then to ten thousand. According to "Big Volume of Business," *HC,* October 21, 1901, 6, col. 1, the version of the ordinance then before the City Council still read "2000" barrels. Andrew Hurley, "Creating Ecological Wastelands: Oil Pollution in New York City, 1870–1900," *Journal of Urban History* 20 (May 1994): 341, noted that petroleum refiners, as an example of enormous centralized and mechanized

production in the late nineteenth century, had to set up shop in "peripheral locations" owing to incompatibility with dense, central-city areas. Further, such activities attracted "ancillary firms" that supplied raw materials or equipment, creating pockets of smoke, terrible odors, and toxic wastes. How these limitations on construction of fuel-oil tanks affected the development of perimeter industrial sites in Houston remains unexplored.

59. "Regular Meeting of the City Council," *HDP,* October 16, 1901, 7, col. 3; "Mayor Woolford and the Council," *HC,* October 16, 1901, 3, col. 1; "Houston Will Use Fuel Oil," *HC,* October 17, 1901, 3, col. 1; "Meeting of City Council," *HDP,* October 22, 1901, 7, col. 3 (quotation 1); "Council To Meet Today," *HC,* October 28, 1901, 6, col. 1; and CCM, book 2L, October 21, 1901, 333; also "A Drilling Company," *HDP,* May 9, 1901, 5, col. 3 (quotation 2); "May Be Another Geyser," *HDP,* May 11, 1901, 3, col. 2; "Big Volume of Business," *HC,* October 21, 1901, 6, col. 1.

60. The second typed version of the fuel-tank ordinance has the notation "Boiler Inspector to report at next meeting," with the date "dec 3 1901" stamped on it. Harrisburg and Reading, Pennsylvania; Wheeling, West Virginia; and Wilmington, Delaware, were also on the list of inquiries. See CCM, book 2L, October 28, 1901, 346; December 3, 1901, 397, and "Meeting of City Council," *HDP,* October 29, 1901, 6, col. 5; December 4, 1901, 6, col. 3. For Miller, who manufactured electrical supplies such as dynamos, motors, and fans, performed electroplating, and cast brass, see *Morrison and Fourmy's General Directory of the City of Houston, 1902–1903,* 310.

61. "Fuel Oil Hazard," *HC,* December 7, 1901, 8, cols. 1–2.

62. Typed letter, Office of City Boiler Inspector, Houston, Texas, December 14, 1901, and endorsed "W. E. Gale/City Boiler Insp" in Crude Oil Ordinance folder (quotations); CCM, book 2L, December 23, 1901, 416; "Meeting of the City Council," *HDP,* December 24, 1901, 7, col. 4; and "Mayor's Veto Sustained," *HC,* December 24, 1901, 6, col. 2. The second typed version of the fuel tank ordinance had the notation "Ordinance Committee," with a stamped date of "dec 23 1901."

63. With regard to fire hazards and to reducing concomitant risks, stated distances in public (regulations) and semipublic (fire-insurance policies) instruments of governance are significant and were subjects of interpretive dispute. See *Michigan Shingle Company* v. *State Investment and Insurance Company of San Francisco,* 53 *N.W.* 945. at 945-51 (1892); and *Mechanics' & Traders' Insurance Co.* v. *Thompson,* 21 *S.W., at 468-69* (1893).

64. CCM, book 2L, January 6, 1902, 428; "An Ordinance to Regulate the Use and Storage of Crude Petroleum and Fuel Oil by Consumers {and the Erection & location of Oil storage tanks} in the City of Houston" (second typed version). As recorded by hand in Office of the City Secretary, Houston, City Ordinance Book, 1901, 631–33, and City Ordinance Book, 1902, 634, the final ordinance has the same title as the second typed version, with the interpolation "{and the Erection & location of Oil storage tanks}" in the final title. See also "Meeting of the City Council," *HDP,* January 7, 1902, 6, cols. 1–2.

65. "Mayor Turned Council Down," *HC,* January 7, 1902, 6, col. 1; "Fuel Oil Tank," *HC,* January 9, 1902, 8, col. 3; "Notes and Personals," *HC,* January 30, 1902, 3, col. 4, reported that bad weather had led to slow progress on the tank.

66. "Will Burn Fuel Oil," *HDP,* August 1, 1901, 7, col. 2; "Popularity of Fuel Oil," *HDP,* August 1, 1901, 7, col. 1; "The S.P. Drilling a Well," *HDP,* September 16, 1901, 5, col. 1; "Southern Pacific Oil Burner," *HDP,* September 22, 1901, 6, col. 3; "Oil Equipment Being Installed," *HC,* December 6, 1901, 6, col. 5; "Division Hospital," *HC,* January 10, 1902, 3, col. 1; "By May 1," *HC,*

February 13, 1902, 8, col. 3; "Oil Field News," *HC,* March 7, 1902, 6, col. 5; "Tank Car Problems," *OIJ* 1 (August 15, 1902): 1.

67. On police powers, road and liability, see Joseph K. Angell and Thomas Durfee, *A Treatise on the Law of Highways* (1857), 15–16; Byron K. Elliott and William F. Elliott, *A Treatise on the Law of Roads and Streets,* 2nd ed. (1900), 1–7, 15; *J. E. Rose* v. *the State,* 19 *Tex. Ct. App.* 470 (1886) at 471; *City of Bloomington* v. *James M. Bay,* 42 *Ill.* 503 (1867) at 506–8; *H. C. Still* v. *City of Houston,* 27 *Tex. Civ. App.* 447 (1901) at 447–49; and CCM, book 2L, June 16, 1902, 43. For Houston's Charter, see "An Act to provide a charter for the City of Houston, Harris County, Texas" (S.B. No. 306), in Gammel, comp., *Laws of Texas,* vol. 10: 1405–6. For City Attorney Joe M. Sam's opinion about the city's unassailable power to "remove any obstructions off of any of the streets or sidewalks of the city of Houston," which he based upon section 29 of the city charter, see "Meeting of City Council," *HDP,* October 29, 1901, 6, col. 2.

68. CCM, book 2L, March 25, 1901, 53 (quotation); and "Teachers to Be Paid," *HDP,* March 26, 1901, 6, col. 1.

69. "Meeting of the City Council," *HDP,* April 2, 1901, 6, col. 2.

70. CCM, book 2L, January 20, 1902, 440; February 3, 1902, 455 (quotation); "Teachers Will Get Salaries," *HDP,* January 21, 1902, 6, col. 1; and "Meeting of the City Council," February 4, 1902, 6, col. 2. On Arthur J. Binz and Jules J. Settegast Jr. operating the Binz Building for Jacob Binz, see Texas State Archives, Records of case no. 34,485 (hereafter TSA, no. 34,485), *National Supply Company* v. *Jacob Binz,* 1906, District Court, Harris County, Texas, 55th JD, Testimony of J. J. Settegast Jr., 23; and *Fourmy's General Directory of the City of Houston, 1902–03,* 64, 393.

71. "An Ordinance/an amendment to section 5, of an ordinance to regulate the use and storage of crude petroleum and fuel oil by consumers and the erection and location of oil storage tanks in the city of houston./be it ordained by the city council of the city of houston," one typed leaf with city-secretary certification pasted at bottom (Crude Oil Ordinance folder). Document endorsements show that the ordinance committee received this February 10 and recommended passage February 17, 1902. See also "An Amendment to Section 5, of an Ordinance to regulate the use and storage of Crude Petroleum and Fuel Oil by Consumers and the erection and location of oil Storage Tanks in the city of Houston," Houston, Office of the City Secretary, City Ordinance Book, 1902, 639. For the final version, Halverton submitted an ordinance entitled "Ordinance amending the Oil Tank Ordinance, adopted Jany. 6, 1902," CCM, book 2L, February 10, 1902, 47; and February 17, 1902, 480–82; "Meeting of the City Council," *HDP,* February 11, 1902, 4, cols. 2–3; and "Meeting of the City Council," *HDP,* February 18, 1902, 6, col. 3.

72. "Fuel Oil in Cities," *HC,* February 20, 1902, 6, col. 1 (quotations). On complaints that making fuel oil an "object of special and stringent legislation" seemed arbitrary, since the city allowed storage of flammable liquids such as "naptha, gasoline, kerosene and whisky" on wharves and in warehouses alongside cotton, see "Fuel Oil Restrictions," *HC,* February 22, 1902, 4, cols. 2–3. On its careless use, see James A. Ruffner, "Two Problems in Fuel Technology," *History of Technology* 3 (1978): 131.

73. The precautionary principle has two variants: relative and absolute. What is discussed here is the *absolute* form, which "urges precaution when the magnitude of the potential adverse event is large or the adverse outcome is severe, even if its probability is small." And one might add, even when its probability is not quantitatively determined. See Paolo F. Ricci et al.,

"Precaution, Uncertainty and Causation in Environmental Decisions," *Environment International* 29 (2003): 1 (quotation 6), 2 (quotations 1–2), 3 (quotations 3, 5), 4–19; and Anthony J. Berry, "Leadership in a New Millennium: The Challenge of the 'Risk Society,'" *Leadership and Organization Development Journal* 21, no. 1 (2000): 5 (quotation 4); and Steve Frosdick, "The Techniques of Risk Analysis Are Insufficient in Themselves," *Disaster Prevention and Management* 6, no. 3 (1997): 165 (quotation 7); see also Hayim Granot, "The Human Factor in Industrial Disaster," *Disaster Prevention and Management* 7, no. 2 (1998): 93.

74. For the letter, see "Where Opposition Comes From: Texas Fire Prevention Association Against Oil Ordinance," *HC,* March 8, 1902, 12, col. 4 (quotations 1, 3–5); CCM, book 2L, February 24, 1902, 492 (quotation 2). On the earlier visit, see "Local Insurance Risks," *HDP,* September 12,1901, 6, col. 3.

75. "Where Opposition Comes From," *HC,* March 8, 1902, 12; for fire accounts, see "Louisville Has a $100,000 Fire," *Lexington* (Ky.) *Leader,* February 18, 1902, 1, col. 2; and "Fire Loss of the Louisville Bolt and Iron Company Will Approximate $100,000," *Lexington Morning Herald,* February 19, 1902, 3, col. 4.

76. "Tear Down Chinese Wall," *HC,* March 4, 1902, 6, col. 3.

77. Willem J. H. Van Groenendaal, "Group Decision Support for Public Policy Planning," *Information and Management* 40 (2003), 371–72.

78. The allegation came to the council's attention on February 3, 1902. See CCM, book 2L, February 3, 1902, 458; February 17, 1902, 478. Also see "To Build City Hall," *HDP,* February 18, 1902, 6, col. 1; and "Meeting of the City Council," *HC,* February 18, 6, col. 3; as well as "Council Adopts Salary List," *HDP,* March 4, 7, col. 1; CCM, book 2L, March 3, 1902, 496.

79. "The Oil Tank Ordinance," *HC,* March 18, 1902, 5, col. 4; "Oil Field News," *HC,* March 18, 1902, 6, col. 5; "Safety of Fuel Oil," *HDP,* April 5, 1902, 5, col. 5.

80. CCM, book 2L, April 3, 1902, 522 (quotations 1–4); also "AN ORDINANCE TO REGULATE THE USE AND STORAGE OF CRUDE PETROLEUM AND FUEL OIL BY CONSUMERS AND THE ERECTION AND LOCATION OF OIL STORAGE TANKS IN THE CITY OF HOUSTON," four unnumbered typed leaves (Crude Oil Ordinance folder), signed by members of the ordinance committee, with the notations "Report Adopted" and "Ordinance Adopted," dated April 8, 1902; see also "Meeting of City Council," *HDP,* April 9, 1902, 8, col. 2 (quotation 5); "What City Council Did," *HC,* April 9, 1902, 5, col. 1 (quotation 6); and CCM, book 2L, April 8, 1902, 535.

81. TSA, no. 34,485; *National Supply Company* v. *Jacob Binz,* 1906, District Court, Harris County, Texas, 55th JD, Transcript of Oral Evidence, "[Contract]," n.d., p. 2, 2nd paragraph (quotation 1); "Proposition for a Fuel Oil Plant," p. 2a (quotation 2); and Specification for Oil Storage Tank/Tank Enclosure," p. 2d. See also Office of the Clerk, 55th JDC, Harris County, Texas, [case no.] 34485, "Conclusions of Law and fact," May 26, 1906; and *Binz* v. *National Supply Co.* et al., 105 *S.W.* 543 (1907).

82. TSA, no. 34,485, "Dispositions of E. E. Billow, taken before Edwin H. Abbott, a Notary Public in and for Cook County, Illinois, on the 23d day of September, A.D. 1905," 3, 7 (quotation 1); and Clerk 55th JDC, no. 34,485, "first amended original answer for answer and cross-bill in lieu of his original answer filed herein September 3rd, 1904," unnumbered leaf 4 (quotations 2–3).

83. Clerk 55th JDC, no. 34,485, "first amended original answer . . . filed herein September 3rd, 1904," unnumbered leaves 4 (quotation 1), 5; and TSA, no. 34,485 "Specification for Oil Storage Tank/Tank Enclosure," p. 2d (quotations 2–3).

84. "Insurance Difficulty," *HC*, March 28, 1902, 6, col. 3, summed up these events and suggested that the only way out of this situation was to set up a "central oil company, with a franchise to pipe the streets."

85. For Billow, see TSA, no. 34,485, "Depositions of E. E. Billow/Direct Interrogatories," 6; for revocation, see "Some Things Council Did," *HC*, March 11, 1902, 5, col. 4; "Meeting of the City Council," *HDP*, March 11, 1902, 7, col. 2. Also CCM, book 2L, March 10, 1902, 505; March 17, 1902, 513–14. The *HC* considered these franchises particularly valuable in light of proposals to build a pipeline from Beaumont to Houston. See "Meeting of the City Council," *HC*, March 17, 5, col. 4.

86. TSA, no. 34,485, "Depositions of E. E. Billow/Direct Interrogatories," 6.

87. Ibid., "Cross Interrogatories of E. E. Billow," 22.

88. "City's Books and Accounts," *HDP*, May 13, 1902, 6, col. 3; and CCM, book 2L, May 12, 1902, 582, and June 9, 1902, 35.

89. "Voted for a New City Hall," *HDP*, July 1, 1902, 6, col. 1; "An Active Council Session," *HDP*, July 8, 1902, 7, col. 2; CCM, book 2M, June 30, 1902, 61, and July 7, 1902, 73. A plea for this rejected container came up first in February 1902. According to "Santa Fe's Oil Tank," *HC*, February 11, 1902, 8, col. 3, GCSFR had petitioned the City Council the day before for permission to start building a thirty-thousand-gallon storage tank for switch engines. See also CCM, book 2L, February 10, 1902, 467. Alderman Halverton introduced an amended fuel-storage ordinance that delayed ordinance consideration for at least a week and in turn put off a decision on the GCSFR request.

90. For the cause of the fire, subsequent damages, and horrific burns suffered by two men, see "Big Oil Tanks in Seething Flames: Thomas Snell, White, and Anthony Cooper, Colored, Burned," *HC*, June 18, 1904, 5, cols. 1–3; "The Injured Doing Well: Thomas Snell, Burned in Saturday's Fire, Somewhat Improved," *HC*, June 20, 1904, 5, col. 5; and "Tells Story of the Fire: Young Snell Says Gasoline Can Did Not Explode," *HC*, June 21, 1904, 5, col. 2. On the instructions to the city engineer, see "Screens Are Sanctioned: Health and Sanitary Matters Considered by Council," *HC*, June 21, 1904, 5, col. 3.

91. For Arto's petition about oil and construction hazards in the opera house, see "After the Opera House: Chief Arto Calls for a Complete Inspection at Once," *HC*, June 20, 1904, 5, col. 4; and "Screens are Sanctioned . . . Fire Precautions," *HC*, June 21, 1904, 5, col. 3. For the political and investigative background to the Houston theater situation, see "New Safe Guards for Theaters: Mayor Holt Has Ordered an Inspection for Next Week," *HC*, January 1, 1904, 10, col. 4; "Will Inspect Theaters: Building Inspector Confers with the Committee Appointed by Mayor," *HC*, January 4, 1904, 6, col. 5; "Inspection Completed: Committee to Look After Safety of Theaters Will Make a Report," *HC*, January 5, 1904, 1, col. 5; "Inspecting Theaters: Building Inspector, Fire Chief and Committee Making the Rounds," *HC*, January 5, 1904, 6, col. 5; "Agreed Not to Make Public: Report from Theater Committee Not Yet Filed With Mayor," *HC*, January 6, 1904, 1, col. 3; "Gallery Was Closed: Sweeney & Coombs' Opera House Will Make Some Changes" and "Inspection of Theaters Today: Council Committee Takes Up Its Work Promptly," *HC*, January 26, 1904, 6, col. 4; "May Close the Theater: City Council Takes a Hand in the Work of Inspection," *HC*, January 26, 1904, 8, cols. 4-5; "Street Paving Is Divided Up: Three Houston Firms Secure Equal Amount of Work," *HC*, January 27, 1904, 7, col. 6; "Theater Improvements: Plans Submitted to Building Inspector Gallagher May Be Changed," *HC*, January 28, 1904,

1, col. 5; and "Gallery to Stay Closed: Council Committee Recommends Changes in Theater Building," *HC*, January 28, 1904, 12, cols. 3–4.

92. "The Injured Doing Well," *HC*, June 20, 1904, 5, col. 5 (quotations 1–2). This article contains the text of Arto's petition against the Waters-Pierce Oil Company.

93. For company plans to move outside the city limits, see "Big Oil Tanks in Seething Flames . . . Were Preparing to Move," *HC*, June 18, 1904, 5, col. 2; for the extra-time appeal, see "Police After Special Fees . . . Petitions," June 28, 1904, 5, col. 3 (quotation).

94. "An Oil Company Would Come In: Anxious to Make Houston State Distributing Center," *HC*, August 11, 1904, 1, col. 5 (quotations 1–3). On the antitrust battles facing Waters-Pierce, see Jonathan W. Singer, *Broken Trusts: The Texas Attorney General Versus the Oil Industry, 1889–1909* (2002).

95. On the positive committee reports, see "Busy Day for City Council . . . Reports," *HC*, August 23, 1904, 5, col. 2 (quotation 1); *HC*, August 23, 1904, 3; "Business Was Not Finished . . . Oil Company Franchises," *HC*, August 30, 1904, 10, cols. 4, 5 (quotations 2–4).

96. See "Call Issue for Meeting," *HC*, August 31, 1904, 5, col. 2; "Demand Money from the City," *HC*, September 1, 1904, 5, col. 2 (quotation 1); and "Bids Changed a Very Little . . . Business Was Blocked," *HC*, September 2, 1904, 5, col. 1.

97. See "Heiser Must Wait Awhile: Park Superintendent Will Not Get Salary Advance Very Soon," *HC*, September 12, 1904, 5, col. 3; and "Where Money Will Be Put. Aldermen Make a Showing for the Board of Public Works . . . Reversed Position," *HC*, September 13, 1904, 5, col. 4.

98. On Gulf Refining's departure, see "Ordinances in the Council: Alderman Had Many Interesting Questions for Study," *HC*, January 30, 1905, 5, col. 4. On the Waters-Pierce petition, see "Oil Company in Fifth Ward: Council's Committee Favorable to Waters-Pierce Company's Petition," *HC*, May 7, 1905, 15, col. 6; and "Sewer Ordered Built: Council Had Little But Routine Matters to Attend To," *HC*, May 9, 1905, 8, col. 7. On the formalization of the hazard-mitigation and fire-prevention measures, which I contend constituted risk management, see "Ordinances Were Passed: City Council Busied With Making of New Laws," *HC*, May 16, 1905, 8, cols. 4–5; and "Waters-Pierce Plant Goes Up: Construction of New Works Will Begin Immediately," *HC*, May 16, 1905, 7, col. 3 (quotation 1).

99. For 1874 charter provisions about restrictions on petroleum storage inside Houston that posed dangers to adjacent property, see note 29, above.

100. "An Ordinance Regulating the Storage and Handling of Gasoline, or other Volatile, Inflammable Liquids within the City of Houston, and Providing Penalties," Crude Oil Ordinance folder.

101. *Charter and Revised Code of Ordinances of the City of Houston: Harris County, Texas, to October 31, 1904* (1904), chap. 38, articles 900–911, 353, 354 (quotation 1 below), 355–56. *The Revised Code of Ordinances of the City of Houston of 1914* (Houston, 1914), chap. 16, article 4, 115, 116 (quotation 2 below), 117–18. The revised sections were identical to articles 900–911 from 1904, except that the change approved in 1913 to amend article 904, which mandated that "in no case will oil tanks be allowed under sidewalks or streets," allowed supply tanks "outside of the buildings or under the sidewalks." See also *The Revised Code of Ordinances of the City of Houston of 1922* (1922), chap. 36, article 4, 800–804. Section 1848 authorized supply tanks up to 250 barrels in size to be "buried under sidewalks" (801).

102. *Revised Code of Ordinances of the City of Houston of 1914*, 116 (quotation in text and first quotation below); "An ordinance amending Article 904 of Chapter 38 of the Revised Code of Ordinances of the City of Houston of 1904, entitled 'Regulations for handling Oil,'" Crude Oil Ordinance folder.

103. For the relationship between environmental hazards and human-capital formation, see Mark D. Agee and Thomas D. Crocker, "Economies, Human Capital, and Natural Assets," *Environmental and Resource Economics* 11 (April 1998): 261, 267.

CHAPTER FOUR

1. Jon C. Teaford, *The Unheralded Triumph: City Government in America, 1870–1900* (1984), 6.

2. For New Orleans and insurance, see "New Orleans as an Insurance Center," *NODS*, June 23, 1891, 3, col. 1. On contracts in Louisiana in 1898, see *The Insurance Year Book, 1900–1901: [Fire and Marine]* . . . (1900), 152; on Low and *L&L&G*, see *New Orleans and the New South*, comp. and ed. Andrew Morrison (1888), 112; and *Hayden's Annual Cyclopedia of Insurance in the United States, 1900–1901* (1901), "Liverpool and London and Globe Insurance Company."

3. For specific police powers, see *The Laws and General Ordinances of the City of New Orleans*, ed. Henry J. Leovy and C. H. Luzenberg (1870), sec. 19, pp. 102–5; *Jewell's Digest Ordinances New Orleans*, 1882, title 6, chap. 1, sec. 2: "Gunpowder," 332–34; *Jewell's Digest of the City Ordinances . . . of the City of New Orleans*, 1887, "Charter of the City of New Orleans," sec. 7–8, pp. 9–11; *Acts Passed by The General Assembly of the State of Louisiana* (1896), no. 45, "An Act To incorporate the city of New Orleans," sec. 14–15, pp. 53–55. On regulating animal slaughtering through municipal police powers, see Norton T. Horr and Alton A. Bemis, *A Treatise on the Power to Enact, Passage, Validity and Enforcement of Municipal Police Ordinances* (1895), secs. 130–31, p. 219. On slaughterhouse litigation, see Ronald M. Labbé and Jonathan Lurie, *The Slaughterhouse Cases: Regulation, Reconstruction, and the Fourteenth Amendment* (2003), chaps. 5 and 6, and pp. 259–60 for cases. On building construction and establishing fire limits inside the city, see *Jewell's Digest Ordinances New Orleans*, 1887, title 14, Ordinance 733 C.S., articles 1–15, pp. 494–497. For city ordinances related to fire-limit restrictions, see W. J. Hardee, City Engineer, *Official Hand-Book: Laws Regulating Construction of Buildings in the City of New Orleans, 1905* (1905). For "managerial society," see Jackson Lears, *Rebirth of a Nation: The Making of Modern America, 1877–1920* (2010), 297.

4. For New Orleans progressivism, see "The Most Meritorious Service to the City," *NODP*, March 13, 1902, 4, col. 1 (quotation 1). See also "The Progressive Union," *NOTD*, November 27, 1901, 4, cols. 2–3; as well as "The Progressive Union," *NOTD*, December 1, 1901, 4, col. 2. On the growing sentiments connecting progressive views, technology, systems management, and expertise, see Thomas P. Hughes, *American Genesis* (2004), 246–47, 298–99; on the economic value of rationalizing and spreading risk through fire insurance see Monkkonen, *America Becomes Urban*, 107, 266n19.

5. On New Orleans's role in fire insurance see W. R. Lyman, "Address of W. R. Lyman on New Orleans as an Insurance Centre," NBFU, *Proceedings of the Twenty-Third Annual Meeting of the National Board of Fire Underwriters*, 53–62. For reinsurance as a technique whereby one insurance company cedes a portion of its contractually acquired liabilities to other companies,

see George A. Clement, *The Law of Fire Insurance*, vol. 2: *Fire Insurance as a Void Contract* (1905), title 6: "Reinsurance," 545–61; and Harvey W. Rubin, *Dictionary of Insurance Terms*, 6th ed. (2013), 41, 437.

6. For fire hazards and city wharves, see T. A. Laughtry who, speaking in behalf of the Marine Underwriters of New Orleans, reminded Mayor Capdevielle in late 1900 that the "financial interests of the Marine Underwriters here are very closely gauged by the value of the cotton export which as you know is very large and they are extremely anxious that all possible precautions [spark arresters] be taken to prevent cotton fires." See NOPL/LD/CA, Record of Mayor Paul Capdevielle, Correspondence, box L-S, folder "M" Miscellaneous, T. A. Laughtry to Paul Capdevielle, December 3, 1900.

7. *The Long Nineteenth Century*, vol. 2 of *The Cambridge Economic History of the United States*, ed. Stanley L. Engerman and Robert E. Gallman (2000), 343, 507, 568–69, 621–23; John G. Clark, "New Orleans and the River: A Study in Attitudes and Responses," *LH* 8, no. 2 (Spring 1967): 132–33; vol. 10, no. 1 (Winter 1969): 52. See also William Cronon, *Nature's Metropolis: Chicago and the Great West* (1991), 296–99, 307; Donald L. Miller, *City of the Century: The Epic of Chicago and the Making of America* (1996), 75, 91–94, 96, 118; and David R. Meyer, "The Industrial Retardation of Southern Cities, 1860–1880," *Explorations in Economic History* 25, no. 4 (October 1988): 366, 373–74, 383. Note, however, that Justin A. Nystrom, *New Orleans After the Civil War: Race, Politics, and a New Birth of Freedom* (2010), 7, describes New Orleans at succession as the center post of the Mississippi Valley and the "unrivaled economic capital of the antebellum South," whose port was third in the nation in "imported goods," making the city a "great international" city.

8. Ari Kelman, *A River and Its City: The Nature of Landscape in New Orleans* (2003): 162; and Walter M. Lowry, "The Engineers and the Mississippi," *LH* 5, no. 3 (Summer 1964): 233–55. On Gilded Age political issues in New Orleans, see Joy Jackson, "Bosses and Businessmen in Gilded Age New Orleans Politics," *LH* 5, no. 4 (Autumn 1964): 387–400. On the observation that an "obsession with race, sectional politics, and the 'Lost Cause' mythology" retarded the city's Reconstruction–Gilded Age economic prospects, see Michael A. Ross, "Resisting the New South: Commercial Crisis and Decline in New Orleans, 1865–85," *American Nineteenth Century History* 4, no. 1 (Spring 2003): 59–71, 72 (quotation above), 73–76.

9. For M. J. Sanders commenting in a speech before the Progressive Union on the need to "take advantage of every opportunity to attract capital" to New Orleans, see "Progressive Union," *NOTD*, July 7, 1901, 3, col. 7; for Houston capitalist John H. Kirby telling a New Orleans audience that "we want eastern and northern capital to help us develop the natural resources of this portion of the South," see "The Progressive Union Has Another Great Night," *NODP*, March 4, 1902, 3, col. 5; see also David R. Meyer, "Midwestern Industrialization and the American Manufacturing Belt in the Nineteenth Century," *Journal of Economic History* 49, no. 4 (December 1989): 937 (quotation 1); and James C. Cobb, "The Sunbelt South: Industrialization in Regional, National, and International Perspective" (1993), 26 (quotation 2).

10. James C. Cobb, "Beyond Planters and Industrialists: A New Perspective on the New South," *JSH* 54, no. 1 (February 1988): 54–55; David L. Carlton and Peter A. Coclanis, *The South, the Nation, and the World: Perspectives on Southern Economic Development* (2003), 42 (quotation 1). For oil prospects, see "A Great Texas Industry," *DMN*, February 25, 1901, 5, cols. 5–6; "Oil in Nacogdoches County: A History of the Search for Petroleum for Fifty Years," *HDP*, March 2,

1901, 2, col. 5; "The New Treasure Field of Texas . . . IX. Vastness of the Field," *HC,* February 10, 1902, 4, cols. 3–5; "Oil Field News," *HC,* March 20, 1902, 6, col. 5.

11. On the power company using fuel oil, see comments by Fred DeLand at a meeting of the NELA in August 1888 in *Proceedings of the National Electric Light Association at Its Eighth Convention, Semi-Annual Meeting . . . 1888* (1888), vol. 4: 77–78, 329–31.

12. "New Business," *NODP,* June 26, 1901, 7, col. 2; and "Streets and Landings," *NODP,* June 29, 1901, 6, col. 1 (quotation).

13. For LSCOC origins and plans, see "Orleans Men Buy Oil Land," *NODS,* April 13, 1901, 1, col. 7; "Star and Crescent," *NODS,* May 4, 1901, 1, cols. 3–4; "Cannot Get Oil," *NODS,* May 5, 1901, 8, col. 3; "L.S. and C. Directors," *NODS,* May 7, 1901, 1, col. 1; "Preferred Stock All Sold," *NODS,* May 10, 1901, 6, col. 6; "Capital is Coming," *HDP,* May 5, 1901, 4, col. 2; "Local Capitalists Secure Oil Lands," *NOTD,* April 14, 1901, 15, col. 1; "At the City Hall," *NOTD,* August 2, 1901, 3, col. 2; "Pipe Line Outlets," *NOTD,* August 2, 1901, 8, col. 2 (Sabine pipeline); "The Corsicana Wells," *NODP,* August 4, 1901, 8, col. 1; "Exporting Beaumont Oil," *NODP,* August 31, 1901, 6, col. 3; and "Lone Star's Oil Tank Steamer," *NODP,* February 5, 1902, 3, col. 4. See also the biographical article for O. B. Greeves, who purchased the original Spindletop property for LSCOC and was first vice-president in 1902, in *Successful American* 5, no. 4 (April 1902): 229–30.

14. On city police powers, see note 3, above. On the city engineer being responsible for superintending "the construction of all public works," see *Acts Passed by The General Assembly of the State of Louisiana* (1896), "An Act To incorporate the city of New Orleans," no. 45, secs. 32–34, 59–60. For Ordinance no. 6533 C.S., July 5, 1892, which required the city engineer to issue a permit "for the erection of any new building, or the alteration of any existing building," see "An Ordinance for the Proper Construction of Buildings," New Orleans *Daily City Item,* July 18, 1892, 2, cols. 5–7.

15. "Oil Tanks," *NODS,* March 12, 1901, 2, col. 5; "Oil Field Notes," *Beaumont Daily Enterprise,* March 14, 1901, 1, col. 3.

16. "Mr. Guffey's Visit," *NODS,* May 14, 1901, 1, col. 3; "Guffey's Big Oil Company," *NODP,* May 17, 1901, 1, cols. 1 (quotation), 2–3; "The Biggest Oil Company," *NODP,* May 17, 1901, 2, col. 4; "The Guffey Company," *NODP,* May 18, 1901, 7, cols. 1–2; "Guffey's Broad View of the Oil Future," *NODP,* May 19, 1901, 3, col. 2; "Guffey Petroleum Company," *NODP,* May 21, 1901, 2, col. 2; and "Plans of the Guffey Company," *NODP,* June 8, 1901, 6, cols. 1–2.

17. Peirce F. Lewis, *New Orleans: The Making of an Urban Landscape* (2003), 46; for canal-traffic statistics, see "Minor Waterways Playing Great Part," *NODP,* September 1, 1901, part 4, p. 2, col. 6.

18. "Oil For Fuel," *NODP,* May 18, 1901, 7, col. 2.

19. See "An Even Break in the City Council," *NODP,* May 15, 1901, 10, col. 1; "Etiquette of the Council Chamber," *NODP,* June 5, 1901, 6, col. 2; "Official Proceedings of the City Council/ Report of Committee No 6. Committee on Streets and Landings," *NODS,* June 6, 1901, 8, col. 3; and "Ordinances on First Reading and Referred under the Rules," col. 4. For the petition, see NOPL/LD/CA, AB300, 1901/May, New Orleans City Council, Official Proceedings (fragments), May 1901, typescript letter on letterhead of Southern Re-Distilling & Rectifying Co. Ltd., dated May 14, 1901, signed by W. A. Gordon, VP/GM.

20. "On Its Dignity Is the City Council," *NODS,* May 29, 1901, 5, col. 4; "Official Proceedings," *NODS,* June 6, 1901, 8, col. 4; "First Oil Burners To Be Put in by the Jackson Brewery,"

NODP, May 21, 1901, 2, cols. 1–2; "Saxon Resignation Fails to Arrive," *NODP*, May 29, 1901, 3, col. 2; "Orleans Franchise Matter Discussed," *NODP*, June 1, 1901, 3, col. 1.

21. "The Council Stops at Telephone Poles," *NODP*, June 26, 1901, 7, col. 1; "Official Proceedings of the City Council," *NODS*, June 28, 1901, 10, col. 1. Subsequently, the Committee on Police and Public Buildings took up the Crescent City Ice petition but decided to forgo action until the new, special oil-committee reported. See "Cheaper Asphalt Again in Sight," July 2, 1901, 9, col. 1; also "Official Proceedings of the City Council," *NODS*, June 28, 1901, 10, cols. 2, 4–6.

22. ABC started with the Dock Board, only to be referred to the City Council. See "The Dock Board Has Oil to Deal With," *NODP*, June 26, 1901, 10, col. 1; "The Dock Board's Weekly Meeting," *NODP*, July 3, part 2, p. 1, cols. 3–4; "Streets and Landings," *NODP*, July 6, 1901, 6, col. 3; "New Business," *NODP*, July 6, 1901, 3, cols. 2–3; and "The I.C. Ordinance Reported Favorably," *NODP*, August 3, 1901, 7, col. 1.

23. "The Mayor Writes Mr. Fish a Letter," *NODP*, July 17, 1901, 3, col. 2; and "Streets and Landings Committee," *NODP*, July 27, 1901, 3, col. 4.

24. For their locations, see *Sanborn Fire Map*, New Orleans, 1908–9, III, 1908, sheet 241; 1908–9, II, 1908, sheet 113.

25. On the delays, see "City Council," *NODS*, June 26, 1901, 2, cols. 4–5; and "The Fight Against Avenue Barrooms," *NODP*, June 29, 1901, 6, col. 1. On indefinite postponement, see "Ordinances on Final Passage," *NODS*, September 8, 1901, 8, cols. 2, for disposition of two ABC calendar ordinances (1067 and 1115), and recommendation that the C. H. Lawrence parties apply to the city engineer under the conditions of Ordinance 805 N.C.S.

26. "The Council Stops at Telephone Poles," *NODP*, June 26, 1901, 7, cols. 1 (quotations 2–3), 2 (quotation 1).

27. "Oil Tank Committee," *NODP*, July 24, 1901, part 2, p. 1, col. 3. Council President William Mehle appointed Moss (chairman) along with notary J. A. Lautenschlager and James McRacken to the committee. See "From the President of the Council," *NODS*, July 5, 1901, 7, col. 1.

28. "John G. Pepper, 70, Is Taken by Death," *NOTP*, April 16, 1934, 2, col. 4. Regarding L&L&G's dominant role in fire insurance, the company between 1893 to 1903 wrote $202,812,945 in fire risks just in Texas. See State of Texas, *Twenty-Ninth Annual Report of the Commissioner of Agriculture . . .* 1903 (1904), 73. For Pepper's comments, see "Devising Rules for Oil Storage," *NODP*, July 28, 1901, part 3, p. 4, col. 3 (quotations). On risk or the possibility of catastrophic outcomes as intrinsic to insurance and loss avoidance, see Paul L. Joskow, "Cartels, Competition and Regulation in the Property-Liability Insurance Industry," *Bell Journal of Economics and Management Science* 4, no. 2 (Autumn 1973): 377.

29. "Canal Street, Oil Storage, Dock Grant," *NODP*, July 31, 1901, 3, cols. 1, 2 (quotation 1); "Wharf Privileges," *NOTD*, July 31, 1901, 3, cols. 6, 7 (quotation 2).

30. "Canal Street, Oil Storage, Dock Grant," *NODP*, July 31, 1901, 3, col. 2. This was a similar but slightly smaller area than those prescribed in city ordinances dating from 1881 (Melpomene, Clairborne, Galvez streets and Napoleon Avenue); for which see *Jewell's Digest Ordinances New Orleans*, 1882, title 16, chap. 6, sec. 1, art. 129, pp. 330–31; *Jewell's Digest Ordinances New Orleans*, 1887, title 15, chap. 1, art. 2; and *Flynn's Digest of the City Ordinances . . . City of New Orleans*, comp. J. Q. Flynn (1896), article 472, p. 238.

31. "Oil and Fire Insurance," *HDP,* February 3, 1901, 28, cols. 5 (quotation 1), 6. For the text of the forms, see Booth, *Liquid Fuel,* 345–46; on L&L&G's rules, see "News and Notables at Local Hotels," *NODP,* May 16, 1901, 9, col. 1.

32. "Canal Street, Oil Storage, Dock Grant," *NODP,* July 31, 1901, 3, col. 2.

33. "Devising Rules for Oil Storage," *NODP,* July 28, 1901, part 3, p. 4, col. 3.

34. "Beaumont Oil Wants a Pipe Line Here," *NODP,* August 1, 1901, 2, col. 6. For McDowell, see D. E. O'Brien, "Jessie C. McDowell," *Successful American* 5, no. 4 (April 1902): 206; an update, "Jesse C. M'Dowell," *Successful American* 10, no. 1 (July 1904): 38; and *Cities Service Co.* v. *McDowell et al.,* 120 Me. 384-387 (1922).

35. For the franchise grant and Policy Jury resolution, see Louisiana, Jefferson Parish, Parish Clerk Office, Mortgage Book, vol. 38, 725, "Act Passed before me John Reuben Langridge, a Notary Public in and for the Parish of Jefferson, on the 29th day of June 1901," filed July 5, 1901.

36. "An Oil Pipe Line: The Guffey Company Granted the Right by Jefferson," *NODP,* June 23, 1901, 8, col. 3; "The Gretna Oil Tank," *NODP,* July 30, 1901, 9, col. 2; "Actual Oil Delivery," *NODS,* August 27, 8, col. 3; "Big Guffey Oil Tank, *NOTD,* October 9, 1901, 10, cols. 3–5. For Cullinan, see University of Houston Libraries, Special Collections, Joseph S. Cullinan Papers, box 1, folder 2 (June 20–October 10, 1902), letterbook carbon, JSC to Arnold Schlaet, New York City, July 14, 1902, 116–19.

37. "Oil Company Asks for a Pipe Line," *NODP,* August 2, 1901, 3, col. 6.

38. For franchise-privilege disputes over telephone service, see "Question of Franchise," *NODP,* January 22, 1902, 3, col. 5; and "Congressman Broussard's Telephone Plan," *NODP,* October 1, 1901, 7, cols. 3–4; "New Telephone Measure," *NODP,* October 1, 1901, 9, cols. 5–6; "Obstructions to the New Telephone," *NODP,* October 2, 1901, 6, col. 1; and "Congressman Broussard Is Puzzled," *NODP,* October 2, 1901, 8, cols. 4–5

39. On troubles and disputes over municipal service franchises, see Joy Jackson, *New Orleans in the Gilded Age: Politics and Urban Progress 1880–1896* (1969), 150–55.

40. See "Franchises," *Cyclopedia of Law and Procedure,* ed. William Mack (1905), vol. 19, §§ 1459–63; and *Black's Law Dictionary,* 6th ed. (1990), 658 (franchise), 1197 (privilege).

41. The Badger ordinance can be followed in "Official Proceedings of the City Council," *NODS,* February 1, 1901, 8, col. 4; "Gossip of the Hall," *NODS,* February 5, 1901, 10, col. 2; "The Fuel Gas Matter," *NODS,* March 2, 1901, 4, col. 2; "Kill the Gas Ordinance," *NODS,* March 21, 1901, 4, col. 3; "Honesty in Public Service," *NODS,* April 19, 1901, 4, col. 2; "The Badger Gas Ordinance," *NODS,* May 6, 1901, 4, col. 2; "The Badger Gas Ordinance," *NODS,* May 14, 1901, 4, col. 2; "City Council," *NODS,* June 26, 1901, 2, cols. 4–5; "Gas Ordinance Withdrawn," *NODS,* June 26, 1901, 4, col. 3; "Committee Differs About That Gas," *NODP,* May 14, 1901, 6, cols. 1–2.

42. "At the City Hall," *NOTD,* August 2, 1901, 3, col. 2 (quotations 1–2); and *1888 La. Acts 135,* §§ 3–4, p. 193 (quotation 3). On Act 135 of 1888 that defined the duties of the New Orleans City Council and city officers in regard to street-railroad franchises, see *Hart* v. *Buckner et al.,* 5 C.C.A. 9 (1892).

43. "The I.C. Ordinance Reported Favorably," *NODP,* August 3, 1901, 7, cols. 1 (quotation 1), 2 (quotation 2).

44. "Joint Meeting," *NODP,* August 3, 1901, 7, col. 2; and "Oil Ordinance Favored," *NOTD,* August 3, 1901, 3, col. 7.

45. "That Pipe Grant," *NODS,* August 5, 1901, 4, col. 5, has an early version of the LSCOC ordinance, in which section 2 allowed a pipeline lateral to Peters Avenue to feed oil to "steamers, barges, etc.," on the river.

46. On the belief that LSCOC pushed for prohibiting landing oil barges on the riverfront to gain an oil-supply monopoly through its pipeline, see NOPL/LD/CA, AB 300, 1901/August, City Council, Official Proceedings [fragment], Higgins Oil & Fuel Company to Hon. Mayor & City Council, New Orleans, letter dated August 5, 1901, signed by C. L. Wallis, general manager.

47. "Councilmen Fell Over Each Other to Vote," *NODP,* August 7, 1901, 3, cols. 1–2.

48. For a slightly modified version of the forms "likely to be used" in New Orleans, see "Fuel Oil Forms and Conditions," *NODS,* August 7, 1901, 3, col. 1; and "Insurance Circles: Rules Governing the Storage of Crude Petroleum," *NODS,* August 7, 1901, 3, col. 1; and "Insurance Circles: Rules for the Safe Burning of Crude Petroleum," *NODS,* August 7, 1901, 7, cols. 1–2. For the text of Ordinance 805 N.C.S., see "Ordinances," *NODS*, August 14, 1901, 8, col. 4. For quotations, see NOPL/LD/CA, New Orleans City Ordinances, microfilm, reel AB311, 1900–1905, NCS 1-3205; Ordinance No. 805, Calendar No. 1134, recorded August 8, 1901 (hereafter NOPL/LD/CA/NOCO). After Egania Street became the upriver limit for handling oil across wharves, LSCOC applied to the Levee Board for permission to construct a wharf "in front of their property" at that point. The board deferred action "until the Dock Commission" acted. See Orleans Parish, Orleans Levee District, Board of Commissioners, "Minutes," book 4, September 11, 1901, 78, communication dated August 29, 1901 (hereafter OLD, BOC, "Minutes").

49. On the first veto and the concessions, see "Mayor May Veto Oil Pipe Grant," *NODP,* August 8, 1901, 3, col. 1; "The Mayor Gives Council a Lesson," *NODP,* August 9, 1901, 3, col. 6; "The Pipe Line," *NODP,* August 11, 1901, part 3, p. 5, col. 2; "City Council Has Short Session," *NODP,* August 28, 1901, 3, col. 1 (quotation 1); and "Oil Pipes Before the Committee," *NODP,* August 31, 1901, 3, col. 2 (quotation 2). For amendments to LSCOC's original ordinance submitted by O'Connor, see "That Pipe Grant," *NODS,* August 5, 1901; and "Unfinished Business," *NODS,* August 8, 1901, 8, col. 5. On the charter requirement to sell franchises, see *Acts Passed by The General Assembly of the State of Louisiana* (1896), "An Act To incorporate the city of New Orleans," no. 45, article 10, secs. 86–87, pp. 71–72.

50. "Public Order After the Public Works," *NODP,* September 14, 1901, 3, col. 6; and "The Bienville Street Suit Compromise," *NODP,* October 5, 1901, 3, col. 5 (quotation); and "To Pay for Peace," *NOTD,* October 5, 1901, 3, col. 5.

51. "A Franchise Worth Millions Veiled Under the Form of a Worthless Privilege," *NODP,* August 8, 1901, 4, col. 1; and "The Oil Pipe Franchise," *NODP,* August 31, 1901, 4, col. 2.

52. The terms "inadequate competition" and "collusion," as well as the notion of exclusivity as a legal privilege, aptly summed up the anti-monopoly critique of special interests seeking franchises from New Orleans officials. See Donna A. Barnes, *The Louisiana Populist Movement, 1881–1900* (2011), 272n1.

53. "Star and Crescent Gets a Pipe Line," *NODP,* September 6, 1901, 3, cols. 4–5; "The Mayor Will Veto Oil Pipe Measure," *NODP,* September 7, 1901, 3, col. 2; and Dennis East, "Health and Wealth: Goals of the New Orleans Public Health Movement, 1879–84," *Louisiana History* 9, no. 3 (Summer 1968): 250–55. See also University of New Orleans, Earl K. Long Library, Louisiana and Special Collections Department, Supreme Court of Louisiana Legal Archives (Mss 106),

New Orleans Auxiliary Sanitary Association v. *The City of New Orleans,* Case no. 13,523; and ordinances no. 6442 (April 20, 1880); no. 6981 (April 12, 1881); and no. 5400 C.S. (June 30, 1891).

54. "The Oil Pipe Job," *NODP,* September 6, 1901, 4, col. 3 (quotation); "The Mayor Will Veto Oil Pipe Measure, *NODP,* September 6, 1901, 3, cols. 1–2; and "Star and Crescent Gets a Pipe Line," *NODP,* September 6, 1901, 3, cols. 4–5.

55. "Other Oil Men," *NODP,* August 10, 1901, part 3, p. 3, col. 7 (quotations); "More Pipe Lines," *NODP,* August 15, 1901, 8, col. 7; and "Council Business. At the Meeting Last Night Some Important Matters were Disposed of," *NODI*, August 28, 1901, 2, col. 3. Word circulated in New Orleans that a third company, possibly Higgins Oil & Fuel Company, would ask for an oil-pipeline privilege. For a list of Franco-American investors see "Texas Oil Notes," *The Age of Steel* (St. Louis) 90, no. 8 (August 24, 1901): 22.

56. "Oil Pipes Before the Committee," *NODP,* August 31, 1901, 3, col. 2; Maginnis-Brunswig Oil," *NODP,* September 4, 1901, 3, col. 2; and "Council Meeting. Many Important Matters Are Considered," *NODI*, September 4, 1901, 2, cols. 2-3. On LSCOC's activities see "Oil for Fuel," *NODP,* August 8, 1901, 7, col. 2, and "Pipe Line Completed to Sabine Pass," *NODP,* October 25, 1901, 2, col. 1. See also "Council Matters," *NODI*, August 31, 1901, 2, col. 5, for a report on S&LC chairman O'Connor and councilman McMahon exchanging "harsh words" over the Maginnis-Brunswig pipeline request. McMahon declared that he would vote against LSCOC's pipeline request that was before the committee if M-B was treated unfairly.

57. "Oil Night Before the City Council," *NODP,* September 4, 1901, 3, col. 1 (quotation 1); "Maginnis-Brunswig Oil," *NODP,* September 4, 1901, 3, col. 3 (quotation 2).

58. "That Pipe Line Franchise," *NOTD,* September 3, 1901, 4, col. 1 (first quotation); and "That Oil Pipe Business," *NOTD,* September 4, 1901, 4, col. 1 (second quotation). The *Daily Item* charged that the M-B request for pipeline privileges was a corrupt project, in light of the fact that city attorney Samuel L. Gilmore was A. A. Maginnis's brother-in-law. Gilmore initially approved of the Badger gas ordinance but reversed his opinion and subsequently claimed that LSCOC, like Badger, should pay the city for pipeline privileges. But Gilmore supported the M-B effort, as did the mayor and a state senator. See "Great Expectations: Backers of the Maginnis-Brunswig Ordinance Have No Oil," *NODI*, September 3, 1901, 1, cols. 3-6; "Great Expectations," cont'd, 8, cols. 5-7; "Mr. Gilmore is Silent," *NODI*, September 3, 1901, 8, cols. 6-7; "The Oil Controversy," *NODI*, September 5, 1901, 4, cols. 1-2; and "The Mayor's Duty," *NODI*, September 5, 1901, 4, cols. 1-2.

59. "The Lone Star Ordinance," *NODP,* September 6, 1901, 3, col. 1.

60. "The Council Voices the City's Sorrow," *NODP,* September 11, 1901, 10 cols. 4 (veto message), 5 (text of new LSCOC ordinance); "Streets and Landings," *NODP,* September 14, 1901, 3, col. 4 (minor amendments); see also "The Oil Pipe Franchise," *NODP,* September 16, 1901, 4, col. 3 (quotation); "Mr. Schabel Objects," *NODP,* September 21, 1901, 6, col. 1.

61. "Moulin Case Goes to City Attorney," *NODP,* September 25, 1901, 3, cols. 1–2; "Obstructions to the New Telephone," *NODP,* October 2, 1901, 6, cols. 1–2; "Council Session Turns Out Tame," *NODP,* October 16, 1901, 3, col. 2; "McMahon Matter All Settled Now," *NODP,* October 23, 1901, 7, col. 2; "Council Meeting/Oil Pipe Line Ordinance Recommitted," *NOTD,* October 2, 1901, 3, cols. 1–2; "Quiet in the Council," *NOTD,* October 16, 1901, 1; and "Report on M'Mahon," *NOTD,* October 23, 1901, 3, col. 2.

62. See "Unfinished Business," *NODP,* October 23, 1901, 7, col. 2, for council approval of LSCOC's tank.

63. LSCOC's large-tank request passed the City Council on October 22. See "Unfinished Business," *NODP,* October 23, 1901, 7, col. 2; and "Council Session Turns Out Tame," *NODP,* October 23, 1901, 3, cols. 1 (quotation 3), 2 (quotations 1–2).

64. "McMahon Matter All Settled Now," *NODP,* October 23, 1901, 7, col. 2; see also "At the Hall," *NODS,* July 24, 1901, 4, col. 4, wherein the City Council charged its special oil committee to fashion a "uniform ordinance that will apply in granting permits for tanks for the storage of fuel oil."

65. "About the Oil Measures," *NODP,* October 20, 1901, 11, col. 6; "Same Old Ordinances," *NOTD,* October 20, 1901, 13, col. 4; "Fire and Lighting," *NOTD,* October 22, 1901, 11, col. 3 (quotation).

66. For the text of Ordinance 918, N.C.S., see NOPL/LD/CA/NOCO, microfilm reel AB311, 1900–1905, City Council, City Ordinances, N.C.S. 1, May 16, 1900–N.C.S. 3205, August 2, 1905.

67. See, for example, "Supreme Court Forfeits the Water Works Charter," *NODP,* November 7, 1901, 3, cols. 1–4; p. 6, cols. 4–7. "The People Win Back Their Water Franchise," *NODP,* November 7, 1901, 4, cols. 1–3; "As to the Proposed Water and Sewerage System," *NODP,* January 26, 1902, 4, cols. 1–2; "The Delay in the Water and Sewerage Work," *NODP,* January 28, 1902, 4, col. 1. "The Problem of City Sewerage," *NODP,* January 30, 1902, 3, cols. 6–7; p. 6, cols. 5–6. "The Obstacles in the Way of Water and Sewerage," *NODP,* January 30, 1902, 4, cols. 1–2. Also "Waterworks Case," *NOTD,* November 7, 1901, 3, cols. 1–2; p. 13, cols. 1–7. "The Waterworks Case," *NOTD,* November 7, 1901, 4, col. 1. "Rehearing Is Asked," November 20, 1901, 3, col. 1; p. 4, cols. 1–2. And "Waterworks Case," *NOTD,* November 22, 1901, 3, cols. 1–2; p. 11, col. 3.

68. On anti-monopoly, republican sentiments about corruption and privilege in the post–Civil War period, see Gretchen Ritter, *Goldbugs and Greenbacks: The Antimonopoly Tradition and the Politics of Finance in America, 1865–1896* (1997), 5–6. For monopolies and trusts in a national context, see Charles R. Geisst, *Monopolies in America: Empire Builders & Their Enemies from Jay Gould to Bill Gates* (2000), 18–62.

69. See "McMahon Matter All Settled Now . . . Unfinished Business," *NODP,* October 23, 1901, 7, col. 2; "Mouline's Case . . . Lone Star Company Gets Tank Privilege," *NODI,* October 23, 1901, 9, col. 5; "City Hall Matters . . . Oil Ordinance Favorably Reported," *NODI,* October 22, 1901, 7, col. 3; and "New Sidewalks . . . Troubles Over a Prospective Ordinance Explained," *NODI,* October 20, 1901, 7, col. 4. For the text of Ordinance 959, N.C.S., see NOPL/LD/CA/NOCO, microfilm reel AB311, 1900–1905, City Council, City Ordinances, N.C.S. 1, May 16, 1900–N.C.S. 3205, August 2, 1905.

70. For the text of Ordinance 1011 N.C.S., see NOPL/LD/CA/NOCO, microfilm reel AB311, 1900–1905, City Council, City Ordinances, N.C.S. 1, May 16, 1900–N.C.S. 3205, August 2, 1905.

71. For the text of Ordinances 1168, 1204 and 1392, N.C.S., see NOPL/LD/CA/NOCO, microfilm reel AB311, 1900–1905, City Council, City Ordinances, N.C.S. 1, May 16, 1900–N.C.S. 3205, August 2, 1905.

72. "Oil Tank Privileges," *NODP,* July 4, 1901, 9, col. 7.

73. "Petitions," *NODP,* October 30, 1901, 10, col. 5; "Fuel Oil Matters," *NODP,* November 3, 1901, part 2, p. 2, col. 4 (quotations); and "At the City Hall," *NOTD,* November 3, 1901, 3, cols. 4–5. Councilman Moss introduced the petitioners' amending ordinance at the October

29 council meeting. See "Petitions and Complaints," *NODS*, November 2, 1901, 8, col. 2; and "Ordinances on First Reading and Referred Under the Rules," col. 3.

74. "Storage Tanks for the Fuel Oil," *NODP*, November 5, 1901, 3, col. 3 (quotation 1); "The Oil Ordinance Must Not be Tampered With," November 5, 1901, 4, cols. 1 (quotations 2–3), 2 (quotation 4); see also "Meeting Not Held/Fire and Lighting," *NOTD*, November 5, 1901, 13, col. 4; and "Fuel Oil Tank," *NOTD*, November 5, 1901, 4, col. 2 (editorial).

75. "Council's Indignation Impressively Made Manifest," *NODP*, November 13, 1901, 7, col. 3; "Insurance Agents Protest," *NODP*, November 14, 1901, 6, col. 3; and "Pipe Lines for Oil," *NOTD*, November 14, 1901, 3, col. 1.

76. "Prof. Caracristi May Locate Here," *NODP*, November 6, 1901, 13, col. 2 (quotations); and "News and Notables at Local Hotels," November 11, 1901, 10, col. 1. See C. F. Z. Caracristi, *Coal and Petroleum in Columbia* (1893); *Coal, Oil, and Asphaltum of the Republic of Columbia* (1895); "Coal Mines in Columbia," in U.S. Department of State, *Consular Reports: Commerce, Manufactures, Etc.*, Report no. 203 (August 1897): 580–81; as well as a critical appraisal of Caracristi's Mexican activities in Charles H. Harris and Louis R. Sadler, *The Texas Rangers and the Mexican Revolution: The Bloodiest Decade, 1910–1920* (2004), 178–80.

77. Underwriters' Restrictions for Fuel Oil," *NODP*, November 25, 1901, 8, cols. 3, 4 (quotation 1). For the fair's power plant and Billow's oil storage and pumping system, see Fred De Land, "The Power Plant," *Electrical Engineering* 1 (January–June 1893): 9–17, 64–66; Clayton O. Billow, "Fuel Oil: Comparative Results in Steam Generation," *Electrical Engineering* 5, no. 4 (April 1895): 202–9; and Booth, *Liquid Fuel*, 182–93. See also George D. Tolbert, "Residual Risk Reduction: Systematically Deciding What Is 'Safe,'" *Professional Safety* 50, no. 11 (November 2005): 25 (quotations 2–3).

78. "No Oil Tanks Under City Sidewalks," *NODP*, November 26, 1901, 7, col. 1; and "Storage Oil Tanks," *NOTD*, November 26, 1901, 3, col. 1. On Moss, see note 73 above.

79. "Insurance Circles: Rules Governing the Storage of Crude Petroleum," *NODS*, August 7, 1901, 3, col. 1; and "Insurance Circles: Rules for the Safe Burning of Crude Petroleum," *NODS*, August 7, 1901, 7, cols. 1–2; see also Filson Historical Society, Louisville, Ky., Louisville Board of Underwriters, Minute Book, vol. 5 (October 29, 1887–March 21, 1891, April 10, 1889): 223. The board secretary pointed out on January 2, 1895, that the Chicago Board, "from whom the present rules governing the instalment of Fuel Oil Systems" had been adopted, had changed the tank distance from seventy-five feet to thirty-five feet, and advised the Louisville Board to amend its rule "to correspond." See Filson Historical Society, Louisville, Ky., Louisville Board of Underwriters, Minute Book, vol. 6 (March 25–May 13, 1896): 357.

80. *The Revised Ordinances of the City of Boston* (1882), chap.6, sec. 6, p. 171, banned storing flammable oil in a "coal-hole or vault under a sidewalk." State storage requirements dictated that any building for crude oil storage "or any of its products" could not be an occupied dwelling and, if less than fifty feet from other buildings, had to be separated from them by a ten-foot-high "stone or brick wall." See State of Massachusetts, *The Public Statutes of the Commonwealth of Massachusetts, Enacted November 19, 1881; to Take Effect February 1, 1882* (1882), title 14, chap. 102, sec. 72, p. 549.

81. On the compromise, see "No Oil Tanks Under City Sidewalks," *NODP*, November 26, 1901, 7, col. 1; "Fuel Oil Measure," *NODI*, December 1, 1901, 9, col. 3; and "The Fuel Oil Measure," *NODP*, December 1, 1901, part 2, p. 7, col. 2.

82. For Wolfe (1837–1904), see "Colonel Udolpho Wolfe Dies of Heart Stroke," *NODP,* March 10, 1904, 3, col. 5–7. For P. S. Morris (1858–1918), veteran oil man and vice-president of SOC of Louisiana, see "The Louisiana Sugar Planters' Association," *Louisiana Planter and Sugar Manufacturer* 26, no. 15 (April 13, 1901): 224, col. 3; and "Orleans Mourns Captain Morris, Called by Death," *NOTP,* September 25, 1918, 16, cols. 1–3. For A. R. Blakely (1841–1908), Irish immigrant and Confederate veteran, who managed the new St. Charles Hotel, see "Colonel Andrew R. Blakely, Prince of Bonifaces," *NODP,* April 11, 1908, 11, cols. 3–7. For P. S. Murphy (1850–1929), manufacturer of sugar-mill and oil-field machinery, see "John H. Murphy Funeral Service To Be Held Today," *NOTP,* August 16, 1929, 1, col. 2; and p. 3, col. 5. On James W. Porch (1858–1921) and Earnest M. Loeb, see "James W. Porch Champion of City's Welfare, Is Dead," *NOTP,* July 21, 1921, 1, col. 2; p. 2, col. 5; and "Earnest M. Loeb, Leader in City, Dies Suddenly," *NOTP,* August 25, 1927, 1, col. 5; and p. 2, col. 5. See also "The Question of Fuel To Be Thoroughly Investigated by the Board of Trade," *NODP,* November 26, 1901, 8, col. 4; and "Business Information," *NODI,* November 26, 1901, 7, col. 2. For the committee, see "Cheap Fuel Investigation," *NODI,* January 2, 1901, 9, col. 2; and New Orleans Board of Trade, Minutes, Board of Directors, vol. 3 (March 6, 1895, to September 26, 1903, December 4, 1901): 271 (hereafter NOBOT, BODM).

83. "Insurance Men Organize," *NOTD,* December 20, 1901, 3, col. 3 (quotations 1–2); and "New Orleans Insurance Men Organize," *The Vindicator* 29, no. 1 (January 1, 1902): 14. For obituaries, see "Clarence F. Low Claimed by Death," *NOTP,* July 13, 1933, 2, col. 2; "A. P. Noll Is Dead at Bay St. Louis," *NOTP,* October 24, 1917, 7, col. 3; "Lucien Monrose's Funeral Conducted," *NOTP,* February 3, 1937; and "Died brand," *NOTP,* February 27, 1926. For Porch, see "The Use of Oil as Factory Fuel," *NODP,* December 20, 1901, 3, col. 1 (quotations 3–6); and "Committees Make Report," *NODI,* December 20, 1901, 3, col. 3.

84. "The Use of Oil as Factory Fuel," *NODP,* December 20, 1901, 3, col. 1 (quotations); "To Make Use Safe," *NOTD,* December 20, 1901, 8, col. 7.

85. "Fuel Oil Experts Discuss Methods," *NODP,* December 22, 1901, part 2, p. 11, col. 1 (quotation 1); and "Discuss Fuel Oil," *NOTD,* December 22, 1901, 11, col. 1 (quotation 2).

86. "Fuel Oil Experts Discuss Methods," *NODP,* December 22, 1901, part 2, p. 11, cols. 1–2; and "All Favor Oil Fuel," *NODI,* December 22, 1901, 12, cols. 3, 4.

87. "Valuable Testimony for Oil as Fuel," *NODP,* December 28, 1901, 9, col. 7 (quotations). On Odenheimer's life, see "Head of Cotton Mills Firm Dies," *NOTP,* March 21, 1945, 1, col. 4. On Guffey Petroleum's plans to build a pipeline to the mills in August 1901, see "Beaumont Oil Wants a Pipe Line Here," *NODP,* August 1, 1901, 2, col. 6. Odenheimer obtained permission from the City Council in October to install an oil tank on-site. See "Council Meeting," *NODP,* October 2, 1901, 3, col. 2; and "Unfinished Business," *NODP,* October 16, 1901, 3, col. 2.

88. "Valuable Testimony for Oil as Fuel, *NODP,* December 28, 1901, 9, col. 7 (quotation). For Frank P. Barbour, a New Orleans engineering contractor, who contrasted the alleged "[c]onservatism . . . disinclination . . . backwardness . . . dilatory actions" of the insurance men with the oil men who were "progressive, up- to-date . . . eager . . . to adopt any modern device," see "Crude Oil as Fuel," *NOTD,* November 17, 1901, 7, col. 7.

89. "Transportation of Fuel Oil," *NODP,* December 29, 1901, 3, col. 5 (quotation); "Fuel Oil Measures," *NODP,* December 31, 1901, 3, col. 3; and NOBOT, BODM, vol. 3 (January 2, 1902):

272. For overlapping superintendence, see the response of a joint meeting of the city, the dock, and the Orleans District Levee boards to the controversial Illinois Central Railroad effort to build city-front wharves ("Dock Board and Levee Board Join," *NODP,* November 16, 1901, 3, cols. 4–5; p. 6, col. 5; and "Important Decisions Rendered in the Courts," *NODP,* December 4, 1901, 3, cols. 4–5).

90. "Fuel Oil Still a Troubling Topic," *NODP,* January 8, 1902, 8, col. 1 (quotation 1), col. 2 (quotations 2–3).

91. "Emergency Work on the Courthouse," *NODP,* November 27, 1901, 3, col. 6; "Oil Pipes, Paving, Fenders and the Spanish Fort Road," *NODP,* November 30, 1901, 3, col. 1; "Sugar Sheds Ask For the Full Amount," *NODP,* January 4, 1902, 7, col. 3; and "Barrooms Can Run on Royal Street," *NODP,* January 11, 1902, 10, col. 4. For LSCOC's successful petition to the Levee Board to run an oil pipe "over and not through the Levee" at the "head of Egania St.," see OLD, BOC, "Minutes," January 15, 1902, 4: 102.

92. "The Fuel Question," *NODP,* January 12, 1902, 3, col. 3. On rational organizations, see W. Richard Scott, *Organizations: Rational, Natural, and Open Systems,* 5th ed. (2002), 5, 33–35, 53. On anti-monopoly movements and ideas up to 1900, see Ritter, *Goldbugs and Greenbacks.*

93. M. Granger Morgan and Max Henrion, *Uncertainty: A Guide to Dealing with Uncertainty in Quantitative Risk and Policy Analysis* (1990), 26 (quotations 1–2), 27.

94. "Insurance Men Organize," *NODI,* January 10, 1902, 3, col. 2.

95. "Insurance Opposition," *NODP,* January 14, 1902, 3, col. 3. The BOT asked its Beaumont counterpart for information about the ignition point of petroleum and received a reply that Beaumont oil was "next to impossible to light . . . with a match flame or torch." See "Particulars about Beaumont Oil," *NODP,* January 23, 1902, 2, col. 3. Opponents to mooring oil vessels could cite the destruction by fire in October 1876 of the German bark *Europa,* an oil- soaked ship "engaged in the petroleum trade" in dry-dock, that damaged the vessel and killed several workers, which in a crowded harbor could have started a conflagration. See "Life and Limb," *DIO,* October 8, 1876, 5, col. 2 (quotations).

96. In mining and petroleum ventures some university and government geologists and chemists acquired marketable knowledge and professional credibility. See Michael Wallis, *Oil Man: The Story of Frank Phillips and the Birth of Phillips Petroleum* (1988), 108; David H. Breen, *Alberta's Petroleum Industry and* the *Conservation Board* (1993), lviii–lix, 33–34, 81; and Lucier, *Scientists and Swindlers,* 4–8, 137–40, 252–55.

97. "Fuel Oil and Insurance," *NODP,* January 14, 1902, 6, cols. 2, 3 (quotations 1–3); and Scott, *Organizations,* 33–35 (quotation 4).

98. "City Council Has Little Business," *NODP,* January 15, 1902, 3, cols. 1 (quotation 1), 2; and "Insurance and Fuel Oil," [editorial] *NODP,* January 15, 1902, 4, col. 4 (quotations 2–4).

99. "The Open Court/Insurance and Fuel Oil," *NODP,* January 15, 1902, 9, col. 3.

100. "Council to Hear Fuel Oil Question," *NODP,* January 23, 1902, 6, col. 1 (quotations); and "The Oil Investigation," *NODP,* January 25, 1902, 3, col. 4.

101. On uncertainty, see Uwe Lubken and Christof Mauch, "Uncertain Environments: Natural Hazards, Risk and Insurance in Historical Perspective," *Environment and History* 17, no. 1 (February 2011): 2–5; T. Aven and V. Kristensen, "Perspectives on Risk: Review and Discussion of the Basis for Establishing a Unified and Holistic Approach," *RE&SS* 90 (2005): 1-5.

102. "A Practical Talk on Oil Dangers," *NODP,* February 3, 1902, 6, cols. 3–4, 5 (quotation). For similar comments from Willis J. Roussel, see "Wants Oil Men to Take Their Own Risks," *NODP,* February 5, 1902, 13, col. 5.

103. "New York Marine Underwriters Come to Investigate," *NODP,* February 4, 1902, 9, col. 2. Chubb & Son, U.S. agent for companies such as London Assurance Marine and the General Marine Insurance Company of Dresden, could through local agents and representatives pressure the city to mitigate oil risks .

104. "The Council Begins Fuel Oil Hearing," *NODP,* February 14, 1902, 15, col. 4.

105. Ibid., cols. 4–5.

106. Ibid., cols. 5 (quotations), 6; for the S&LC meeting, see "No Oil Tanks under City Sidewalks," *NODP,* November 26, 1901, 7, col. 1.

107. On objective risk, see Terje Aven and Ortwin Renn, "On risk defined as an event where the outcome is uncertain," *Journal of Risk Research* 12, no. 1 (January 2009): 5–6.

108. "Exchanges Heard About Fuel Oil," *NODP,* November 26, 1901, 10, cols. 1 (quotations 1–3), 2 (quotations 4–5).

109. For quotations, see Claude Henry and Marc Henry, "Formalization and applications of the Precautionary Principle," CECO-965, hal-00243001 (2002, 2008), 2.

110. "Fuel Oil's Advance in the New Orleans Harbor," *NODP,* February 21, 1902, 3, cols. 6, 7 (quotations). On Coyle's company, Louisiana Petroleum, see "Latest News from Louisiana," *NODP,* March 3, 1902, 12, col. 3; for its charter see *NODS,* January 27, 1902, 9, col. 2. The charter was filed with the state March 6; see State of Louisiana, Secretary of State, Commercial Division, coraweb.sos.la.gov/commercialsearch.

111. On the report submission, distribution, and resolutions, see NOBOT, BODM, vol.. 3, February 22, 1902, 283; vol. 3, February 26, 1902, 284 (quotations); and "Fuel Oil/The Board of Trade Committee Completes Its Report," *NODP,* February 23, 1902, 7, col. 7.

112. NOBOT, BODM, vol. 3, February 26, 1902, 285.

113. "The Report on Fuel Oil," *NODP,* February 27, 1902, 10, col. 2.

114. Ibid., cols. 2, 3 (quotations); on McDowell's claim that oil had not contributed to any great fires, see "Oil-Works at Cleveland Burned—Loss $15,000, No Insurance," *CTrib,* January 3, 1884, 2, col. 6; "A Sea of Fire," *CTrib,* January 16, 1884, 2, cols. 5–6; "Conflagrations," *CTrib,* February 5, 1884, 5, col. 6; "New York: Destructive Fire at Long Island City, Caused by an Exploding Oil-Tank," *CTrib,* February 9, 1884, 3, col. 5; "Conflagrations," *CTrib,* March 3, 1884, 2, col. 5.

115. "The Report on Fuel Oil," *NODP,* February 27, 1902, 10, col. 4; on risk and technical systems, see Kathleen J. Tierney, "Toward a Critical Sociology of Risk," *Sociological Forum,* 14, no. 2 (June 1999): 226–27 (quotation 2).

116. "The Report on Fuel Oil," *NODP,* February 27, 1902, 10, cols. 5 (quotation 1), 6 (quotations 2-4); see also NOBOT, BODM, special meeting, vol. 3 (February 26, 1902): 285.

117. "The Report on Fuel Oil," *NODP,* February 27, 1902, 10, col. 6 (quotation 2); p. 11, col. 7. "The Fuel Oil Hearing Ends," *NODP,* February 27, 1902, 10, col. 7 (quotations 3–5). On the withdrawal, see "Fire Protection Notes/New Orleans," *Insurance Press* 23, no. 568 (July 18, 1906): 9 (quotation 1).

118. "The Fuel Oil Hearing Ends," *NODP,* February 27, 1902, 10, col. 7 (quotations 1–2); p. 11, col. 1 (quotation 3). Conflagrations raised the issue of "reinsurability of catastrophic risk."

American insurance firms sometimes sold part of their risk portfolio to foreign insurance companies to spread the "costs of catastrophes across a wider financing base." See Philip D. Bougen, "Catastrophe Risk," *Economy and Society* 32, no. 2 (May 2003): 253–54.

119. "The Fuel Oil Hearing Ends," *NODP*, February 27, 1902, 11, col. 2. For a concise statement of the insurance position see "Fuel Oil and Insurance," *NODP*, February 26, 1902, 9, cols. 5–6.

120. "Another Hearing on the City Fuel Oil Question," *NODP*, March 4, 1902, 3, col. 1.

121. Ibid., cols. 1 (quotations 1, 3–4), 2 (quotation 5). For risk as "disadvantages that may or may not materialize" while benefits were advantages "of which one is sure," see Sven Ove Hansson, "Weighing Risks and Benefits," *Topoi* 23, no. 2 (September 2004): 145 (quotation 2), 146.

122. "Another Hearing on the City Fuel Oil Question," *NODP*, March 4, 1902, 3, col. 3.

123. Ibid. See also Moore, *Fire Insurance and How to Build* (1903), pointing to a large fire in New York City in April 1889 that destroyed a lard-oil refinery releasing "burning oil . . . floating on harbor water . . . ignited piers from beneath." Burning petroleum oil constitutes an identical hazard (70).

124. "Another Hearing on the City Fuel Oil Question," *NODP*, March 4, 1902, 3, col. 3.

125. Ibid., 9, col. 5.

126. See NOPL/LD/CA/NOCO, microfilm reel AB311, 1900–1905, City Council, City Ordinances N.C.S. 1, May 16, 1900–N.C.S. 3205, August 2, 1905, for text of Ordinance 1159.

127. "Fuel Oil Committee," *NODP*, May 16, 1902, 7, col. 2 (quotation 1); May 20, p. 7, col. 7; May 21, p. 6, col. 4; and "Powers of Chairmen to be Determined," *NODP*, May 28, 1902, 10, col. 1.

128. On the meaning of risk analysis and assessment, see Lawrence B. Gratt, "Risk Analysis or Risk Assessment: A Proposal for Consistent Definitions," in *Uncertainty in Risk Assessment, Risk Management, and Decision Making*, ed. Vincent T. Covello et al. (1984), 242, 244.

129. See *Assessing Microbial Safety of Drinking Water: Improving Approaches and Methods* (2003), P.R. Hunter et al., chap. 3: Assessment of Risk, sec. 2, "What is risk?," 80, in which risk is defined as "'the possibility of loss, harm or injury'."

130. John Tabak, *Probability and Statistics: The Science of Uncertainty* (2005), 97.

131. Vincent T. Covello and Joshua Menkes, "Issues in Risk Analysis," in *Risk in the Technological Society*, ed. Christoph Hohenemser and Jeanne X. Kasperson (1982), 289–93; Robert W. Kates and Jeanne X. Kasperson, "Comparative Risk Analysis of Technological Hazards," *Proceedings of the National Academy of Sciences of the United States of America* 80, no. 22, part 2: "Physical Sciences" (November 15, 1983): 7029; and Bernard L. Cohen, "Probabilistic Risk Analysis for a High-Level Radioactive Waste Repository," *Risk Analysis* 23, no. 5 (October 2003): 909–10.

132. S. Yearley, "Sociology and Politics of Risk," in *International Encyclopedia of the Social & Behavioral Sciences*, ed. Neil J. Smelser and Paul B. Baltes (2001), vol. 20: 13361.

133. "Oil Tank Committee," *NODP*, July 24, 1901, 1, col. 3; "Canal Street, Oil Storage Dock, Dock Grant," *NODP*, July 31, 1901, 3, col. 1. On Hardee's revisions and his duties in Ordinance 918, see "Joint Meeting," *NODP*, August 3, 1901, 7, col. 2; "About the Oil Measures," *NODP*, October 20, 1901, 11, col. 6; "Fire and Lighting and Streets and Landings," *NODP*, October 22, 1901, 6, col. 2.

134. "The Council Begins Fuel Oil Hearing," *NODP*, February 14, 1902, 15, col. 5.

135. "Many Plants Apply to Use Fuel Oil," *NODP*, April 23, 1902, 11, col. 6; "A List of Permits for Fuel Oil Tanks," *NODP*, May 20, 1902, 7, col. 5; and NOPL/LD/CA/NOCO, microfilm

reel AB311, 1900–1905, City Council, City Ordinances, N.C.S. 1, May 16, 1900–N.C.S. 3205, August 2, 1905. For a list of permits issued to city businesses through May 6, see NOPL/LD/CA, CE, OC, vol. 18: 295, "List of Permits issue to construct Oil Tanks."

136. NOPL/LD/CA, CE, OC, vol. 18: January 28, 1902–August 11, 1902: W. J. Hardee, City Engineer, to J. T. Whittspoon, Manager, American Sugar Refining Co., City, March 25, 1902, p. 173; W. J. Hardee, City Engineer, to Central Ice and Cold Storage Co., Ltd., 1125 Julia St., City, March 25, p. 175; W. J. Hardee, City Engineer, to Messrs. Bacher Brothers, City, April 3, 1902, p. 202; W. J. Hardee, City Engineer, to Leigh Carroll, President, Algiers Waterworks & Electric Co., 222 Elmire St., City, April 7, 1902, p. 214.

137. For plans, see NOPL/LD/CA, CE, OC, W. J. Hardee, City Engineer, to New Orleans Brewing Ass'n, City, March 29, 1902, vol. 18:182; W. J. Hardee, City Engineer, to New Orleans Cypress Co., 507 Hennen Building, City, April 1, 1902, vol. 18: 195; W. J. Hardee, City Engineer, to Louisiana Distillery Co., Ltd., April 2, 1902, vol. 18: 200; for details, see W. J. Hardee to W. A. Gordon, Vice-President and General Manager, Louisiana Distillery Co, Ltd., 313–315 Tchoupitoulas Street, City, n.d., vol. 18: 195; on Brown's application, see W. J. Hardee, City Engineer, to L. W. Brown, Civil Engineer, City, April 8, 1902, vol. 18: 219.

138. For quotation 1, see NOPL/LD/CA, CE, OC, Hardee to Charles E. Rich, April 7, 1902, vol. 18: 26; for quotations 2–3, see W. J. Hardee, City Engineer, to Frances R. Cogswell Sugar Refining Co., April 8, 1902, vol. 18: 220; for quotation 4, see W. J. Hardee, City Engineer, to Charles [M.] Schenck, Secretary-Treasurer, Standard Brewing, City, May 10, 1902, vol. 18: 315. Hardee told Schenck that if he buried tanks less than ten feet from a building, they had to be "encased in masonry."

139. NOPL/LD/CA, CE, OC, W. J. Hardee, City Engineer, to W. T. Maginnis, Manager, Maginnis Cotton Mills, April 8, 1902; W. J. Hardee, City Engineer, to Lane Cotton Mills, City, May 5, 1902, vol. 18: 301. See also J. O. Hansam, Acting City Engineer, to John Otnett, President, Home Ice & Distilled Water Co., June 11, 1902, warning Otnett that a recently received application "in regard to [an] additional oil tank" had to be accompanied by another blueprint "showing the location of the new tank" (vol. 18: 376); and W. J. Hardee, City Engineer, to Henry Kahn, President, National Rice Mill Co., City, July 25, 1902, vol. 18: 450.

140. On Jones, see NOPL/LD/CA, CE, OC, W. J. Hardee, City Engineer, to the Riverside Ice Company, City, March 25, 1902, vol. 18: 172; W. J. Hardee to Messrs Tate, Jones & Co., City, April 17, 1902, vol. 18: 256; W. J. Hardee, City Engineer, to F. G. Ernst, Manager, Southern Rice Milling Co, May 22, 1902, vol. 18: 341; W. J. Hardee to J. W. Joachim, President, Washington Ice Co., May 27, 1902; and W. J. Hardee, City Engineer, to Ben A. Michel, Secretary, New Orleans Butchers' Co-operative Abattoir Co., Ltd., June 30, 1902, vol. 18: 353; and J. O. Hansam, Acting City Engineer, to John G. Woods, General Manager, NO Railways Co., July 7, 1902, vol. 18: 417; on the amended permits, see W. J. Hardee to Louisiana Distillery Co., City, April 17, 1902, vol. 18: 257; on a successful permit, see W. J. Hardee, City Engineer, to S. Odenheimer, Lane Cotton Mills, City, May 7, 1902, vol. 18: 310.

141. NOPL/LD/CA, CE, OC, W. J. Hardee to Charles E. Rice, Morris Building & Land Improvement Co., April 10, 1902, vol. 18: 230.

142. Ibid., May 5, 1902, vol. 18: 289.

143. NOPL/LD/CA, CE, OC, W. J. Hardee to Mssrs. Tate, Jones & Co., of La., Ltd., 541 Carondelet St., City, March 28, 1902, vol. 18: 181.

144. NOPL/LD/CA, CE, OC, W. J. Hardee, City Engineer, to J. K. Newman, President, NO & Carrolton R.R. Co. & L. & P. Co., City, April 9, 1902, vol. 18: 227; Von Phul presented the original plan, dated March 26, 1902, to the city engineer, vol. 18: 229; W. J. Hardee, City Engineer, to A. G. Phelps, President, St. Charles St. RR Co., City, April 9, 1902, vol. 18: 227; W. J. Hardee, City Engineer, to Hon. A. T. Moss, Councilman, City, May 5, 1902, vol. 18: 295.

145. NOPL/LD/CA, CE, OC, W. J. Hardee to Alonzo Church, April 7, 1902, vol. 18: 212; for a list of fifteen companies including square locations, see "Permits Issued by the City Engineer For the Construction of Oil Tanks," vol. 18: 243.

146. See W. J. Hardee to N. O. Butcher's Co-operative and Abbatoir Co. Ltd., April 22, 1902, vol. 18: 270; W. J. Hardee, City Engineer, to American Brewing Co., City, May 1, 1902, vol. 18: 282; W. J. Hardee, City Engineer, to Messrs Wogan Brothers, Port & Decatur Streets, City, May 5, 1902, vol. 18: 298; W. J. Hardee to St. Louis Rice Mill Co, City, May 5, 1902, vol. 18: 302; W. J. Hardee, City Engineer, to Security Brewing Co., 341–345 N. Diamond St, City, May 19, 1902, vol. 18: 327; and W. J. Hardee, City Engineer, to Robert E. Fallis, 905 Howard Ave, City, May 28, 1902, vol. 18: 357.

147. For Ordinance 3273 C.S., October 22, 1888, "An Ordinance Establishing Fire Limits for the City of New Orleans" see NOPL/LD/CA/NOCO, microfilm reel AB311, CS 1584–CS 16231, December 28, 1885–May 2, 1900.

148. Jane Holder and Donald McGillivray, "Introduction," *Locality and Identity: Environmental Issues in Law and Society,* ed. Jane Holder and Donald McGillivray (1999), 1; Leon A. G. Oerlemans et al., "Innovation and Space: Theoretical Perspectives," Eindhoven Centre for Innovation Studies, *Working Paper* 99.3 (February 1999): 9–10; and Jens Rasmussen, "Risk Management in a Dynamic Society: A Modelling Problem," *Safety Science* 27, no. 2–3 (November–December 1997): 2 (quotation 1).

149. Tom L. McKnight and Darrel Hess, *Physical Geography: A Landscape Appreciation* (2002), 1, col. 2 (quotation 1); Alan Strahler and Arthur Strahler, *Physical Geography: Science and Systems of the Human Environment* (2002), 5–6.

150. Arthur Getis et al., *Introduction to Geography* (1996), 6 (quotation 1); Strahler and Strahler, *Physical Geography,* 6. On the U.S. grid system, see William D. Pattison, *Beginnings of the American Rectangular Land Survey System, 1784–1800* (1964); Lisa C. Tolbert, *Connecting Townscapes: Space and Society in Antebellum Tennessee* (1999), 23–34; and Dan Stanislawski, "The Origin and Spread of the Grid-Pattern Town," *Geographical Review* 36, no. 1 (January 1946): 106, 108.

151. See "An Act To Incorporate the City of New-Orleans," *Louisiana Gazette,* February 22, 1805, 1–2; and "An Act to Incorporate the City of New-Orleans" (approved February 17, 1805), in *A Digest of the Ordinances, Resolutions, By-Laws and Regulations of the Corporation of New-Orleans* (1836), 282–94. For the use of "squares as a commonly understood term for bound urban spaces, see "Insurance Men Voice Complaints," *NODP,* August 22, 1901, 3, col. 3.

152. For Ordinance 2209, see NOPL/LD/CA/NOCO, microfilm reel AB311, CCS 1-11029, 1911–1928, reel 1241; Ordinance 2700, adopted October 18, 1904; no. 2209, adopted March 2, 1915; Ordinance 2488, adopted June 15, 1915; Ordinance 2789, adopted October 5, 1915.

153. See NOPL/LD/CA/NOCO, microfilm reel AB311, CCS 1-11029, 1911–1928, reel 1241; Ordinance 5843, adopted August 17, 1920.

154. See NOPL/LD/CA/NOCO, microfilm reel AB 311, 1928–1950, reels no. 1241–42.

155. Ibid.
156. Ibid.

CHAPTER FIVE

1. See the "Chicago Fire Underwriters' Association: Rules Governing the Use of Crude Petroleum for Fuel, Chicago, January 26th, 1892," in *Ordinances of the Building and Health Departments of the City of Chicago,* comp. D. O. Wilkie (1892), 107–10.

2. See Olivier Zunz, *Why the American Century?* (1998), 4–13; for a contemporary description of the web of fire experts, risk knowledge, and the "problems of the fire hazard," see Edward F. Croker, *Fire Prevention* (1917), 310 (quotation), 311.

3. For "hazard" see *Global Trends in Water-Related Disasters: an Insight for Policy Makers, World Water Assessment Program* (UNESCO 2009), Appendix, Important UN/ISDR Definitions, "Hazard", 22 (quotations 1-2). On hurricanes, see National Weather Service, National Hurricane Center, Miami, Eric S. Blake, Edward N. Rappaport, and Christopher W. Landsea, NOAA Technical Memorandum NWS TPC-5, *The Deadliest, Costliest, and Most Intense United States Tropical Cyclones from 1851 to 2006,* April 2007, table 1, "Saffir/Simpson Hurricane Scale" (2), and appendix A (38) (2007). On lack of warning, see John Edward Weems, *A Weekend in September* (1957, 2000), 12, 15. On risks and natural hazards, see W. Neil Adger, "Vulnerability," *Global Environmental Change* 16, no. 3 (2006): 268–69; on vulnerability and community decisions, see Greg Bankoff, "Constructing Vulnerability: The Historical, Natural and Social Generation of Flooding in Metropolitan Manila," *Disasters* 27, no. 3 (September 2003): 225; on the formal meaning of risk, see V. Kristensen, T. Aven, and D. Ford, "A New Perspective on Renn and Klinke's Approach to Risk Evaluation and Management," *RE&SS* 91, no. 4 (April 2006): 424.

4. Robert B. Cumming, "Is Risk Assessment a Science?" *Risk Analysis* 1, no. 1 (March 1981): 1; for the prohibition against refining or storing flammable liquids or petroleum on wharves, streets, sidewalks or alleys, see City of Galveston, *Charter and Revised Ordinances* (1875), chap. 26, article 1, secs. 1–3, 5, pp. 204–6; rpt. in City of Galveston, *Charter and Revised Ordinances of the City of Galveston* (1883), chap. 2, article 3, secs. 1–5, pp. 215–16; and City of Galveston, *Charter of the City of Galveston with the Amendments Thereto and the Revised Ordinances,* digested and rev. Henry W. Rhodes (1893), title 6, chap. 5, articles 235, 239, pp. 160–62 [hereafter CCG&RO 1893]. For fire limits and available water around stored cotton, see *Charter of the City of Galveston and the Revised Ordinances,* digested and rev. George P. Finlay (1902), chap. 3, articles 238–40, pp. 134–35; and article 248, pp. 137–38.

5. On the risks of introducing a new technology, see Sally H. Clarke, "Unmanageable Risks: MacPherson v. Buick and the Emergence of a Mass Consumer Market," *Law & History Review* 23, no. 1 (Spring 2005): 2–3, 14–15, 18.

6. Cumming, "Is Risk Assessment a Science?" 2; on subjective or qualitative risk assessment, see Lucia Mitello and Fabrizio Rufo, "Individual Risk and Collective Fears: Towards a Governance of Medical Uncertainty," *Topoi* 23, no. 2 (September 2004): 222; and P. K. Marhavilas, D. Koulouriotis, and V. Gemeni, "Risk Analysis and Assessment Methodologies in the Work Sites: On a Review, Classification and Comparative Study of the Scientific Literature of the Period 2000–2009," *Journal of Loss Prevention in the Process Industries* 24, no. 5 (September 2011): 490, table 8.

7. The University of Texas's Perry-Castaneda Map Collection in the university library has ninety-four sheets of the Sanborn Fire Maps for Galveston dated 1899, in addition separate sheets for the title page, index, key and a view of the Cotton Compress and Warehouse District. (www.lib.utexas.edu/maps/sanborn/g.html).

8. See C. Pinney, *Atlas of the City of Boston* (1861), and John E. Whiting, *A Schedule of the Buildings and Their Occupancy on the Principal Streets and Wharves in the City of Boston* (1877), title page (quotation 1); for essays on fire-insurance maps and their use, see Walter W. Ristow, "United States Fire Insurance and Underwriters Maps, 1852–1968," *Quarterly Journal of the Library of Congress* 25, no. 3 (July 1968): 194–218; and Helena Wright, "Insurance Mapping and Industrial Archeology," *IA: The Journal of the Society for Industrial Archeology* 9, no. 1 (1983): 1–18.

9. "And What of Galveston?" *GDN*, January 10, 1901, 6, col. 1; "The Labor Council," *GDN*, January 29, 1901, 12, col. 4; "Report of the Mayor," *GDN*, February 5, 1901, 10, col. 2; "The City Council," *GDN*, February 5, 1901, cols. 3–6; "Bond Situation Summed Up," *GDN*, February 20, 1901, 10, col. 4; "To Be, Or Not to Be," *GDN*, March 13, 1901, 1, cols. 2–3; p. 6, cols. 4–5. "Galveston Charter Bill," *GDN*, April 14, 1901, 13, cols. 5–6; for city-bond-ordinance texts, 1873–97, see *Charter of the City of Galveston and the Revised Ordinances*, appendix, chap. 2, "Galveston Thirty-Year Limited Bonds—1876," 247–72; for city indebtedness, see "The Needs of Galveston," *GDN*, January 15, 1901, 6, cols. 6–7; "Bonds of Galveston," *GDN*, February 3, 1901, 7, col. 1; "County's Finance," *GDN*, February 8, 1901, 10, cols. 1–7; "Refunding Galveston's Bonded Indebtedness," *GDN*, February 22, 1901, 6, col. 6; "Galveston Bonds," *GDN*, February 24, 1901, 8, col. 2; "Obstructionists," *GDN*, March 10, 1901, 1, cols. 6–7; "City Is Enjoined," *GDN*, September 5, 1901, 10, cols. 3–4; "A Budget Adopted," *GDN*, October 17, 1901, 4, col. 3; "As To City Bonds," *GDN*, November 20, 1901, 4, cols. 1–5; untitled paragraph, *GDN*, November 21, 1901, 6, col. 1; "City Has Defaulted," *GDN*, December 7, 1901, 7, col. 2; "Default on Bonds," *GDN*, December 13, 1901, 10, col. 6; "Galveston Bondholders," *GDN*, December 31, 1901, 10, col. 4; for sewer debts, see "City Sewer Bonds," *GDN*, January 4, 1901, 10, cols. 3–6; "The Sewer Company Plant," *GDN*, January 11, 1901, 6, col. 1; "Galveston Sewer Company," *GDN*, January 13, 1901, 8, cols. 1–2; "The Sewer System," *GDN*, February 3, 1901, 3, col. 7; "Refunding Galveston's Bonded Indebtedness," *GDN*, February 22, 1901, 6, col. 6; "Not an April Fool," *GDN*, March 10, 1901, 8, col. 3; "No One Fare Round Trip," *GDN*, March 11, 1901, 8, cols. 2–3; and "The Sewer Plant," *GDN*, March 13, 1901, 10, cols. 3–4.

10. "Galveston Legislation," *GDN*, February 5, 1901, 2, col. 2; "Report to the Mayor," *GDN*, February 5, 1901, 10, col. 2; "The Senate," *GDN*, February 6, 1901, 2, col. 1; "Bill for Galveston," *GDN*, February 6, 1901, 4, col. 5; "Bills in the House," *GDN*, February 7, 1901, 2, cols. 1–2; "Towns & Cities," *GDN*, February 8, 1901, 2, cols. 2–3; "The Galveston Bills," *GDN*, February 8, 1901, 2, col. 7; "Galveston Court," *GDN*, February 8, 1901, 4, col. 2; "Galveston Denied," *GDN*, April 3, 1901, 2, cols. 1–3; "The Legislature," *GDN*, April 4, 1901, 2, cols. 1–4; "How Does Texas Stand To-day in the Eyes of the World," *GDN*, April 5, 1901, 10, col. 2; "The City Council," *GDN*, April 16, 1901, 3, cols. 1–4; "Texas' Cruelty to Galveston," *GDN*, April 16, 1901, 4, cols. 3–4; "Shame Upon Them," *GDN*, April 17, 1901, 4, cols. 5–6; "Galveston's Case," *GDN*, April 17, 1901, 4, col. 7; "The Constitution," *GDN*, April 20, 1901, 2, col. 1; "State Capitol News," *GDN*, April 20, 1901, 3, col. 1; "Commission Bill," *GDN*, August 4, 1901, 24, cols. 2–3; "Dr. Trueheart Right/And Question Arises," *GDN*, August 4, 1901, 24, col. 3; "Proposition to Aid Galveston," *GDN*, August 10, 1901, 7, cols. 2–3; "The Grade Bill," *GDN*, August 12, 1901, 8, cols. 3–4; "Galveston:

Its Importance to the State," *GDN,* August 19, 1901, 4, cols. 6–7; "The Governor's Message," *GDN,* August 21, 1901, 1, col. 5; "Something to Do," *GDN,* August 21, 1901, 1, cols. 6–7; "Mr. Davidson's Bill," *GDN,* August 21, 1901, 2, col. 6; and "State Tax for City, *GDN,* September 26, 1901, 10, col. 2.

11. "Galveston in 1900," *GDN,* January 1, 1901, 4, cols. 1–2; "Summary of Losses and Rehabilitation," *GDN,* January 1, 1901, part 2, p. 1, cols. 2–3; "Work of Relief Committee," *GDN,* January 1, 1901, part 2, p. 3, col. 7; "Relief Provisions," *GDN,* January 9, 1901, 10, cols. 3–4; "Homes Constructed," *GDN,* January 10, 1901, 3, col. 5; "The Needs of Galveston: A Synopsis of Proposed Legislation For Its Relief," *GDN,* January 15, 1901, 6, cols. 6–7; "New York Balance," *GDN,* January 18, 1901, 10, col. 5; "The Building Fund," *GDN,* January 21, 1901, 8, col. 3; "Helped Galveston," *GDN,* January 25, 1901, 10, col. 5; "Transport of M'Pherson," *GDN,* January 27, 1901, 8, col. 2; "Rockefeller Party," *GDN,* January 27, 1901, 8, col. 3; "A Relief Measure," *GDN,* January 30, 1901, 1, cols. 6–7; "Storm Relief Fund," *GDN,* January 31, 1901, 7, col. 3; "The Relief Work," *GDN,* February 9, 1901, 4, cols. 1–2; "The Galveston Commission Bill," *GDN,* February 10, 1901, 16, cols. 1–2; "The Work of Relief," *GDN,* February 12, 1901, 10, col. 3.

12. "Hays Was Pleased," *GDN,* January 2, 1901, 1, cols. 6–7; "West End Wharves," *GDN,* January 7, 1901, 8, col. 1; "Galveston Harbor," *GDN,* January 7, 1901, 6, cols. 6–7; "Report of the Mayor," *GDN,* February 5, 1901, 10, col. 2; "A Party of Solons," *GDN,* February 10, 1901, 8, cols. 1–2; West End Wharves," *GDN,* February 16, 1901, 7, col. 1; "Bulkhead Channel," *GDN,* February 27, 1901, 4, col. 1; "The Sunset Officials," *GDN,* April 21, 1901, 4, col. 4; and "Wharves Finished," *GDN,* October 21, 1901, 5, cols. 1–2.

13. Companies with Galveston connections included Galveston-Beaumont Oil, Galveston Oil & Gas, Galveston Land, Oil & Gas, Galveston Oil, Bonanza Oil, Enterprise Oil, and Manhattan Oil, based on information taken from the Web site of the Texas secretary of state and in advertisements in the *GDN,* April–July 1901. See also "Organization of Oil Companies in Texas," *GDN,* May 21, 1901, 4, cols. 1–4; for Beatty's real estate dealings in Galveston County, see *D. R. Beatty et al.* v. *Charles W. Bulger,* 28 *Tex. Civ. App.* 117–23 (1903); for his oil career, see "Oil Geyser Sold," *DMN,* April 10, 1901, 8, col. 1 (quotation); "Plans of the Western," *GDN,* March 28, 1901, 4, col. 3; "Echoes of Streets," *GDN,* March 30, 1901, 10, col. 4; "Mr. Beatty On Oil," *GDN,* March 31, 1901, 8, col. 3; "Texas Western Directors," *GDN,* April 7, 1901, 6, col. 2; "Big Oil Well Sold," *GDN,* April 10, 1901, 4, cols. 4–5; "Texas Petroleum," *GDN,* August 24, 1901, 6, col. 7; "Thinks It Only a Starter," *DMN,* April 11, 1901, 2, col. 5; "Quality of Beaumont Oil," *DMN,* April 13, 1901, 2, col. 3; "Met Oil Magnates," *DMN,* April 13, 1901, 2, col. 2; "Money Is Common," *DMN,* April 14, 1901, 4, col. 2; "Excitement Runs High," *NOTD,* April 14, 1901, 14, col. 1; and "Injunction Granted," *NOTD,* February 14, 1901, 10, col. 2.

14. For Sealy, see "A Fourth Spouter," *GDN,* April 4, 1901, 7, col. 1; and "Texas Western Directors," *GDN,* April 4, 1901, 6, col. 2; on Moody, see the advertisements for High Island and National Oil in *GDN,* April 14, 1901, 26.

15. See advertisement for Roby Oil Company in *GDN,* April 14, 1901, 26, cols. 5–7.

16. "Oil Leases Filed," *GDN,* May 11, 1901, 10, col. 5.

17. See Brearley, *History NBFU,* 85–86, 88–89; and Brearley, Introduction, notes 34, 54; and chapter 3, notes 57, 84.

18. "Oil Fuel Contract," *GDN,* April 12, 1901, 10, cols. 2–4; "From Coal to Oil," *GDN,* April 13, 1901, 12, cols. 4–7; "Galveston and Oil," *GDN,* November 23, 1901, 7, cols. 1–2; for Brush, see

"Galveston Will Burn Oil," *DMN*, April 12, 1901, 2, col. 2 (oil contract); "Oil at Brush Power House," *GDN*, August 19, 1901, 8, col. 5 (oil tank); "A $27,500 Offer," *GDN*, July 2, 1901, 7, cols. 1–2 (two-inch pipeline; also Merchants & Planters oil tank); and "Round About Town," *GDN*, August 26, 1901, 8, col. 3. See also Galveston, Texas, Office of City Secretary, Galveston City Council, City Council Records, "Minutes of City Council," 1898–1901, large folio vol., July 1, 1901, Brush Electric, 735–36 (pipeline); Merchants & Planters, 735 (oil tank under sidewalk); hereafter GCC, CCR, "Minutes."

19. "The First Oil Plant," *GDN*, May 3, 1901, 10, col. 3. In January 1902, Texas Star successfully petitioned the city to run a pipeline from the flour mill to its rice mill. See "City Commission," *GDN*, January 15, 1902, 10, cols. 3–4; "Sand Ordinance," *GDN*, February 5, 1902, 10, cols. 2–3 (approved). GCC, CCR, "Minutes," vol. 2, "Proceedings of the Board of Commissioners of the City of Galveston, September 18, 1901 thru May 5, 1903," January 14, 1902, 107; January 28, 1902, 137 (hereafter GCC, CCM, PBOCCG, "Minutes"). On Red Snapper, see GCC, CCR, "Minutes," May 27, 1901, 717 (tank); June 3, 1901, 721 (pipeline); "City Council Meets," *GDN*, May 28, 1901, 10, col. 3; "The City Council," *GDN*, June 4, 1901, 10, col. 3. For Texas Ice, see GCC, CCR, "Minutes," June 17, 1901, 728; and "Going to Use Oil," August 20, 1901, 7, cols. 2–3. On burning fuel oil, see "Oil Is a Cheap Fuel," *GDN*, December 7, 1901, 7, col. 3; on the Waterworks, see "Water Commissioners," *GDN*, February 8, 1901, 7, col. 2; "Oil for Fuel," *GDN*, April 12, 1901, 12, col. 4; and "From Coal to Oil," *GDN*, April 13, 1901, 12, col. 6.

20. "A Pipe Line to the Oil Fields," *GDN*, April 4, 1901, 6, cols. 2–3; "An Oil Pipe Line," *GDN*, April 4, 1901, 10, col. 5; "Some Oil Gossip," *GDN*, April 5, 1901, 10, cols. 2–3; "The Impending Revolution," *GDN*, April 16, 1901, 4, col. 1; "Oil for Steamships," *GDN*, April 27, 1901, 10, col. 4; and "Galveston as a Fuel Station," *GDN*, November 3, 1901, 8, cols. 1–2.

21. For railroads and Galveston, see Earle B. Young, *The Tracks to the Sea: Galveston and Western Railroad Development, 1866–1900* (1999), 101–21, 123–30; for supply plans, see "Oil Fuel Contract," *GDN*, April 12, 1901, 10, col. 2; and "From Coal to Oil," *GDN*, April 13, 1901, 12, col. 4.

22. On the connection between cheap fuel oil in Galveston and manufacturing promotion and recruitment, see "No June Election," *GDN*, May 7, 1901, 2, col. 3; on the derivation of the term "rock oil," see Samuel Harries Daddow and Benjamin Bannan, *Coal, Iron, and Oil; or, the Practical American Miner: A Plain and Popular Work on Our Mines and Mineral Resources* (1866), 651–52; for the text of the 1871 ordinance, see City of Galveston, *Charter and Revised Ordinances* (1875), chap. 26, article 1, sec. 1, pp. 204 (quotations 1–4), 205. See also City of Galveston, *Charter and Revised Ordinances* (1883) article 3, pp. 215–17; and *CCG&RO* (1893), articles 235–41, contents of which closely follow the 1871 ordinance.

23. For fuel-oil adoptions see "A New Power House," *GDN*, March 6, 1901, 10, col. 2; "The City Council," *GDN*, August 6, 1901, 10, col. 5; "Going to Use Oil," *GDN*, August 20, 1901, 7, cols. 2–3; "Street Railway Improvements," *GDN*, August 22, 1901, 10, col. 3; "Galveston Is Busy," *GDN*, October 5, 1901, 10, col. 3; "The City Council," *GDN*, October 5, 1901, 10, col. 3; GCC, CCR, "Minutes," August 5, 1901, 741; "Fuel Oil for Elevators," *GDN*, May 19, 1901, 22, col. 4; "Change to Fuel Oil," *GDN*, May 26, 1901, 22, cols. 1–2; "Galveston and Oil," *GDN*, May 26, 1901, 7, col. 1; "A Dog Ordinance," *GDN*, August 27, 1902, 10, col. 4; "City Waterworks," *GDN*, July 23, 1902, 12, cols. 2–3. On Clarke, see "It Is Businesslike," *GDN*, May 3, 1901, 2, col. 1; "Oil for Galveston," *GDN*, May 11, 1901, 10, col. 4; "No June Election," *GDN*, May 7, 1901, 2, cols. 1–5; and Morrison, ed., *The Port of Galveston and the State of Texas*, 67.

24. "A Called Meeting," *GDN,* June 19, 1901, 7, col. 1 (Stewart); and "The City Council," *GDN,* May 10, 1901, 12, col. 3; "The City Council," *GDN,* May 21, 1901, 10, cols. 3–5; "The City Council," *GDN,* June 4, 1901, 10, cols. 3–5; "Reports Received," *GDN,* June 18, 1901, 7, cols. 1–3; "The Oil Ordinances," *GDN,* June 20, 1901, 10, col. 4; "Lacked a Quorum," *GDN,* June 27, 1901, 10, col. 4; "A $27,500 Offer," *GDN,* July 2, 1901, 7, cols. 1–2; "The City Council," *GDN,* August 6, 1901, 10, cols. 3–6.

25. "Round About Town," *GDN,* January 3, 1902, 8, col. 3; "City Commission," *GDN,* January 8, 1902, 10, col. 3; "City Commission," *GDN,* January 22, 1902, 10, col. 4; "Protection Bonds," *GDN,* January 29, 1902, 10, col. 3; "Tax Relief Cases," *GDN,* January 31, 1902, 10, col. 2; "News of the Shipping," *GDN,* March 11, 1902, 8, col. 1.

26. "No June Election," *GDN,* May 7, 1902, 2, cols. 1, 3; "The City Council," *GDN,* May 10, 1902; "The City Council," *GDN,* May 21, 1902, 10, cols. 3–5.

27. "The City Council," *GDN,* May 21, 1902, 10, col. 6; "City Council Meets," *GDN,* May 28, 1902, 10, col. 4; "The City Council," *GDN,* June 4, 1902, 10, col. 4 (quotations).

28. "The City Council," *GDN,* June 4, 1902, 10, cols. 4 (quotation 1), 5 (quotations 2–4), 6.

29. "Reports Received," *GDN,* June 18, 1901, 7, col. 2; for the grant to the Galveston and Bolivar Steam Ferry Company, see "In the City Council," *GDN,* March 22, 1898, 8, col. 3. For Stubbs's reports, see GCC, CCR, "Minutes," 729–30, dated June 7, 1901, and recorded June 17, 1901.

30. "Reports Received," *GDN,* June 18, 1901, 7, col. 2; for charter-based powers, see City of Galveston, *Charter and Revised Ordinances* (1875), Charter, title 4, article 3, sec. 9, p. 28, granting the City Council the power to "exclusively regulate, direct and control the laying and repairing of the gas-pipes and gas fixtures in the streets, alleys, sidewalks and elsewhere"; and *CCG&RO 1893,* title 4, article 3, sec. 27, p. 21; sec. 34, p. 23.

31. "A Called Meeting," *GDN,* June 19, 1901, 7, cols. 2 (quotations 1–2), 3 (quotations 3–4). Critics called the GWC the "Octopus of the Gulf" and a "hydra-headed monster." See Susan Wiley Hardwick, *Mythic Galveston: Re-Inventing America's Third Coast* (2002), 90, citing Marilyn McAdams Sibley, *The Port of Houston* (1968), 85, 87, 97.

32. "A $27,500 Offer," *GDN,* July 2, 1901, 7, col. 2.

33. Ibid.

34. "The City Council," August 6, p. 10, cols. 5 (quotations), 6.

35. Ibid., col. 5, letter, dated July 6, 1901, from Mayor Walter C. Jones to the Galveston City Council (quotations 1–3). See also GCC, CCR, "Minutes," Texas Ice & Cold Storage, June 17, 1901, 728; and Merchants' & Planters' Compress and Warehouse, July 1, 735; *CCG&RO 1893,* title 3, article 9, sec. 19; chap. 3, articles 494–510; and *Charter of the City of Galveston and the Revised Ordinances,* "An Act to incorporate the City of Galveston and grant it a new Charter; and to repeal all pre-existing Charters," approved April 19, 1901 (in effect October 16), sec. 19i, "Engineer," p. 18. For the Clarke ordinance, see GCC, CCR, "Minutes," May 9, 1901, 709 (quotations 4–5).

36. On Jones as a target of the commission forces, see McComb, *Galveston: A History,* 135–36.

37. The Santa Fe system used 243,382 and 408,894 tons of fuel oil respectively in 1902–3. See the *Eighth Annual Report of the Atchison, Topeka, & Santa Fe Railway Company . . . 1903* (1903), 38; on adopting fuel oil, see "Going to Use Oil," *GDN,* August 20, 1901, 7, cols. 2, 3

(quotation 1); "Santa Fe Oil Fuel," *GDN,* August 22, 1901, 5, col. 1; "Oil Storage Tanks / Santa Fe Experimenting With Oil and a Dust Layer," *GDN,* August 24, 1901, 5, col. 1; and "New Oil Reservoirs, *GDN,* October 12, 1901, 7, col. 1. For GCSFR plans to put up "three monster tanks" in Temple, Texas, each 90 feet in diameter, 15 feet high, holding 2,225,000 gallons, see "Santa Fe Oil Tanks," *GDN,* November 17, 1901, 7, col. 2.

38. "City Commission," *GDN,* October 2, 1901, 12, cols. 2, 3 (quotation).

39. Ibid., 8, cols. 3, 4 (quotation); "City Commission," *GDN,* October 16, 1901, 4, col. 2; and *CCG&RO 1893,* title 6, chap. 5, article 239. See also GCC, CCM, PBOCCG, "Minutes," vol. 2, October 1, 1901, 13; October 8, 1901, 20.

40. "City Commission," *GDN,* October 16, 1901, 4, col. 4.

41. Ibid.

42. Ibid.

43. "Proposed Oil Law," *GDN,* October 18, 1901, 7, col. 1 (quotations 1–2, 4); and Klein and Cork, "Approach to Technical Risk Assessment," 345 (quotation 3).

44. "Proposed Oil Law," *GDN,* October 18, 1901, 7, cols. 1 (quotations 2–4), 2 (quotation 1).

45. Ibid., October 19, 1901, 7, col. 1.

46. "Proposed Oil Law," *GDN,* October 19, 1901, 7, cols. 1 (quotations 1–3), 2 (quotation 4). For a clear understanding of construction, hazard-mitigation, and engineering expertise in dealing with storing or mixing oil in railroad operations, using compiled information similar to what insurance companies and standards associations offered to fire underwriters and municipal officials, see Walter G. Berg, *Buildings and Structures of American Railroads: A Reference Book* (1893), chap. 12, "Oil-Storage Houses," 81–92; and chap. 13, "Oil-Mixing Houses," 93–112.

47. Erik Vanem, "Ethics and Fundamental Principles of Risk Acceptance Criteria," *Safety Science* 50, no. 4 (April 2012): 958.

48. A civil engineer advised the city to steer clear of a "uniform law to regulate all oil plants." Instead, it should authorize the city engineer, an insurance inspector, the chief of the Fire Department, or "other competent authority" to "pass upon all plans and specifications" submitted by permit applicants, addressing the "special features" of each installation while allowing the work to proceed. See "Proposed Oil Law," *GDN,* October 19, 1901, 7, col. 1.

49. Colleen Murphy and Paolo Gardoni, "The Acceptability and the Tolerability of Societal Risks: A Capabilities-based Approach," *Science and Engineering Ethics* 14 (2008): 78; Gerhard Ersdal and Terje Aven, "Risk Informed Decision-Making and Its Ethical Basis," *RE&SS* 93, no. 2 (February 2008): 197, 200; and Edward M. Marszal, "Tolerable Risk Guidelines," *ISA Transactions* 40, no. 4 (September 2001): 391–95.

"50. "Fuel Oil Storage," *GDN,* October 24, 1901, 4, col. 1; and "City Commission," *GDN,* October 16, 1901, 4, col. 3 (quotation).

51. "City Commission," *GDN,* October 16, 1901, 4, col. 4; and "Fuel Oil Storage," *GDN,* October 24, 1901, 4, col. 1 (quotation; substitute ordinance). See CCG&RO 1893, title 6, chap. five, articles 235–36; "City Commission," *GDN,* October 16, 1901, 4, col. 4; and "Fuel Oil Storage," *GDN,* October 24, 1901, 4, col. 1 (quotation).

52. "City Commission," *GDN,* October 16, 1901, 4, col. 4; and "Fuel Oil Storage," *GDN,* October 24, 1901, 4, col. 1 (quotations).

53. "Fuel Oil Storage," *GDN,* October 24, 1901, 4, col. 2.

54. Ibid.; and "Santa Fe Tanks," *GDN,* October 29, 1901, 10, col. 4; for Hill, see *Galveston in Nineteen Hundred,* ed. Clarence Ousley (1900), 158; and "J. H. Hill on Embargoes," *Traffic World* 15, no. 17 (April 24, 1915): 911–14.

55. For Chicago, see City of Chicago, *Laws and Ordinances Governing the City of Chicago, From April 2, 1890, to July 10, 1894,* comp. Henry Binmore (1894), secs. 4940–41, pp. 83–84. On Brown and his famous father, J. M. Brown, see *Historical Review of South-East Texas and the Founders, Leaders and Representative Men of Its Commerce, Industry and Civic Affairs,* ed. Dermont H. Hardy and Ingham S. Roberts (1910), vol. 2: 556–59; for the discussion of oil safety, see "Fuel Oil Storage," *GDN,* October 24, 1901, 4, col. 3 (quotations 1–3). New Orleans insurance men charged that the substitute ordinance paved "the way to all kinds of negligence and unsafe methods which in time would endanger the entire city." See "Scheme of the Oil Men," *The Vindicator* 28, no. 9 (November 1, 1901): 140.

56. "New Oil Ordinance," *GDN,* October 25, 1901, 10, col. 3.

57. Ibid.

58. Ibid., cols. 3–4; and "Santa Fe Tanks," *GDN,* October 29, 1901, 10, col. 4.

59. "New Oil Ordinance," *GDN,* October 25, 1901, 10, col. 4.

60. "City Commission," *GDN,* October 30, 1901, 4, cols. 1–4.

61. For McClure Kelly as representative for St. Paul Fire and Marine Insurance Company in Dallas, see Texas, Department of Agriculture, *Twenty-Fifth Annual Report of the Commissioner of Agriculture . . . 1899* (1900), 69; *The Insurance Year Book, 1902–1903: [Fire and Marine]* (1902), 214 (TFPA executive-committee member); and "Kelly and Dexter Join," *The Insurance Field,* 18, no. 3 (July 16, 1908): 17.

62. "City Commission," *GDN,* October 30, 1901, 4, col. 3.

63. Ibid., November 6, 1901, 5, col. 1. For the TFPA activities, see "Oil Storage Too Hazardous," *DMN,* January 31, 1902, 7, col. 2; "Locating Oil Storage Tanks," *DMN,* February 1, 1902, 2, col. 2; and "Movement Against Oil Tanks," *DMN,* February 1, 1902, 8, col. 2; "Fighting Oil Tanks, *GDN,* January 26, 1902, 1, col. 4; "Texas Fire Prevention Association," *GDN,* January 31, 1902, 1, cols. 4–5, "Insurance People," *GDN,* February 1, 1902, 1, cols. 6–7; "Locating Oil Tank," *GDN,* February 1, 1902, 1, col. 7; "Oil Is All Right," *GDN,* February 4, 1902, 11, cols. 6–7; and "Railroad Pipe Line," *GDN,* February 5, 1902, 4, col. 4; "Reforms Needed at Houston," *The Vindicator* 28, no. 8 (October 15, 1901): 124; "Houston Fire Protection," *The Vindicator* 28, no. 11 (December 1, 1901): 169; "Fuel Oil Hazard in Texas"; *The Vindicator* 29, no. 3 (February 1, 1902): 40; "Watching the Fuel Oil Hazard," *The Vindicator* 29, no. 4 (February 15, 1902): 56; "Texas Municipalities Asked to Act," *The Vindicator* 29, no. 4 (February 15, 1902): 56–57; "Feeding Fuel Oil by Gravity Pressure," *The Vindicator* 29, no. 6 (March 15, 1902): 87; "Dangerous Ordinance at Houston," *The Vindicator* 29, no. 7 (April 1, 1902): 101; "Houston Gets Back in Line," *The Vindicator* 29, no. 8 (April 15, 1902): 122; and "Two Sides to a Dangerous Proposition," *The Vindicator* 29, no. 10 (May 15, 1902): 149–50.

64. "City Commission," *GDN,* November 6, 1901, 5, col. 1 (quotations), 2–3 (ordinance). For the official text of the ordinance, see GCC, CCM, PBOCG, "Minutes," vol. 2 (November 5, 1901): 49–51, "Unfinished Business," text of "Ordinance Regulating the Handling, Storing, Manufacture, Etc., of Oil, Etc., and Regulating the Building, Erection and Location of Oil Tanks." McClure Kelly was chairman of the Construction and Combustible Committee of the TFPA. See "Houston Gets Back in Line," *The Vindicator* 29, no. 8 (April 15, 1902): 122.

65. See "Swayne's Project," *GDN*, February 1, 1902, 4, col. 1, wherein it is claimed that these "three grand divisions" were given in order of importance.

66. "Fuel Oil on Dredges," *GDN*, July 23, 1901, 8, col. 5; "Clarke to Burn Oil," *GDN*, November 11, 1901, 5, col. 1; "Clarke Burning Oil," *GDN*, November 15, 1901, 8, col. 3; "Economy in Fuel," *GDN*, November 26, 1901, 3, col. 1; and "City Commission," *GDN*, January 8, 1902, 10, col. 3.

67. "Illinois Road Sold Under Foreclosure," *Electric Railway Journal* 54, no. 3 (July 19, 1919): 140; "Franchise Rights Questioned," *Electric Railway Journal* 54, no. 5 (August 2, 1919): 252; "Sale Under Foreclosure Confirmed," *Electric Railway Journal* 54, no. 8 (August 23, 1919): 410–11; "Foreclosure Sale Approved," *Electric Railway Journal* 54, no. 9 (August 30, 1919): 453; "Purchase Payments Being Made," *Electric Railway Journal* 54, no. 14 (October 4, 1919): 700; "More Time Allowed for Payment," *Electric Railway Journal* 54, no. 17 (October 25, 1919): 806; "Promoter Behind in Payment," *Electric Railway Journal* 54, no. 20 (December 13, 1919): 952; and "Sale Under Foreclosure Confirmed," *Electric Railway Journal* 54, no. 21 (December 20, 1919): 1018.

68. "The Lone Star Injunction," *GDN*, February 14, 1898, 5, col. 2; "Bolivar Pipe Line," *GDN*, October 30, 1901, 10, col. 2 (quotations); and "Galveston as a Fuel Station," *GDN*, November 3, 1901, 8, cols. 1–2.

69. On tank-car scarcity, see "Scarcity of Cars," *GDN*, November 13, 1901, 2, col. 1; "Rumors of a Deal," *GDN*, November 22, 1901, 3, col. 1; "Oil Handling Facilities," *GDN*, November 22, 1901, 3, col. 2; "Delay in Deliveries," *GDN*, November 24, 1901, 6, col. 4; "Shortage of Cars," *GDN*, November 26, 1901, 11, col. 3; "Scarcity of Cars," *GDN*, November 29, 1901, 7, col. 2; "No Service Made," *GDN*, September 7, 1902, 7, col. 1; "Oil Car Movement," *GDN*, October 17, 1902, 7, col. 2; "Tank Car Hearing," *GDN*, October 18, 1902, 1, col. 7; p. 2, cols. 1–3; "A Tank Car Suit," *GDN*, November 12, 1902, 7, col. 1; for waterworks, see "Will Burn Fuel Oil," *GDN*, November 16, 1901, 7, col. 3; and "What Fuel Oil Costs," *GDN*, November 17, 1901, 8, col. 2.

70. "Santa Fe Oil Tanks," *GDN*, November 17, 1901, 4, col. 1; and "As to City Bonds," *GDN*, November 20, 1901, 4, col. 2.

71. "City Commission," *GDN*, November 27, 1901, 4, col. 2.

72. "City Commissioner," *GDN*, December 4, 1901, 8, cols. 1 (quotation 5), 3 (quotations 1, 4).

73. "City Commission," *GDN*, December 31, 1901, 4, cols. 1–4.

74. "Oil Tank Franchise," *GDN*, January 18, 1902, 4, col. 5 (quotation); "City Commission," *GDN*, January 22, 1902, 10, col. 3; "On the Santa Fe," *GDN*, January 24, 1902, 10, col. 2.

75. For the quotation, see the U.S. Nuclear Regulatory Agency, Glossary, definition of "performance-based regulation," www.nrc.gov/reading-rm/basic-ref/glossary.html.

76. For Clarke & Courts, see Denise Alexander, *Images of America: Galveston's Historic Downtown and Strand District* (2010), 75, 117; for Henchman, see *Municipal Engineering*, vol. 18, no. 3 (March 1900): 214; and *Directory of American Cement Industries*, ed. Charles Carroll Brown, 4th ed. (1906), 440; for Model Laundry, see "City Commission," *GDN*, April 2, 1902, 10, col. 3; "Oil Tank Permit," *GDN*, April 9, 1902, 10, col. 4; Clarke & Courts: "Pipe Line for Oil," *GDN*, April 30, 1902, 7, col. 1; White: "The French Wells," *GDN*, August 6, 1902, 10, col. 3 (quotation 1); and Henchman: "A Dog Ordinance," *GDN*, August 27, 1902, 10, col. 3; and "The Fire Limit Law," *GDN*, September 10, 1902, 10, col. 3. See also GCC, CCM, PBOCCG, "Minutes," vol. 2 (April 1, 1902): 177; vol. 2 (April 8, 1902): 190 (Model); vol. 2 (April 22, 1902): 201; vol. 2 (April 29, 1902): 206 (Clarke & Courts); vol. 2 (August 5, 1902): 286 (White); and vol. 2 (August 26, 1902): 301; vol. 2 (September 9, 1902): 310 (quotation 2) (Henchman).

77. For Wisrodt, see "The Scavenger Act," *GDN,* May 28, 1902, 10, col. 3 (quotations); and "A Scavenger Law," *GDN,* June 4, 1902, 10, col. 3; for St. Mary, see "Must Use Sewer," *GDN,* November 12, 1902, 6, col. 3. See also GCC, CCM, PBOCCG, "Minutes," vol. 2 (May 27, 1902): 224; vol. 2 (June 3, 1902): 228 (Wisrodt); and vol. 2 (November 11, 1902): 345 (Sisters).

78. "City Commissioners," *GDN,* November 5, 1902, 10, col. 4; "Plenty Business," *GDN,* November 19, 1902, 8, col. 3 (quotation 3); "A Special Meeting," *GDN,* November 26, 1902, 9, col. 2 (quotation 2); and "The City Commissioners," *GDN,* December 3, 1902, 8, col. 3. See also GCC, CCM, PBOCCG, "Minutes," November 4, 1902, 342; November 18, 1902, 353; November 25, 1902, 357, 358 [text of ordinance & quotation 1]; December 2, 1902, 365. For article 279, see *Charter of the City of Galveston and the Revised Ordinances* (1902), 147–48.

79. For GHH, see "City Commission," *GDN,* December 31, 1901, 4, cols. 2–3 (original ordinance); "City Commission," *GDN,* January 22, 1902, 10, col. 4 (quotations); "Scavenger Law," *GDN,* March 26, 1902, 10, col. 4; and "New Machine Shop," *GDN,* April 15, 1902, 10, col. 5. See also GCC, CCR, "Minutes," December 30, 1901, 100; and GCC, CCM, PBOCCG, "Minutes," January 21, 1902, 111–12; March 11, 1902, 164–65; March 25, 1902, 173–74 (GHH).

80. "The City Commissioners," *GDN,* September 30, 1903, 8, col. 1 (quotation 1); "The City Commissioners," *GDN,* October 7, 1903, 12, col. 2 (quotation 2); and "The City Commissioners," *GDN,* October 14, 1903, 8, col. 1 (quotations 3–4). See also GCC, CCM, PBOCCG, "Minutes," vol. 2 (September 15, 1903): 117; vol. 2 (September 29, 1903): 129; vol. 2 (October 6, 1903): 134; and vol. 2 (October 13, 1903): 137–38.

81. "The City Commissioners," *GDN,* November 18, 1904, 12, col. 5; "Special Notices," *GDN,* November 20, 1904, 3, col. 7; "The City Commissioners," *GDN,* December 2, 1904, 5, col. 4; "The City Commissioners," *GDN,* December 9, 1904, 10, col. 4; "Four Miles of Line," *GDN,* January 13, 1905, 12, col. 4; and "The City Commissioners," *GDN,* February 17, 1905, 10, col. 5 (quotations 2–3; oil tank approved). See also GCC, CCM, PBOCCG, "Minutes," vol. 2 (November 10, 1904): 423; vol. 2 (November 17, 1904): 428; vol. 2 (December 1, 1904): 438–39 (text of pipeline ordinance); vol. 2 (December 8, 1904): 443; vol. 2 (December 15, 1904): 445, 447; vol. 2 (February 9, 1905): 475; vol. 2 (and February 16, 1905): 485 (quotations 1, 4). Guyton subsequently incorporated to "go into the business of selling fuel oils much more extensively than in the past." However, the company dissolved in 1911 and sold its oil tank at Forty-Second Street and Avenue A and approximately eight miles of pipeline that serviced "local concerns" to the Gulf Pipe Line Co. See "New Corporations," *Paint, Oil and Drug Review,* vol. 39, no. 26 (June 28, 1905): 37 (quotation 1 above); and "Competition in Texas," *The Petroleum Gazette* 16, no. 6 (September 1911): 8 (quotation 2 above).

82. On licensing see "The City Commissioners," *GDN,* December 23, 1903, 3, col. 5; "The City Commissioners," *GDN,* January 13, 1904, 8, cols. 2–3; "City Railroad's Pipe Line," *GDN,* January 27, 1904, 7, col. 2; "The City Commissioners," *GDN,* February 3, 1904, 8, col. 4; "Oil Tank Petition," *GDN,* January 27, 1905, 10, col. 4; "Pumping Engine Bids Submitted," *GDN,* June 23, 1911, 12, cols. 1–2; "Commission Makes City Improvements," *GDN,* July 7, 1911, 12, col. 1; "7000-Gallon Tank," *GDN,* September 4, 1911, 12, col. 3; "City Commission Puts in Overtime, *GDN,* February 2, 1912, 5, col. 1; "Tenth Ward Club Protests Delay," *GDN,* November 9, 1923, 4, col. 1; "More Improvement Bonds to be Issued," *GDN,* January 15, 1925, 3, cols. 1–2; "Would Change Gas Tank Sidewalk Ban," *GDN,* August 14, 1925, 12, col. 1; "Industrial Site Creation

and Dump Removal Covered in New Grade Raising Plan," *GDN,* September 18, 1925, 3, col. 5; "Receipts and Expenditures of the City of Galveston, Texas, for the Fiscal Year Ending June 30, 1926," Oil Tank Licenses $950.00, *GDN,* July 11, 1926, 22, cols. 5–6; "Action by Health Officer Is Urged," *GDN,* August 13, 1926, 9, col. 6; "Sidewalk Displays Curtailed by City," *GDN,* August 20, 1926, 5, col. 3; "Seek Bond Issue and Paving Bids," *GDN,* September 17, 1926, 3, cols. 4–5; "City Tax Rate Raised 8c as New Budget Is Tentatively Adopted," *GDN,* July 22, 1927, 2, col. 4; "Board Will Create Zoning Commission Here in Line with Legislative Action," *GDN,* December 16, 1927, 8, col. 5; "Committee Reports Pelican District Plan and Filling of East End to Commission," *GDN,* March 9, 1928, 9, col. 3; "Gas Rate Ordinance Will Be Drafted," *GDN,* June 15, 1928, 9, col. 6; "Santa Fe Seeks Track Franchise," *GDN,* August 17, 1928, 1, col. 3; "Board Considers Revoking Franchise of Gas Company," *GDN,* August 24, 1928, p. 10, col. 4; "Commissioners Open Bids for 1,500 Feet Fire Hose," *GDN,* December 21, 1928, 9, col. 4; "City Gives Use of Street for Mardi Gras Festival," *GDN,* January 25, 1929, 4, col. 8; "Tribute Paid to Memory of Cone," *GDN,* January 24, 1930, 5, cols. 4–6; "Election Order Is Adopted by Board," *GDN,* March 7, 1930, 12, col. 3; "Building Ordinance Hearing Scheduled," *GDN,* September 1, 1933, 12, col. 2; "Quarterly Finance Report Submitted," *GDN,* January 18, 1935, 10, col. 2; "Insurance Change Ordered by Board," *GDN,* April 19, 1935, 3, col. 1; and "Building Code Revision Awaits End of War," *GDN,* August 25, 1943, 12, col. 4. For Karbach, see GCC, CCM, PBOCCG, "Minutes," vol. 27 (January 8, 1942, to March 4, 1943), "Karbach, C. W. Refund on Oil Tank Authorized," 316 (quotation 2), 317.

83. GCC, CCM, PBOCCG, "Minutes," vol. 15 (October 18, 1928): 233. For the "Texas Form" and the "Chicago Form-amended," see Booth, *Liquid Fuel and Its Combustion,* 344–46. Under the Texas form, fuel-oil tanks cannot be closer than fifty feet from the building serviced ("the risk described," 345), while in the Chicago form, an "authorized inspector" (346) determines the acceptable location of a tank. Tank capacity regulates its distance from "any property" (346). "Forms" contain the rules and regulations with which one must comply to gain a permit from insurance underwriters.

84. See *Charter of the City of Galveston and the Revised Ordinances* (1902), 144–46, 147 (quotations 1–2), 148; and City of Galveston, *Revised Ordinances of the City of Galveston To June 1st, 1917* (1917), chap. 5, "Oil and Oil Storage," articles 239–47, pp. 98–100, 101 (table), 103–4.

85. "Replace cracked concrete floor with steel floor in fuel oil storage tank—Galveston 1929," Letters, Superintendent [A. P. Hill] to W. E. Maxson, March 14, 1929; W. E. Maxson to A. P. Hill, March 15; [A. P. Hill] to W. E. Maxson, March 18; F. Merritt, Chief Engineer, to W. E. Maxson, March 27; W. E. Maxson to A. P. Hill, April 15. On repairs, see Z. A. Green to A. P. Hall, April 18; A. P. Hill to Z. A. Green (Division Engineer), April 20 and April 25; A. P. Hill to W. H. Kluever (General Foreman), May 7. See Houston Public Library, HMRC, RG D 004, Santa Fe Railroad Collection, Records of the Gulf, Colorado, and Santa Fe Railway Company, 1897–1968, part 5, box 312, folder 20, Operational File #1626 (part 1).

86. On the crisis and how to proceed, see the telegrams and letters: A. P. Hall to F. G. Pettibone and W. E. Maxson, MEMO, May 18, 1929; A. P. Hall to W. H. Kluever (quotation 2); A. P. Hall to Mr. Jim Kluever (quotation 1); A. P. Hall to Z. A. Green; A. P. Hall to Z. A. Green, MEMO, May 20; F. G. Pettibone to A. P. Hall; A. P. Hall to F. G. Pettibone (Vice-President and General Manager), MEMO; A. P. Hall to Z. A. Green and W. H. Kluever, May 21; A. P. Hall to Z. A. Green,

MEMO; A. P. Hall to F. G. Pettibone, May 22; A. P. Hall to Z. A. Green, June 13, with handwritten notation dated June 14. See HMRC, RG D 004, GCSFR Records, box 312, folder 20.

87. W. A. Hudson to K. B. Duncan, June 6, 1929; A. P. Hall to W. E. Maxson, June 19; "Revised Ordinances of the City of Galveston, June 1st, 1917," with annotation, "copied June 6, 1929"; F. Merritt to [A. P.] Hall, July 24, Office of Chief Engineer, Form 1575, "Recommended Expenditure, $6,430.00, re: replacing concrete with steel floor, moving tank to original location"; Z. A. Green to A. P. Hall, August 31; and A. P. Hall to G. G. Moore, September 3. See HMRC, RG D 004, GCSFR Records, box 312, folder 20.

88. Hudson to Duncan, June 6, 1929 (quotations 1–2); Z. A. Green to A. P. Hall, June 19 (quotation 3); A. P. Hall to G. G. Moore, June 22 (quotation 5); [G. G. Moore] to A. P. Hall, June 25; A. P. Hall to W. E. Maxson, July 5; F. Merritt to W. E. Maxson, July 24 (quotation 4); W. E. Maxson to F. Merritt and A. P. Hall, August 6; and G. G. Moore to A. P. Hall, August 27 (quotations 6–10). See HMRC, RG D 004, GCSFR Records, box 312, folder 20.

CONCLUSION

1. Saleh and Marais, "Highlights from the Early (and Pre-) History of Reliability Engineering," *RE&SS* 91: 250–53, 254 (quotation); Keller and Modarres, "A Historical Overview of Probabilistic Risk Assessment Development," 271–85; T. A. Kletz, "The Origins and History of Loss Prevention," *Transactions of the Institution of Chemical Engineers* 77, part B (May 1999): 109–16; Marc Mangel and Francisco J. Samaniego, "Abraham Wald's Work on Aircraft Survivability," *Journal of the American Statistical Society* 79, no. 386 (June 1984): 259–67; W. Allen Wallis, "The Statistical Research Group, 1942–1945," Journal of the American Statistical Society 75, no. 370 (June 1980): 320–30; Joseph R. Fragola, "Reliability and Risk Analysis Data Base Development: An Historical Perspective," *RE&SS* 51, no. 2 (February 1996): 126–28; William Denson, "The History of Reliability Prediction," *IEEE Transactions on Reliability* 47, no. 3 (September 1998): 321–28; Anthony Coppola, "Reliability Engineering of Electronic Equipment: A Historical Perspective," IEEE Transactions on Reliability, vol. R-33, no. 1 (April 1984): 29–35.

2. On southern industrialization, see Richard H. Steckel and Roderick Floud, eds., *Health and Welfare during Industrialization* (1997), 2 (quotation 1); and Robert A. Margo, "The South as an Economic Problem: Fact or Fiction?" in *The South as an American Problem*, ed. Larry J. Griffin and Don H. Doyle (1995), 173–74; and David L. Carlton, "Unbalanced Growth and Industrialization: The Case of South Carolina," in *Developing Dixie: Modernization in a Traditional Society*, ed. Winfred B. Moore Jr., Joseph F. Tripp, and Lyon G. Taylor Jr. (1988), 112. On specific industries and areas, see David L. Carlton and Peter A. Colcanis, "Capital Mobilization and Southern Industry, 1880–1905: The Case of the Carolina Piedmont," *Journal of Economic History*, 49, no. 1 (March 1989): 73–94; and David L. Carlton and Peter A. Colcanis, "The Uninventive South? A Quantitative Look at Region and American Inventiveness," in *The South, the Nation, and the World*, 120; and Pamela C. Edwards, "Entrepreneurial Networks and the Textile Industry: Technology, Innovation, and Labor in the American Southeast, 1890–1925," in *Technology, Innovation, and Southern Industrialization: From the Antebellum Era to the Computer Age*, ed. Susanna Delfino and Michele Gillespie (2008), 125–62. On Houston and industrialization, see chapter 3, above.

EPILOGUE

1. City of Mobile, *The Charter and the Code of Ordinances of the City of Mobile, Alabama, With Appendix* (1907), chap. 61, sec. 589, p. 283; and *Ordinances, Grants and Franchises of the City of Mobile Since the Code of 1907* (1921), pp. 15–16, 233–35, 326–28, 416–19, 432–34.

2. City of Mobile, Board of Commissioners, *The Commission Government Charter and the Code of Ordinances of the City of Mobile, Alabama, with Appendix,* prepared by Shelton H. Hendrix (1934), chap. 86, article 1, secs. 722–24. See also Board of Commissioners, *Charter and the Code of Ordinances . . . Mobile* (1907), sec. 590, p. 283; and Board of Commissioners, *Ordinances, Grants and Franchises* (1921), p. 327.

3. City of Mobile, Board of Commissioners, *Charter and the Code of Ordinances . . . Mobile* (1907), chap. 61, sec. 592, pp. 284–85; Board of Commissioners, *Charter and the Code of Ordinances . . . Mobile* (1934), chap. 86, article 3, sec. 729, pp. 177–78; Board of Commissioners, *The Commission Charter and the Code of Ordinances of the City of Mobile, Alabama with Appendix,* prepared by Shelton H. Hendrix (1947), chap. 106, article 2, sec. 969, p. 213; chap. 108, sec. 978, p. 216; and Board of Commissioners, *The Code of the City of Mobile, Alabama 1955: The General Ordinances of the City,* sec. 19–31, pp. 150–51; sec. 19–36, p. 153.

4. City of Mobile, Board of Commissioners, *Charter and the Code of Ordinances . . . Mobile* (1907), sec. 591, pp. 283–84; Board of Commissioners, *Charter and the Code of Ordinances . . . Mobile* (1934), chap. 86, article 3, secs. 728–29, 177–78; chap. 86, article 2, secs. 725–27, 176–77; Board of Commissioners, *Charter and the Code of Ordinances . . . Mobile* (1947), chap. 107, secs. 974–76, 215; chap. 108, sec. 977, pp. 215–16; and Board of Commissioners, *The Code of the City of Mobile, Alabama 1955,* sec. 19–35, pp. 152–53. On municipal power to devise ordinances to "provide for the safety, preserve the health, promote the prosperity" of its residents, see *The Code of Alabama, Adopted by Act of the Legislature of Alabama,* 3 vols. (1907), vol. 1, article 18, sec. 1251, p. 609.

5. "City Adopts Material Standards Fire Code," *MDR,* March 8, 1972, 1-D; "Flurry of Legislation Proposed by City of Mobile," *MDR,* January 22, 1978, 7-C; "Section of Home Rule Bill 'Terrifies' Mims," *MDR,* January 25, 1978, 1-B; "Notice of Proposal By the Council of the City of Mobile To Adopt a Fire Prevention Code," *MDR,* February 3, 1986 , 5-B; "Council Adopts Fire Prevention Code Without Life-Safety Section," *MDR,* February 26, 1986, 1-B; "19-063-1986, An Ordinance to Adopt a Fire Prevention Code for the City of Mobile, Alabama," *MDR,* February 27, 1986, 4-F; "Notice of Proposal by the City Council of the City of Mobile to Adopt a Fire Prevention Code," *MDR,* March 27, 1992, 8-C, col. 1; "Mobile City Council Agenda for Noon, Tuesday, April 14," *MDR,* April 11, 1992, 2-B; and "City Council Postpones Decision on 5-Month NDC Contract Extension," *MDR,* April 15, 1992, 1-B.

6. On the legal history of the Office of State Fire Marshal, see Alabama Bureau of Insurance, *Insurance Laws of the State of Alabama,* comp. Russel P. Coleman (1928), article 24, sec. 333; Alabama Department of Insurance, State Fire Marshal's Office, Alabama State Fire Marshal, 2009 *Annual Report* (2009) (I thank Alesha Womble, Alabama State Fire Marshal's Office, for providing me with a digital copy of this item), 5; "An Act to Create the Office of State Fire Marshal, and to Provide for the Appointment of Deputy Marshals and Assistants to Define Their Powers and Duties and to Fix Their Compensation," *General Laws (and Joint Resolutions) of the Legislature of Alabama Passed at the Session of 1919* (1919), sec. 8, 1017 (quotation); and

Code of Alabama, title 36, chap. 19, article 1, sec. 36-19-9; see also "Alignment of City, State Fire Codes Eyed by Commission," *MR,* February 4, 1981, 1-B; "Fire Code Problems," March 1, 2-B; and *NFPA 101 Life Safety Code* (2000), "Origin and Development of NFPA *101,*" 1–2.

7. International Code Council, *IFC: International Fire Code 2009* (2009), preface, iii–v; chapter 34, "Flammable and Combustible Liquids," 319–58; see also Kevin A. Cassidy, *Fire Safety and Loss Prevention* (1992), 30 (quotation). Flammable and combustible liquids are covered in *International Fire Code 2012* in chapter 57, and fuel oil is addressed in section 603.

8. "An Ordinance to Adopt a Fire Code for the City of Mobile, Alabama," sec. 1, adopting the International Fire Code and cited appendices as the "Fire Code of the City of Mobile"; 11 pp., dated December 2, 2010. For adoption of the *International Fire Code 2012 Edition* (2012), as well as amendments and appendices, see Mobile, Alabama, *Code of Ordinances,* chap. 22, Article I.—Fire Prevention Code, sec. 22-1.—Code and appendices; sec. 22-2.—Amendments. For the State Fire Marshal, see *Alabama Administrative Code,* Alabama Department of Insurance, Fire Marshal Regulation, chap. 482-2-101, Fire Codes; Building Code, 482-2-101-.01 "Adoption of National Fire Codes (2003 ed.); exceptions"; and 483-2-101-.02 "Adoption of International Fire Codes (2009 ed.); exceptions." For discretion to use "later editions of fire or building codes promulgated by the NFPA® or the International Code Council® than those specified in this chapter," see 482-2-101-.05 "Use of later code editions" in chap. 482-2-101 noted above.

9. See *International Fire Code 2009* (2009), chap. 6, 54–55; chap. 34, "Flammable and Combustible Liquids," 319–58; and *International Fire Code 2012* (2012), chap. 6, sec. 603, "Fuel-Fired Appliances," 71–74; capacities are delimited in section 603.3.1–3.2.

10. City of Houston, *The Revised Code of Ordinances of the City of Houston of 1914* (1914), chap. 16, article 2, secs. 265 (quotation 1), 266–72; and *The Revised Code of Ordinances of the City of Houston of 1922,* comp. J. G. Hautier (1922), chap. 36, article 3, secs. 1835–37, 1838 (quotation 2), 1839–43.

11. City of Houston, *The Revised Code of Ordinances . . . Houston* (1914), chap. 16, article 4, secs. 278–92; see sec. 283 amending Article 904 of 1904 edition; *The Revised Code of Ordinances . . . Houston* (1922), sec. 1853; and *The Code of the City of Houston Texas* (1943), chap. 20, article 6, "Fuel Oil," secs. 649–61; see sec. 658 for inter-tank distance.

12. For Houston codes issued in loose-leaf binders that allowed for revisions by means of new-page insertions under numbered supplements, see *Code of Ordinances: City of Houston, Texas,* 2 vols. (1958), vol. 1, part 2, "Code of Ordinances," chap. 15A, "Fire Prevention," articles 1–8, supp. no. 6. See also *Code of Ordinances: City of Houston, Texas: General Ordinances of the City* , 3 vols. (1968), vol. 1, chap. 17, "Fire Code," articles 1–40, supp. No. 58; and *Code of Ordinances City of Houston,* Adoption Ordinance No. 85-1380, "An Ordinance Adopting a Codification of the General and Criminal Ordinances of the City of Houston, Texas," sec. 2, (a), 5 (the adoption of the fire code, chapter 17, of the 1968 *Code of Ordinances*, in 1973, and still in force); accompanied by explanations of the code's numbering system and the system of loose-leaf supplements. The adoption was effective August 21, 1985.

13. *Fire Code: City of Houston,* 1974.

14. Ibid., pp. viii–viii, iii, 959–62, supplement no. 21.

15. *Fire Code: City of Houston, Texas* (1974), Division 1, "General Provisions," 17-5.1(c) (quotation), 17-5.2–3.

16. Ibid., Division 2, "Tank Storage," 17-15.12–15, 19; Division 3, "Pumps, Piping, Valves and Fittings," 17-15.22.

17. City of Houston, *Houston Amendments to the 2006 International Fire Code, Adopted by Ord. No. 2010–871* (2010); according to ICC, *International Codes—Adoption by Jurisdiction (Updated 06-01-2011),* For Houston's adoptions of the IFC, 1973-2016, see https://edocs.publicworks.houstontx.gov/documents/divisions/planning/enforcement/1204_code_adoption_history_for_houston_codes.pdf

18. "Committee Named For Revision of Building Code," *GDN,* December 3, 1941, 1, col. 8; "Revised Building Code Expected To Be Ready March 1," *GDN,* February 2, 1942, 5, col. 1 (quotation); and "Board to Submit Library Tax Levy With Other City Amendments; Adopt Code," *GDN,* March 22, 1946, 1, cols. 4–5; "Increase Shown in Building Permits," *GDN,* March 26, 1946, 3, col. 4.

19. "City Still Waits For Decision on Building Code," *GDN,* October 22, 1942, 12, cols. 4–5; "Building Code Revision Awaits End of War," *GDN,* August 25, 1943, 12, cols. 3 (quotation 2), 4 (quotation 1); and City of Galveston, *Building Code, City of Galveston: Adopted October Twenty-Ninth, Nineteen Fourteen* (1914); *Building Code, City of Galveston 1946: Officially Adopted July 16, 1946* (1946).

20. National Board of Fire Underwriters, Committee on Fire Prevention, *Report on the City of Galveston, Texas: National Board of Fire Underwriters, Committee on Fire Prevention* (1910); *Report on the City of Galveston, Texas: (superseding that of 1925): National Board of Fire Underwriters, Committee on Fire Prevention and Engineering Standards* (1946); "Improvement in Fire Fighting Facilities Urged," *GDN,* August 27, 1946, 3, col. 1 (quotations 1–2); "City Officials Agree With Report's Finding," *GDN,* August 31, 1946, 12, col. 4; "50 New Members Elected to CC," *GDN,* September 7, 1946, 12, col. 5.

21. "Bond Issue Is Favored Among Candidates for Improvement of Fire Fighting Equipment," *GDN,* May 6, 1947, 1, cols. 5–6; "City Commissioners are Urged to Speedily Cut Fire Hazards," *GDN,* February 27, 1948, 8, cols. 1 (quotation 1), 2 (quotation 2).

22. City of Galveston, *Building Code, City of Galveston 1946* (1946), article 20, secs. 2000–2004, 99–100.

23. See "Standard Fire Prevention Code for Galveston Okayed," *GDN,* December 4, 1953, 21, cols. 4–6; "City Asked to Adopt Fire Code," *GDN,* October 29, 1954, 30, cols. 1–3; "City to Be Asked to Adopt Model Fire Prevention Code," *GDN,* October 10, 1956, 1, cols. 5–6; p. 6, col. 6; "New Fire Code Sought for City," *GDN,* October 11, 1956, 10, cols. 1–3; "Fire Losses Cited in Drive For Insurance Code Here," *GDN,* October 24, 1956, 8, cols. 4–5; "Revised Model Fire Prevention Code Recommended for Island," *GDN,* December 22, 1956, 10, cols. 1–2; "McCray's Formula: 'Positive Thinking, Unlimited Action,'" *GDN,* January 1, 1957, S2–15, col. 2; "Standard Code Asked For Fire Prevention," *GDN,* June 14, 1957, 1, cols. 1–2; "Fire Heads Inspecting Buildings Here," *GDN,* June 18, 1957, 2, col. 7; "Causeway Work of Chamber Is Cited," *GDN,* January 2, 1958, 2, col. 3; "1958 Galveston Chamber of Commerce Calendar Sets 150 Projects," *GDN,* January 17, 1958, 13, col. 4; and "Legal Notices, Ordinance no. 66-[8], An Ordinance Adopting a Fire Prevention Code," *GDN,* March 4, 1966, 7-B, cols. 5–6; "Legals, Ordinance No. 66-36, An Ordinance Amending the Fire Prevention Code of 1965 as Adopted by Ordinance No. 66-8," *GDN,* June 7, 1966, 4-B, col. 1.

24. "Ordinance No. 71-10, An Ordinance Amending Chapter 8 of the Code of the City of Galveston, Texas, 1950," *GDN,* March 4, 1971, 9-B, cols. 6 (quotation 1), 7; "Ordinance No. [75-40], An Ordinance Designating the Fire Marshall as Chief Inspector and Chief of the Bureau of Fire Prevention," *GDN,* July 22, 1975, 11, col. 8; cont'd as "Inspector Cites Deficiencies," *GDN,* April 1, 1977, 1, cols. 1–4; p. 2-A, col. 1; "All Downtown Public Buildings To Be Inspected Soon," *GDN,* July 14, 1977, 1, cols. 1–8; p. 2-A, col. 4. "Look Into $497,147 Street Program Asked by Council," *GDN,* May 5, 1978, 1, cols. 5–8; cont'd as "Council Agenda," p. 2-A, col. 5. "Iran Group Allowed To Hold March," *GDN,* June 2, 1978, 2-A, col. 4; "Ordinance no. 78-61, An Ordinance Amending Chapter 8 (Fire Protection) of the Code of the City of Galveston, Texas, 1960," *GDN,* August 4, 1978, 7-B, col. 8; "Business Fire Checks Begun," *GDN,* October 16, 1978, 3-A, cols. 1–2 (quotation 2).

25. "In Wake of Fires: Additional City Inspections Due In Strand Area," *GDN,* December 31, 1978, 1, col. 5; "Restaurant Owners Warned of Dangers: Fire Marshal Urges Compliance," *GDN,* September 5, 1, 1979, col. 5; cont'd p. 2-A, cols. 6–7. "Fire Inspections Aimed at Public Safety, Not at Picking on People," *GDN,* October 14, 1979, 1, cols. 5–8; cont'd p. 2-A, cols. 1–2. "Most Galveston Buildings Comply with Fire Codes," *GDN,* December 5, 1980, 1, cols. 1–5; "Legal Notice, An Ordinance of the City of Galveston . . . Adopting the 1994 Standard Fire Prevention Code," *GDN,* November 19, 1995; "Ordinance No. 10-036 . . . Adopting the 2009 of the 'International Fire Code' . . . Adding the 2009 'NFPA 101 Life Safety Code' as Chapter 48 of the 2009 Edition of the 'International Fire Code,'" www.galvestonfire.com/pdfs/cogfireord/.

26. For the proposed authority to be held by the commissioner of the Fire Department to enter premises to ascertain violations of the NBFU building code regarding "unlawful quantities of explosives . . . petroleum and coal oils" and chemicals, see National Board of Fire Underwriters, *Building Code Recommended by The National Board of Fire Underwriters* (1905), pt. 34, sec. 162, pp. 250–51.

27. Powell Evans, "Fire Waste and Prevention," in *A Five Years' Fight against Fire Waste and Its Possible Control in the United States by Fire Prevention and Protection,* 171 (quotations 1–2) (1912). See also Alan Waring and A. Ian Glendon, *Managing Risk* (1998), who observed that insurance markets traditionally have been in the "business of risk transfer" (101).

28. See National Board of Fire Underwriters, *Building Code Recommended by The National Board of Fire Underwriters,* 1905 ed., and the *Building Code Recommended by The National Board of Fire Underwriters,* 4th ed. (1915).

29. On the Louisiana Fire Prevention Bureau and the Office of Fire Marshal, see "The Senate," *NODP,* July 7, 1904, 2, col. 3; "Governor's Opinion," *NODP,* July 10, 1904, 26, col. 4; "Insurance Laws," *NODP,* July 10, 1904, col. 7; "Saw Mill Insurance," *NODP,* April 20, 1906, 2, col. 3; "Hughes Explains Electric Bill," *NODP,* June 11, 1906, 5, cols. 3-4; "Fire Prevention Bureau Hearing," *NODP,* July 3, 1906, 5, cols. 6–7; "Special Investigations," *NODP,* July 15, 1906, 26, col. 2; "Review of the Work Done by the Louisiana Legislature," Act 152, *NODP,* July 15, 1906, col. 7; "Fire Prevention Bureau Suggests," *NODP,* September 7, 1907, 4, col. 4; "Legislative Day Full of Activity," *NODP,* May 22, 1908, 1, cols. 2–3.

30. On building fire-code deficiencies and conflagrations in New Orleans, see "Fire Prevention Bureau Suggests," *NODP,* September 6, 1907, 4, col. 4; "Fire Prevention Bureau Meeting," *NODP,* October 16, 1907, 2, cols. 5–6; "Cause of Dreyfous Fire Very Plain," *NODP,* May 12, 1908, 4, cols. 3–4; on updating the city building code and fire-fighting shortcomings with the city's

response, see "A Building Department Vital Necessity Here," May 14, 1908, 5, cols. 3–4; "Fire Department Condemned/Insurance Rates are Raised," August 5, 1908, 1, cols. 1–2; "Defense of Advanced Insurance Rates," August 7, 1908, 4, cols. 1–2; "Hardee Answers Bloodworth Some More," August 12, 1908, 5, cols. 4–5.

31. "Building Code Ready," *NODP*, November 21, 1908, 12, col. 3; and "Insurance Hearing By Board of Trade," June 6, 1909, 11, cols. 4, 5 (quotation 1), 6.

32. On electrical issues see "City Electricity/New Building Code Rules Have Improved Installation," *NODP*, September 1, 1912, 12, col. 6; "City Electricity," *NODP*, September 12, 1912, 12, col. 6; "Electrical Code Proposed as Aid to Householders," *NODP*, March 5, 1932, 2, col. 2; on the LFPB, see "Does Not Pass on Details of Work," *NOI*, August 24, 1913, S1-7, col. 1; on fire prevention, see "Fire Prevention in New Orleans Wins Recognition," *NOTP*, March 18, 1927; on the association, see Anthony J. Stanonis, *The Big Easy: New Orleans and the Emergence of Modern Tourism, 1918–1945* (2006), 33, 34 (quotation 1); for the association's commitment to expertise regarding a proposed harbor-terminal project, see John H. Bernhard, *New Orleans: As the World's Greatest Port* (1921), 103.

33. On the mayor, see "Walmsley Acts for Lower Basic Fire Rate in City," *NOTP*, October 18, 1931, [13], cols. 4–5; on Strong, see "Funeral Held for Strong, 96," *NOTP*, March 2, 1975, S1–20, cols. 1–2; on Stone, see "City Government Reforms Needed, Declares Stone, *NOTP*, January 31, 1934, part 2, p. 11, col. 2; and "Traffic Hazard Remedy Sought by A. of C. Bureau," *NOTP*, February 2, 1934, 18, col. 4. On the NBFU's 1934 report on New Orleans, see "Fire Prevention Study Opens Here," *NOTP*, April 4, 1946, 2, col. 7; on the 1931 (5th) edition of the NBFU's building code as reprinted in 1934, see NBFU, *Building Code Recommended by The National Board of Fire Underwriters*, 5th ed. (1934), viii.

34. "Compulsory Fire Checks Asked for Rooming Houses," *NOTP*, November 30, 1944, 1, col. 1; cont'd "Fire Checks," p. 2, cols. 1–2, 3 (quotation 1), 4; also "Fire Prevention Study Opens Here," *NOTP*, April 4, 1946.

35. "Fire Prevention Ordinance Ready for Council Action," *NOTP*, January 2, 1945, 1, cols. 2–3; cont'd "Fire Prevention," p. 2, cols. 3 (quotation 1), 4. For postwar calls for municipalities to upgrade their building codes and fire rules, see "Outmoded Codes Factor in Higher Building Costs," *NOTP*, February 20, 1944, S4-01, col. 8.

36. "Fire Prevention Discussions Set," *NOTP*, December 11, 1946, 4, col. 3; "Stairwells Offer Threat of Fire, Engineer Warns," *NOTP*, December 12, 1946, 1, cols. 2–3; cont'd "Fire Protection," p. 2, col. 5. "Committee Reviews Draft of Fire Prevention Code," *NOTP*, December 18, 1946, 17, cols. 2–3; and "Statewide Fire Meeting Mapped," *NOTP*, December 25, 1946, part 2, p. 15, cols. 5 (quotation 1), 6.

37. "Experts Chosen for City Survey," *NOTP*, February 28, 1946, 1, col. 2 (quotation 1); cont'd "Experts Named for City Survey," p. 2, cols. 3 (quotation 2), 4. For suggested reorganization of city services, based in the case of fire protection on the NBFU survey, see "Revisions in City Setup are Urged," *NOTP*, August 29, 1946, 18, cols. 3–5.

38. "Fire Conference Termed Success," *NOTP*, May 11, 1947, 18, col. 4; and "New Code Ready on City Building," *NOTP*, November 25, 1948, 4, col. 1.

39. For a discussion of the relationship between fire-prevention and building codes, see International City Managers' Association, Institute for Training in Municipal Administration, *Municipal Fire Administration*, 7th ed. (1967), 228–32; and J. E. Bihr, "Building and Fire Codes:

The Regulatory Process," in Bihr, *Fire Standards and Safety* (1976): 209–18. See also the 1934 reprint noted above of the NBFU *Building Code,* sec. 1212, on oil-burning equipment that referred readers to the NBFU's regulations on oil consumption (NFPA 31).

40. On code changes and conflicts, see "Abbreviated Fire Code Now Ready," *NOTP,* January 22, 1950, S15-5, col. 3; "Adoption of New Fire Code Asked," *NOTP,* February 1, 1955, 16, col. 5; "Report Urges 16 Fire Inspectors," *NOTP,* February 2, 1955, 7, col. 1; "Clasen Elected to Head Council," *NOTP,* April 29, 1955, 18, cols. 3–5; "Oil Stove Ban," *NOTP,* April 30, 1958, 10, col. 1; "Voids, Overlappings Noted by Fire Prevention Group," *NOTP,* April 14, 1966, S2-09, cols. 3–4. See also the NBFU, *Fire Prevention Code, Abbreviated Edition* (1965); *Fire Prevention Code: A Code Prescribing Regulations Governing Conditions Hazardous to Life and Property* (1960); and AIA, *Fire Prevention Code: A Code Prescribing Regulations Governing Conditions Hazardous to Life and Property from Fire, also, a Suggested Ordinance Adopting the Fire Prevention Code and Establishing a Bureau of Fire Prevention* (1965), 227.

41. "Revision of Fire Code Submitted," *NOTP,* February 3, 1970, S1-08, cols. 3, (quotations 1–2), 4.

42. "Second Phase of Fire Code Hearings Begins," *NOTP,* October 2, 1973, S1-04, cols. 1–5; on sprinklers, see French, *1860–Fifty Years–1910 Arkwright Mutual Fire Insurance Company,* 42–53, 59; Steeb, *Agents' and Inspectors Pocket-Book of Fire Protection* (1902), 107–21; Brearley, *History NBFU,* 166–67; Wermiel, *Fireproof Building,* 112–13.

43. "Sprinklers Use Benefits Seen," *NOTP,* February 7, 1974, S1-16, col. 4; "Grisbaum Seeks New Fire Office," *NOTP,* May 2, 1974, S1-13, col. 1.

44. "Better Fire Safety Laws Eyed in Legislature," *NOTP,* June 16, 1974, S3-06, cols. 1–8; "House Passes Sprinkler Bill," *NOTP,* June 29, 1974, S3-06, cols. 1–4; "Fire Code Debate On," *NOTP,* July 4, 1974, S3-06, cols. 1–4; "Fire Code Board Bill," *NOTP,* July 5, 1974, 1, col. 2; "Unit Okays Fire Code Bill," *NOTP,* July 5, 1974, S1-12, cols. 1–2.

45. On the *SSFPC* and New Orleans, see New Orleans, City Council Minutes, Calendar Ordinance 20,125, adopted September 17, 1995, "An Ordinance to Amend and Reordain Ordinance 4912 (known as the Fire Prevention Code of the City of New Orleans) Relative to the Standard Fire Prevention Code"; Ordinance 18,822 M.C.S., adopted August 6, 1998, "An Ordinance to Amend and Reordain Ordinance No. 11,625, M.C.S., as Amended to Conform to the Standard Building Code, 1997 Edition," subsection 906.2: "Detail Requirements," 1; Ordinance 21145, adopted June 19, 2003, "An Ordinance to Amend and Reordain Ordinance No. 11,625, M.C.S., as Amended, to Conform to the International Building Code, 2000 Edition, and Appendix Thereto," subsection 101.4.6: "Fire Prevention," www.library1.municode/minutes/home.htm. On adoption of the *IFC,* see City of New Orleans, Fire Prevention Code, "Amendments to the 2006 International Fire Code," Ordinance 22,958 M.C.S, approved January 10, 2008; to adopt and amend the *IFC,* 2006 edition; and chapter 34, "Flammable and Combustible Liquids," subsection 3404.2.9.5.1, which excepted underground tanks "legally installed prior to adoption" (text and note quotations).

46. On technological rationality, see Denis Goulet, "Three Rationalities in Development Decision-making," *World Development* 14, no. 2 (February 1986): 302 (quotation).

47. On risk in the United States, see Richard D. Brown, "Modernization and the Modern Personality in Early America, 1600–1865: A Sketch of a Synthesis," *Journal of Interdisciplinary History* 2, no. 3 (Winter 1972): 211.

BIBLIOGRAPHY

MANUSCRIPTS

Alabama Department of Archives and History. Private Records. Zebulon Montgomery Pike Inge Papers, 1833–1918.

Filson Historical Society. Louisville, Ky. Louisville Board of Underwriters. Minute Books, vols. 5–6.

Galveston, Tex. Galveston City Council. City Council Records, 1898–1901. "Minutes of City Council."

Hagley Museum and Library. George H. Gilbert & Co. Collection. Accession No. 2178.

Harris County, Texas. Office of the Clerk. *Harris County Deed Records*. Vol. R.

———. 55th Judicial District Court. [Case No.] 34485.

The Historical Society of Pennsylvania. Am 35812. Philadelphia Board of Fire Underwriters, "Minutes, 1853–1881."

Houston Public Library. Houston Metropolitan Research Center. RG D 004. Santa Fe Railroad Collection. Records of the Gulf, Colorado, and Santa Fe Railway Co., 1897–1968.

Jefferson Parish, Louisiana. Parish Clerk Office. Mortgage Book, vol. 38.

John Ripley Freeman Papers. MIT. Cambridge, Mass.

Joseph Cullinan Papers. Special Collections. University of Houston Libraries.

Mobile County Chancery Court Records. McCall Rare Books & Manuscript Library. University of South Alabama.

Mobile Municipal Archives. Mobile, Ala.

New Orleans Board of Trade. Minutes, Board of Directors. Vol. 3: March 6, 1895, to September 26, 1903, December 4, 1901.

New Orleans City Council. Official Proceedings Office of City Secretary.

New Orleans Public Library. City Archives. Capdevielle Records. Correspondence City Engineer. Outgoing Correspondence.

Sanborn Fire Insurance Maps. New Orleans. 1908–9.

Office of City Secretary. Galveston, Tex. Galveston City Council. City Council Minutes. Vol. 2. "Proceedings of the Board of Commissioners of the City of Galveston."

Office of the City Secretary. Houston, Texas.
City Council Minutes.
City Ordinance Book.
Crude Oil Ordinance Folder.
Miscellaneous Papers 1900–1903.
Orleans Parish, Louisiana. Records of the Levee Board.
Philadelphia, Pennsylvania. Ace Insurance Company Archives (now CHUBB), Records of the Insurance Company of North America.
Directors Minutes, February 1803–February 1812, book 2.
Directors Minutes, January 1877–September 1887, vol. 7.
Directors Minutes, January 1895–April 1904, vol. 9.
Texas State Archives, Austin. Records of case no. 34,485.
University of New Orleans. Special Collections. Historical Archive of the State Supreme Court of Louisiana.
University of Texas at Austin. University of Texas Libraries. Perry-Castañeda Library Map Collection. Sanborn Fire Insurance Maps. Texas (1877–1922).

NEWSPAPERS

Argus and Patriot (Montpelier, Vt.).
Baltimore Sun.
Boston Daily Advertiser.
Boston Evening Journal.
Boston Herald.
Chicago Herald.
Cincinnati Commercial Gazette.
Cincinnati Daily Gazette.
Cleveland Daily Leader.
Cleveland Leader and Morning Herald.
Cleveland Morning Leader.
Cleveland Plain Dealer.
Daily Evening Telegraph (Philadelphia).
Daily Evening Traveller (Boston).
Daily Inter Ocean (Chicago).
The Daily Messenger (St. Albans, Vt.).
Galveston Daily News Houston Chronicle Houston Daily Post.
Lexington Leader (Lexington, Ky.).
Manufacturers' Record.
Mobile Daily Commercial Register & Patriot.
Mobile Daily Register.
Mobile Register.

Montgomery Advertiser & Gazette.
New Orleans Daily Item.
New Orleans Daily Picayune.
New Orleans Times-Democrat.
New Orleans Times-Picayune.
New York Commercial Advertiser.
New York Daily Graphic.
New York Daily Tribune.
New York Herald.
New York Sun.
New York Times.
New York Tribune.
North American (Philadelphia).
The North Star (Danville, Vt.).
Petroleum Centre Daily Record (Petroleum Centre, Pa.).
Philadelphia Daily Evening Telegraph.
Philadelphia Inquirer.
Press and Horticulturist (Riverside, Calif.).
San Francisco Chronicle.
San Francisco Evening Bulletin.
Springfield Daily Republican.
Springfield Republican.
St. Johnsbury Caledonian (St. Johnsbury, Vt.).
Weekly Portage Sentinel (Ravenna, Ohio).
Western Reserve Chronicle (Warren, Ohio).

PERIODICALS AND SERIALS

The American Chemist.
The American Engineer: An Illustrated Weekly Journal.
The American Exchange and Review .
American Gas Light Journal.
American Manufacturer and Iron World.
The Bankers' Insurance Managers' & Agents Magazine.
Bulletin of the Executive Committee of the National Board of Fire Underwriters. Volume 1 November 1873–April 1874; Volume 2 January–December, 1875. Bound with Index to Bulletin. Library of Insurance Library Association. Boston.
The Chemical Gazette.
The Chemical News and Journal of Physical Science.
The Chemical News and Journal of Physical Science. American Supplement.
The Chronicle A Journal Devoted to the Interests of Insurance.

Electrical Engineering: An Illustrated Monthly Magazine.
Electric Railway Journal.
The Electric World.
The Electrical Engineer: A Weekly Review of Theoretical and Applied Electricity.
The Engineering and Mining Journal.
Gas Age-Record.
Hunt's Merchants' Magazine and Commercial Review.
Insurance Engineering.
The Insurance Journal (Hartford).
The Insurance Monitor.
The Insurance Press.
The Insurance Times: A Journal Solely Devoted to Life, Fire and Marine Insurance.
The Iron Age.
The Iron Trade Review.
The Journal of the Franklin Institute.
The Mechanic's Magazine and Journal of Engineering, Agricultural Machinery, Manufactures, and Shipbuilding.
The Mining Magazine: Devoted to Mines, Mining Operations, Metallurgy.
National Cotton Manufactures' Association. Proceedings.
National Cotton Manufactures' Association. Transactions.
The Oil Investors' Journal.
Paint, Oil and Drug Review.
Proceedings of the American Gas Institute.
Proceedings of the American Gas-Light Association.
Proceedings of the American Philosophical Society.
Popular Science Monthly.
Quarterly of the National Fire Protection Association.
The Spectator.
The Standard.
Stevens Institute Indicator.
United States Gazette & Magazine of Useful Knowledge.
The Vindicator (Atlanta-New Orleans).
The Weekly Underwriter: An Insurance Newspaper.

LEGAL

"An Act further to provide for the Safety of the Lives of Passengers on Board of Vessels propelled in Whole or in Part by Steam, to regulate the Salaries of Steamboat Inspectors, and for other Purposes." 14 Stat. 234 at 227–29 (1866).
"An Act To amend an act entitled an act to incorporate the port of Mobile, and provide for the government thereof. Approved February 11th, 1879." *Acts of the*

General Assembly of Alabama, Passed at the Session of 1886–87. Montgomery, Ala.: W.D. Brown, 1887.

"An Act to Create the Office of State Fire Marshal, and to Provide for the Appointment of Deputy Marshals and Assistants to Define Their Powers and Duties and to Fix Their Compensation," *General Laws (and Joint Resolutions) of the Legislature of Alabama Passed at the Session of 1919*. Montgomery, Alabama: Brown Printing Co., 1919.

"An Act to provide for the better Security of Life on board of Vessels propelled in Whole or in Part by Steam, and for other Purposes." 16 Stat. 100, at 440–459 (1871).

Cities Service Co. v. McDowell et al. 120 Me., at 384–87 (1922).

City of Bloomington v. James M. Bay. 42 Ill. 503, at 506–8 (1867).

D. R. Beatty et al. v. Charles W. Bulger. 28 Tex. Civ. App., at 117–23 (1903).

Hart v. Buckner et al. 5 C.C.A., at 9 (1892).

H. C. Still v. City of Houston. 27 Tex. Civ. App. 447, at 447–49 (1901).

J. E. Rose v. the State. 19 Tex. Ct. App., at 470–71 (1886).

Mechanics' & Traders' Insurance Co. v. Thompson. 21 S.W., at 468–69 (1893).

Michigan Shingle Company v. State Investment and Insurance Company of San Francisco. 53 N.W., at 945–51 (1892).

Patterson v. Kentucky. 97 U.S., at 501–9. In 1 A.R.C., at 1378–81 (1912).

United States. 42 U.S.C. 85 (2013).

BOOKS

Abbott, Carl. *Boosters and Businessmen: Popular Economic Thought and Urban Growth in the Antebellum Middle West*. Westport, Conn.: Greenwood Press, 1981.

Alabama, State of. *Alabama Administrative Code*. Alabama Department of Insurance. Fire Marshal Regulation, chap. 482-2-101, Fire Codes; Building Code. www.firemarshal.alabama.gov/PDF/FM-101r-2010.pdf.

———. *Acts of the General Assembly of Alabama Passed at the Session of 1894–5*. Montgomery: State of Alabama, 1895.

———. *Annual Report of the Insurance Commission of the State of Alabama . . . December 31, 1901*. Fire, Life and Miscellaneous Insurance. Montgomery, Ala.: Brown Printing Co., 1902.

———. *Alabama Bureau of Insurance. Insurance Laws of the State of Alabama*. Compiled by Russel P. Coleman. Birmingham, Ala.: Brown Printing Co., 1928.

———. *The Code of Alabama 1896*. Prepared by William L. Martin. Atlanta: Foote & Davies Co., 1897.

———. *The Code of Alabama*. codes.findlaw.com/al/title-36-public-officers-and-employees/al-code-sect-36-19-9.html.

———. *The Code of Alabama, Adopted by Act of the Legislature of Alabama*. 3 vols. Prepared by James J. Mayfield. Nashville: Marshall & Bruce Co., 1907.

Alexander, Denise. *Galveston's Historic Downtown and Strand District*. Charleston, S.C.: Arcadia Publishing, 2010.

Alabama Department of Insurance. State Fire Marshal's Office. Alabama State Fire Marshal. 2009 *Annual Report*. 2009.

American Insurance Association. *Fire Prevention Code: A Code Prescribing Regulations Governing Conditions Hazardous to Life and Property from Fire, also, a Suggested Ordinance Adopting the Fire Prevention Code and Establishing a Bureau of Fire Prevention*. New York: The Association, 1965.

———. *Fire Prevention Code: A Code Prescribing Regulations Governing Conditions Hazardous to Life and Property from Fire, also A Suggested Ordinance Adopting the Fire Prevention Code and Establishing a Bureau of Fire Prevention*. New York: Engineering and Safety Service, 1976. Supersedes 1970 edition with 1972, 1973, 1974 revisions.

The American Ruling Cases as Determined by the Courts. 5 vols. Chicago: National Law Book Co., 1912.

America's Successful Men of Affairs: An Encyclopedia of Contemporaneous Biography. Ed. Henry Hall. 2 vols. New York: Tribune Association, 1895–96.

Amos, Harriet E. *Cotton City: Urban Development in Antebellum Mobile*. Tuscaloosa: University of Alabama Press, 1985.

Andreas, A. T. *History of Chicago From the Earliest Period to the Present Time*. Vol. 3: *From the Fire of 1871 Until 1885*. Chicago: A. T. Andreas Co., 1886.

Andros, Stephen O. Fuel Oil in Industry. Chicago: Shaw Publishing Co., 1920.

Angell, Joseph K., and Thomas Durfee. *A Treatise on the Law of Highways*. Boston: Little, Brown and Co., 1857.

Ash, Mark. *The New York City Consolidation Act, as in Force in 1891, With Notes Indicating the Statutory Sources, References to Judicial Decisions, and All Laws Relating to New York City, Passed Since January 1, 1882*. Albany, N.Y.: Weed, Parsons and Co., 1891.

Atchison, Topeka, and Santa Fe Railway Co. *Eighth Annual Report of the Atchison, Topeka, & Santa Fe Railway Company for the Fiscal Year Ending June 30, 1903*. New York: Atchison, Topeka, and Santa Fe Railway Co., 1903.

Atkinson, Edward. *The Prevention of Loss by Fire and the System of Factory Mutual Insurance: An Address by Edward Atkinson, of Boston*. Chicago: John Morris Co., 1885.

———. *The Prevention of Loss by Fire: Fifty Years' Record of Factory Mutual Insurance Based on Thirty-Five Years' Personal Experience with Suggestions for the Protection of Cities from Conflagrations*. Boston: Damrell & Upham, 1900.

Banks, Erik. *Catastrophic Risk: Analysis and Management*. Chichester, West Sussex, U.K.: John Wiley & Sons, 2005.

Barnard, Frederick A. P., and Arnold Guyot, eds. *Johnson's (Revised) Universal Cyclopædia: A Scientific and Popular Treasury of Useful Knowledge*. 16 vols. New York, 1886.

Barnes, Donna A. *The Louisiana Populist Movement, 1881–1900*. Baton Rouge: Louisiana State University Press, 2011.

Barnes, L. Diane, Brian Schoen, and Frank Towers. *The Old South's Modern Worlds: Slavery, Region, and Nation in the Age of Progress*. New York: Oxford University Press, 2011.

Battle, John Rome. *The Handbook of Industrial Oil Engineering*. Philadelphia: J. B. Lippincott Co., 1920.

Beach, Lee Roy. *The Human Element: Understanding and Managing Employee Behavior*. Armonk, New York: M. E. Sharpe, 2007.

Beckmann, Martin J. "An Economic Model of Urban Growth." In *Cities and Their Vital Systems: Infrastructure Past, Present, and Future*, ed. Jesse H. Ausubel and Robert Herman. Washington, D.C.: National Academy Press, 1988.

Berg, Walter G. *Buildings and Structures of American Railroads: A Reference Book for Railroad Managers, Superintendents, Master Mechanics, Engineers, Architects, and Students*. New York: John Wiley & Sons, 1893.

Bernard, John H. *New Orleans: As the World's Greatest Port*. [New Orleans]: New Orleans Association of Commerce, 1921.

Best's Insurance Reports upon all American and Foreign Joint-Stock Companies, and American Mutual Companies and "Lloyds" Associations. 2nd annual ed. New York: Alfred M. Best Co., 1901.

Bihr, J. E. "Building and Fire Codes: The Regulatory Process." In *Fire Standards and Safety: A Symposium Presented at National Bureau of Standards. Gaithersburg, Md., April 5–6, 1976*. Ed. A. F. Robertson. Washington, D.C.: American Society for Testing and Materials, 1976. 209–18.

Birdseye, Charles F. *The Revised Statutes, Codes and General Laws of the State of New York. 3 vols. 2nd ed.* New York: Baker, Voorhis & Co., 1896.

Bixel, Patricia Bellis, and Elizabeth Hays Turner. *Galveston and the 1900 Storm: Catastrophe and Catalyst*. Austin: University of Texas Press, 2000.

Black's Law Dictionary. 6th ed. St. Paul, Minn.: West Publishing Co., 1990.

Blue Book of American Shipping: Marine and Naval Directory of the United States . . . 1901. Cleveland, Ohio: Marine Review Publishing Co., 1901.

Booth, James C., and Campbell Morfit. *On Recent Improvements in the Chemical Arts. In Smithsonian Miscellaneous Collections*. Vol. 2. Washington, [D.C.], Smithsonian Institution, 1862.

Booth, William H. *Liquid Fuel and Its Combustion*. Westminster, U.K.: Archibald Constable & Co., 1903.

Boston Manufacturers Mutual Fire Insurance Co. *Reports*. Vol. 19. HG8018 .I59. Cornell University Library.

Brearley, Harry Chase. *The History of the National Board of Fire Underwriters: Fifty Years of a Civilizing Force*. New York: Frederick A. Stokes Co., 1916.

Breen, David H. *Alberta's Petroleum Industry and the Conservation Board*. Edmonton: University of Alberta Press, 1993.

Brightly, Frank F. *Laws and Ordinances of the City of Philadelphia, From the Year One Thousand Seven Hundred and One to the Twenty-First Day of June, One Thousand Eight Hundred and Eighty-Seven*. Philadelphia: Kay & Brother, 1887.

Brightly, Frederick C. *Annual Digest of the Laws of Pennsylvania for the Years 1862 to 1868: From 21 May 1861 to 1 August 1868, Together with Some Laws of Older Date inadvertently omitted in Purdon's Digest, Completing Brightly's Purdon's Digest To The Present Date*. Philadelphia: Kay & Brother, 1868.

———. *Annual Digest of the Laws of Pennsylvania for The Years 1873 to 1878. Together with Some Laws of Older Date inadvertently omitted in Purdon's Digest. Completing Brightly's Purdon's Digest To The Present Date*. Philadelphia: Key & Brother, 1878.

Brown, Charles Carroll, ed. *Directory of American Cement Industries. 4th ed*. Indianapolis: Municipal Engineering Co., 1906.

Brownell, Blaine A., and David R. Goldfield, eds. *The City in Southern History: The Growth of Urban Civilization in the South*. Port Washington, N.Y.: Kennikat Press, 1977.

Bugbee, Percy. *Men against Fire: The Story of the National Fire Protection Association 1896–1971*. Boston: National Fire Protection Association, 1971.

Calvert, Monte A. *The Mechanical Engineer in America, 1830–1910: Professional Cultures in Conflict*. Baltimore: Johns Hopkins University Press, 1967.

The Cambridge Economic History of the United States. Vol. 2: *The Long Nineteenth Century*. Ed. Stanley L. Engerman and Robert E. Gallman. New York: Cambridge University Press, 2000.

Caracristi, C. F. Z. *Coal and Petroleum in Columbia: Commercial Information*. Washington, D.C.: Bureau of the American Republics, 1893.

———. *Coal, Oil, and Asphaltum of the Republic of Colombia*. Washington, D.C.: [Parks & Co.], 1895.

Caranci, Paul F. *North Providence: A History and the People Who Shaped It*. Charleston, S.C.: History Press, 2012.

Carlton, David L., and Peter A. Coclanis. *The South, the Nation, and the World: Perspectives on Southern Economic Development*. Charlottesville: University of Virginia Press, 2003.

Cassidy, Kevin A. *Fire Safety and Loss Prevention*. Stoneham, Mass.: Butterworth-Heinemann, 1992.

Chamberlain, John. *Farewell to Reform: the Rise, Life and Decay of the Progressive Mind in America*. 2nd ed. 1933. Rpt. Gloucester, Mass.: Peter Smith, 1958.

Chandler, Charles F. *Report on Petroleum as an Illuminator, and the Advantages and Perils Which Attend Its Use; with Special Reference to the Prevention of the Traffic in Dangerous Kerosene and Naphtha*. New York: New York Printing Co., 1871.

The Chicago Fire and the Fire Insurance Companies: An Exhibit of the Capital, Assets, and Losses of the Companies, Together with a Graphic Account of the Great Disaster, Accompanied by Maps of Chicago Showing the Burned District. New York: J. H. and C. M. Goodsell, 1871.

Chubb, John E. *Interest Groups and the Bureaucracy: The Politics of Energy*. Stanford, Calif.: Stanford University Press, 1983.

City of Boston. *The Charter and Ordinances of the City of Boston, Together with the Acts of the Legislature Relating to the City, and an Appendix*. Boston: Boston City Council, 1856.

———. *Documents of the City of Boston, for the Year 1866*. Vol. 2: *Containing Documents from No. 67 to No. 138, Inclusive*. Boston: Alfred Mudge & Son, 1867.

———. *Documents of the City of Boston, For the Year 1872*. In Three Volumes. Boston: Rockwell & Churchill, 1873.

———. *Ordinances and Rules and Orders of the City of Boston, Together with the General and Special Statutes of the Massachusetts Legislature Relating to the City*. Boston: Alfred Mudge & Son, 1869.

———. *The Revised Ordinances of the City of Boston*. 8th rev. Boston: Rockwell and Churchill, 1882.

City of Chicago. *Laws and Ordinances Governing the City of Chicago, From April 2, 1890, to July 10, 1894. With Notes and Cross References to All Amended and Repealed Sections of Prior Ordinances. Being a Supplement to the Compilation of April 2, 1890*. Comp. Henry Binmore. Chicago: E. B. Myers and Co., 1894.

———. *The Municipal Code of Chicago: Comprising the Laws of Illinois Relating to the City of Chicago, and the Ordinances of the City Council*. Codified and rev. Egbert Jamieson and Francis Adams. Chicago: Beach, Barnard & Co., 1881.

———. *Ordinances of the Building and Health Departments of the City of Chicago and the Annexed Towns of Lake View, Hyde Park and the Town of Lake, in Force, March 1, 1892*. 3rd ed. Comp. D. O. Wilkie. Chicago: A. L. Swift & Co., 1892.

———. *Report of the Board of Health of the City of Chicago, for 1867, 1868 and 1869; and a Sanitary History of Chicago; From 1833 to 1870*. Chicago: Lakeside Publishing and Printing Co., 1871.

City of Galveston. *Building Code, City of Galveston: Adopted October Twenty-Ninth, Nineteen Fourteen*. Galveston: F. F. Hunter, 1914.

———. *Building Code, City of Galveston 1946: Officially Adopted July 16, 1946*. Galveston: Pfeifer Agency, 1946.

———. *Charter and Revised Ordinances of the City of Galveston, and All Ordinances in Force to December 1, 1875, The Standing Rules of the City Council, and Catalogue

of the City Government From 1839 to 1875, Inclusive. Galveston: "News" Book and Job Establishment, 1875.
———. *Charter and Revised Ordinances of the City of Galveston, and All Ordinances in Force to October 1883; The Standing Rules of the City Council; Rules and Regulations Governing the Police Force; Certain Acts of the Legislature of the State of Texas Affecting the City of Galveston; and Catalogue of the City Government From March, 1839, to October, 1883, Inclusive*. Galveston: Clarke & Courts, 1883.
———. *Charter of the City of Galveston and the Revised Ordinances as Ordained and Published by the Board of Commissioners of the City of Galveston*. Digested and rev. George P. Finlay. Galveston: Clarke & Courts, 1902.
———. *Charter of the City of Galveston with the Amendments Thereto and the Revised Ordinances As Ordained and Published by the City Council of the City of Galveston*. Digested and rev. Henry W. Rhodes. Galveston: Strickland Printing Co., 1893.
———. *Code of Ordinances City of Galveston, Texas: Charter and General Ordinances*. Tallahassee, Fla.: Municipal Code Corporation, 1960.
———. *The Code of the City of Galveston, Texas, 1948*. Charlottesville, Va.: Michie City Publications Co.
———. *Revised Ordinances of the City of Galveston to June 1st, 1917 as Ordained by the Board of Commissioners*. Rev. John D. Kelley. Galveston: Oscar Springer, 1917.
City of Houston. *Charter and Revised Code of Ordinances of the City of Houston: Harris County, Texas, to October 31, 1904*. Houston: City Council of the City of Houston, 1904.
———. *Charter of the City of Houston and General Ordinances Passed by the City Council of the City of Houston from the 31st Day of October, 1904, to and Including October 31st, 1910*. Houston: W. H. Coyle, 1910.
———. *Charters and Rev. Code of Ordinances of the City of Houston 1897–1905*. Bound volume of pamphlets. Office of the City Attorney, Houston, Texas. 2003.
———. *Code of Ordinances City of Houston*. Adoption Ordinance No. 85-1380, "An Ordinance Adopting a Codification of the General and Criminal Ordinances of the City of Houston, Texas." Effective August 21, 1985. library.municode.com/tx/houston/codes/codeof_ordinances?nodeId=COORHOTE.
———. *Code of Ordinances: City of Houston, Texas*. 2 vols. Tallahassee, Fla.: Municipal Code Corporation, 1958.
———. *Fire Code: City of Houston*. Houston, 1974.
———. Harris County. *City of Houston Map*. Ed. William E. Wood. Houston: W. E. Wood, 1869.
———. *Houston Amendments to the 2006 International Fire Code*. Adopted by Ord. No. 2010-871. Passed November 11, 2010. Effective December 31, 2010. www.houstontx.gov/fire/business/HoustonAdmendtoIFC.pdf.

———. *The Revised Code of Ordinances of the City of Houston of 1914 Passed May 6, 1914, Approved by the Mayor May 6, 1914*. Comp. and indexed by E. P. Phelps, 1914.

———. *The Revised Code of Ordinances of the City of Houston of 1922*. Comp. and Indexed by J. G. Hautier. Houston: Houston City Council, 1922.

City of Mobile. *The Charter and Code of Ordinances of the City of Mobile*. Comp. Reuben A. Lewis. Mobile, Ala.: City of Mobile, 1866.

———. *The Charter and Code of Ordinances of the City of Mobile*. Mobile, Ala.: George Matzenger, 1889.

———. Board of Commissioners. *The Code of the City of Mobile, Alabama 1955. The General Ordinances of the City*. Charlottesville, Va.: Michie City Publications Co., 1956.

———. Board of Commissioners. *The Commission Government Charter and the Code of Ordinances of the City of Mobile, Alabama, with Appendix*. Prepared by Shelton H. Hendrix. Mobile, Ala.: City of Mobile, 1934.

———. *The Charter and the Code of Ordinances of 1897 of the City of Mobile, With an appendix*. Compiled by Peter Joseph Hamilton. Mobile: City of Mobile, 1897.

———. *The Charter and the Code of Ordinances of the City of Mobile, Alabama, With Appendix*. Prepared by Saffold Berney. Mobile, Ala.: City of Mobile, 1907.

———. *Code of Ordinances*. Chap. 22. Article I.—Fire Prevention Code. library.municode.com/al/mobile/codes/code_of_ordinances?nodeId=CICO_CH22FIPR. Accessed March 31, 2017.

———. *The Code of Ordinances of the City of Mobile, With the Charter, And An Appendix*. By Alexander McKinstry. Mobile: S. H. Goetzel & Co., 1859.

———. *The Commission Charter and the Code of Ordinances of the City of Mobile, Alabama with Appendix*. Prepared by Shelton H. Hendrix. Mobile, Ala.: City of Mobile, 1947.

———. "An Ordinance to Adopt a Fire Code for the City of Mobile, Alabama." December 2, 2010. Access http://www.cityofmobile.org/fire/pdfs/Fire_Code_Ord.pdf

———. *Ordinances, Grants and Franchises of the City of Mobile Since the Code of 1907*. N.p., 1921.

City of New Orleans. *Manual of the City of New Orleans*. Comp. T. W. Campbell. New Orleans: City of New Orleans, 1903.

City of New York. Fire Department. Municipal Explosives Commission. *Regulations Governing the Manufacturing, Storage, Sale and Use of Explosive and Combustible Substances*. New York: n.p., 1903.

———. Health Department. Board of Health. *First Annual Report of the Board of Health of the Health Department of the City of New York. April 11, 1870, to April 10, 1871*. New York: New York Printing Co., 1871.

———. Metropolitan Board of Health. Manual of the Metropolitan Board of Health, and the Metropolitan Board of Excise. June, 1869. New York: D. Appleton & Co., 1869.

Clement, George A. The Law of Fire Insurance. Vol. 2: *Fire Insurance as a Void Contract, and in Both Volumes the Conditions of the Contract as Affected by Construction, Waiver, or Estoppel*. New York: Baker, Voorhis & Co., 1905.

Cobb, James C. *Industrialization and Southern Society 1877–1984*. Lexington: University Press of Kentucky, 1984.

———. *The Selling of the South: the Southern Crusade for Industrial Development, 1936–1990*. 2nd ed. Urbana: University of Illinois Press, 1993.

———. "The Sunbelt South: Industrialization in Regional, National, and International Perspective." In *Searching for the Sunbelt: Historical Perspectives on a Region*, ed. Raymond A. Mohl. Athens: University of Georgia Press, 1993. 25–46.

Coleman, W. H. *Historical Sketch Book and Guide to New Orleans and Environs, With Map*. New York: Will H. Coleman, 1885.

Colten, Craig E., ed. *Transforming New Orleans and Its Environs: Centuries of Change*. Pittsburgh: University of Pittsburgh Press, 2000.

Cote, Arthur, and Percy Bugbee. *Principles of Fire Protection*. Quincy, Mass.: National Fire Protection Association, 1988.

Covello, Vincent T., and Joshua Menkes. "Issues in Risk Analysis." In *Risk in the Technological Society*, ed. Hohenemser and Kasperson. 287–301.

Croker, Edward F. *Fire Prevention*. New York: Dodd Mead & Co., 1917.

Cronon, William. *Nature's Metropolis: Chicago and the Great West*. New York: W. W. Norton & Co., 1991.

Crosby, Everett U., Henry A. Fiske, and H. Walter Forster. *Hand-Book of Fire Protection*. 6th ed. New York: D. Van Nostrand Co., 1919.

Cross, Roy. *A Handbook of Petroleum, Asphalt and Natural Gas: Methods of Analysis, Specifications, Properties, Refining Processes, Statistics, Tables and Bibliography*. Kansas City, Mo.: Kansas City Testing Laboratory, 1919.

Daddow, Samuel Harries, and Benjamin Bannan. *Coal, Iron, and Oil; or, the Practical American Miner: A Plain and Popular Work on Our Mines and Mineral Resources, and a Text-Book or Guide to their Economical Development*. Philadelphia: J. B. Lippincott & Co., 1866.

Daly, Charles P., ed. *Reports of Cases Argued and Determined in the Court of Common Pleas of the City and County of New York*. Vol. 3. New York: Baker, Voorhis & Co., 1872.

Dana, David D. *The Fireman: The Fire Departments of the United States with a Full Account of All Large Fires, Statistics of Losses and Expenses, Theatres Destroyed by Fire and Accidents, Anecdotes, and Incidents*. Boston: James French and Co., 1858.

Davis, Thomas Frederick. *History of Jacksonville, Florida and Vicinity, 1513–1924*. St. Augustine, Fla.: Record Co., 1925.

Day, Thomas Henry. Practical Instruction on "Electrical Fire Hazards". Hartford, Conn.: Insurance Institute of Hartford, 1918.

Deitch, Guilford A. *The Standard Fire Policy: Lectures Before the Fire Insurance Club of Chicago.* Indianapolis: Rough Notes Co., 1905.

De Leon, T. C. *Belles Beaux and Brains of the 60's.* New York: G. W. Dillingham Co., 1907.

Delfino, Susanna, and Michele Gillespie, eds. *Global Perspectives on Industrial Transformation in the American South.* Columbia: University of Missouri Press, 2005.

———, eds. *Technology, Innovation, and Southern Industrialization: From the Antebellum Era to the Computer Age.* Columbia: University of Missouri Press, 2008.

The Derrick's Hand-Book of Petroleum: A Complete Chronological and Statistical Review of Petroleum Developments During 1898 and 1899. Vol. 2. Oil City, Pa.: Derrick Publishing Co., 1900.

Diamantes, David. *Principles of Fire Prevention.* Clifton Park, N.Y.: Thomson Delmar Learning, 2005.

A Digest of the Laws of Pennsylvania, from the Year One Thousand Seven Hundred to the Sixth Day of June, One Thousand Eight Hundred and Eighty-Three. Comp. John Purdon. 11th ed. Vol. 2. Rev. and ed. Frederick C. Brightly. Philadelphia: Kay and Brother, 1885.

A Digest of the Ordinances, Resolutions, By-Laws and Regulations of the Corporation of New-Orleans. And a Collection of the Laws of the Legislature Relative to the Said City. New Orleans: New Orleans City Council, 1836.

A Digest of the Statute Law of the State of Pennsylvania From the Year 1700 to 1894. Comp. John Purdon. 12th ed. Vol. 2. Rev. and ed. Frank F. Brightly. Philadelphia: Kay and Brother, 1895.

Dillon, John F. *Commentaries on the Law of Municipal Corporations.* 4th ed. Vol. 1. Boston: Little, Brown, and Co., 1890.

———. *Commentaries on the Law of Municipal Corporations.* 5th ed. 5 vols. Boston: Little, Brown, and Co., 1911.

Directory of American Cement Industries. Ed. Charles Carroll Brown. 4th ed. Indianapolis: Municipal Engineering Co., 1906.

Doyle, Don H. *New Men, New Cities, New South: Atlanta, Nashville, Charleston, Mobile, 1860–1910.* Chapel Hill: University of North Carolina Press, 1990.

Dunn, F. B. *Industrial Uses of Fuel Oil.* San Francisco: Technical Publishing Co., 1916.

Dunn, Joe P., and Howard L. Preston, eds. *The Future South: A Historical Perspective for the Twenty-first Century.* Champaign: University of Illinois Press, 1991.

Edmonds, John W., ed. *Statutes at Large of the State of New York Containing the General Statutes Passed in the Years 1863, 1864, 1865 & 1866.* 6 vols. Albany, N.Y.: Weed, Parsons & Co., 1868.

Ekbladh, David. *The Great American Mission: Modernization and the Construction of an American World Order.* Princeton, N.J.: Princeton University Press, 2010.

Elliott, Byron K., and William F. Elliott. *A Treatise on the Law of Roads and Streets.* 2nd ed. Indianapolis: Bowen-Merrill Co., 1900.

Elliott, William F. *A Treatise on the Law of Roads and Streets.* 2nd ed. 1900.

Encyclopedia Americana: A Library of Universal Knowledge. 30 vols. New York: Encyclopedia Americana Corporation, 1919.

Encyclopedia of the United States in the Nineteenth Century. Ed. Paul Finkelman. 3 vols. New York: Charles Scribner's Sons, 2001.

Engineering Facts and Figures for 1866: An Annual Register of Progress in Mechanical Engineering and Construction. London: A. Fullerton & Co., 1867.

Ericson, Richard V., and Aaron Doyle, eds. *Risk and Morality.* Toronto: University of Toronto Press, 2003.

Ericson, Richard V., Aaron Doyle, and Dean Barry. *Insurance as Governance.* Toronto: University of Toronto Press, 2003.

Fire Underwriters' Association of the Northwest. *Proceedings of the Thirty-Fifth Annual Meeting of the Fire Underwriters' Association of the Northwest. Chicago, Ill., September 28–29, 1904.* Milwaukee: Wright & Joys Co., 1904.

———. *Proceedings of the Twelfth Regular Meeting Fire Underwriters' Association of the Northwest. Chicago, Ill., September 14 and 15, 1881.* Milwaukee: Cramer, Aikens & Cramer, 1881.

———. *Proceedings of the Twenty-Eighth Annual Meeting of the Fire Underwriters' Association of the Northwest. Chicago, Ill., September 29–30, 1897.* Chicago: Fire Underwriters' Association of the Northwest, 1897.

———. *Proceedings of the Twenty-Second Annual Meeting of the Fire Underwriters' Association of the Northwest. Chicago, Ill., October 7, 8 and 9, 1891.* Milwaukee: King, Fowle & Co., 1891.

Fire Underwriters' Association of the Pacific. *Proceedings of the Seventeenth Annual Meeting of the Fire Underwriters' Association of the Pacific, San Francisco, February 20th and 21st, 1893.* San Francisco: FUAP, 1893.

———. *Proceedings of the Sixteenth Annual Meeting of the Fire Underwriters' Association of the Pacific, San Francisco, February 16th and 17th, 1892.* San Francisco: FUAP, 1892.

———. *Proceedings of the Twelfth Annual Meeting of the Fire Underwriters' Association of the Pacific, San Francisco, February 20th and 21st, 1888.* [San Francisco]: FUAP, 1888.

———. *Proceedings of the Twenty-Eighth Annual Meeting of the Fire Underwriters' Association of the Pacific, San Francisco, California, January 12–13, 1904.* [San Francisco]: FUAP, 1904.

———. *Proceedings of the Twenty-Fourth Annual Meeting of the Fire Underwriters' Association of the Pacific San Francisco, Cal., February 20th and 21st, 1900.* N.p., 1900.

———. *Proceedings of the Twenty-Seventh Annual Meeting of the Fire Underwriters' Association of the Pacific San Francisco, Cal., February 17–18, 1903*. N.p., 1903.

Flynn's Digest of the City Ordinances Together with the Constitutional Provisions, Acts of the General Assembly, and Decisions of the Courts Relative to the Government of the City of New Orleans. Comp. J. Q. Flynn. New Orleans: City Council of New Orleans, 1896.

Folmer, Henk, and Tom Tietenberg, eds. *The International Yearbook of Environmental and Resource Economics 2003/2004: A Survey of Current Issues*. Northampton, Mass.: Edward Elgar Publishing, Inc., 2003.

Forrest, Clarence H. *Official History of the Fire Department of the City of Baltimore Together With Biographies and Portraits of Eminent Citizens of Baltimore*. Baltimore: n.p., 1898.

Francis, C. K. *Manual Issued by the Home Insurance Company of New York, Containing General Rules and Instructions, with Forms of Policies and Endorsements, for the Use of Its Agents*. [New York]: Home Insurance Co., 1878.

Franks, Kenny A., and Paul F. Lambert. *Early California Oil: A Photographic History, 1865–1940*. College Station: Texas A&M University Press, 1985.

———. Early Louisiana and Arkansas Oil: A Photographic History, 1901–1946. College Station: Texas A&M University Press, 1982.

Freedley, Edwin T. *Philadelphia and Its Manufactures: A Hand-Book Exhibiting the Development, Variety, and Statistics of the Manufacturing Industry of Philadelphia in 1857, Together with Sketches of Remarkable Manufactories; and a List of Articles Now Made in Philadelphia*. Philadelphia: Edward Young, 1859.

[Freeman, Hovey T., and Norris G. Abbott Jr.]. *The Factory Mutuals 1835–1935: Being Primarily a History of the Manufacturers Mutual Fire Insurance Company the Original of the Factory Mutual Companies during its Fire One Hundred Years*. Providence, R.I.: Manufacturers Mutual Fire Insurance Co., 1935.

Frehner, Brian. Finding Oil. *The Nature of Petroleum Geology, 1859–1920*. Lincoln: University of Nebraska Press, 2011.

Freitag, Joseph Kendall. *Fire Prevention and Fire Protection As Applied to Building Construction: A Handbook of Theory and Practice*. New York: John Wiley & Sons, 1912.

French, Edward V. *1860–Fifty Years–1910 Arkwright Mutual Fire Insurance Company: The Development of the Prevention Idea*. Boston: Arkwright Mutual Fire Insurance Co., 1912.

Freudenburg, William R., and Robert Gramling. *Blowout in the Gulf: The BP Oil Spill Disaster and the Future of Energy in America*. Cambridge, Mass.: MIT Press, 2011.

Frothingham, F. E. *The Boston Fire, November 9th and 10th, 1872: Its History, Together with the Losses in Detail of Both Real and Personal Estate; Also, a Complete List of*

Insurance Losses and an Appendix Containing the City Loan, Insurance, and Building Acts. Boston: Lee & Shepard, 1873.

Fryer, William J., Jr., ed. *Laws Relating to Buildings, in the City of New York, with Marginal Notes, a Complete Index, and Colored Engravings*. New York: Record and Guide, 1887.

Gaston, Paul M. *The New South Creed: a Study in Southern Mythmaking*. New York: Knopf, 1970.

Geisst, Charles R. *Monopolies in America: Empire Builders and Their Enemies from Jay Gould to Bill Gates*. New York: Oxford University Press, 2000.

George Matzenger's Mobile Directory for 1895. Vol. 30. Mobile: George Matzenger, 1895.

George Matzenger's Mobile Directory for 1901. Vol. 36. Mobile: George Matzenger, 1901.

Getis, Arthur, Judith Getis, and Jerome D. Fellman. *Introduction to Geography*. 5th ed. Boston: William C. Brown Publishers, 1996.

Gesner, Abraham. *A Practical Treatise on Coal, Petroleum, and Other Distilled Oils*. 2nd ed. Revised by George Weltden Gesner. New York: Bailliere Brothers, 1865.

Giddens, Anthony. *The Consequences of Modernity*. Stanford, Calif.: Stanford University Press, 1990.

Giddens, Paul H. *Standard Oil Company (Indiana): Oil Pioneer of the Middle West*. New York: Appleton-Century-Crofts, 1955.

Gigerenzer, Gerd, Zeno Swijtink, Theodore Parker, Lorraine Daston, John Beatty, and Lorenz Krüger. *The Empire of Chance: How Probability Changed Science and Everyday Life*. Cambridge, U.K.: Cambridge University Press, 1989.

Gildemeister, Eduard, and Frederick Hoffmann. *The Volatile Oils*. 2nd ed. 3 vols. New York: John Wiley and Sons, 1913.

Goldfield, David. *Black, White, and Southern: Race Relations and Southern Culture 1940 to the Present*. Baton Rouge: Louisiana State University Press, 1990.

———. *Region Race and Cities: Interpreting the Urban South*. Baton Rouge: Louisiana State University Press, 1997.

———. *Still Fighting the Civil War: The American South and Southern History*. Baton Rouge: Louisiana State University Press, 2002.

———. *Urban Growth in the Age of Sectionalism: Virginia, 1847–1861*. Baton Rouge: Louisiana State University Press, 1977.

Gordon, Dane R., and Józef Niżnik, eds. *Criticism and Defense of Rationality in Contemporary Philosophy*. Amsterdam: Rodopi, 1984.

Gostin, Lawrence O. *Public Health Law: Power, Duty, Restraint*. Berkeley: University of California Press, 2000.

Grant, H. Roger. *Insurance Reform: Consumer Action in the Progressive Era*. Ames: Iowa State University Press, 1979.

Gratt, Lawrence B. "Risk Analysis or Risk Assessment: A Proposal for Consistent Definitions." In *Uncertainty in Risk Assessment, Risk Management, and Decision*

Making, ed. Vincent T. Covello, Lester B. Lave, Alan Moghissi and V. R. R. Uppuluir. New York: Plenum Press, 1984. 241–49.

Greene, Casey Edward, and Shelly Henley Kelly, eds. *Through a Night of Horrors: Voices from the 1900 Galveston Storm.* College Station: Texas A&M University Press, 2000.

Griffin, Larry J., and Don H. Doyle, eds. *The South as an American Problem.* Athens: University of Georgia Press, 1995.

Griswold, Jeremiah. *The Fire Underwriters' Text-Book.* New York: Insurance Monitor Office, 1872.

———. *Hand-Book of Adjustments of Loss or Damage by Fire, for the Use of Fire Underwriters.* 2nd ed. New York: Insurance Monitor Office, 1877.

Haber, Samuel. *Efficiency and Uplift: Scientific Management in the Progressive Era 1890–1920.* 1964. Rpt. Chicago: University of Chicago Press, 1973.

Hacking, Ian. *The Emergence of Probability: A Philosophical Study of Early Ideas about Probability, Induction and Statistical Inference.* Cambridge, U.K.: Cambridge University Press, 1975, 1984.

Hanchett, Thomas W. *Sorting Out the New South: City, Race, Class, and Urban Development in Charlotte, 1875–1975.* Chapel Hill: University of North Carolina Press, 1998.

Haque, C. Emdad. *Hazards in a Fickle Environment: Bangladesh.* Dordrecht, Netherlands: Kluwer Academic Publishers, 1997.

[Hardee, W. J.] *Official Hand-Book: Laws Regulating Construction of Buildings in the City of New Orleans, 1905.* New Orleans: L. Graham Co., 1905.

Hardwick, Susan Wiley. *Mythic Galveston: Re-Inventing America's Third Coast.* Baltimore: Johns Hopkins University Press, 2002.

Hardy, Dermot H., and Major Ingham S. Roberts, eds. *Historical Review of South-East Texas and the Founders, Leaders and Representative Men of Its Commerce, Industry and Civic Affairs.* Vol. 2. Chicago: Lewis Publishing Co., 1910.

Harris, Charles H., and Louis R. Sadler. *The Texas Rangers and the Mexican Revolution: The Bloodiest Decade, 1910–1920.* Albuquerque: University of New Mexico Press, 2004.

Harris, Richard A., and Sidney M. Milkis. *The Politics of Regulatory Change: A Tale of Two Agencies.* 2nd ed. New York: Oxford University Press, 1996.

Harrison, Robert. *Congress, Progressive Reform, and the New American State.* New York: Cambridge University Press, 2004.

Hayden, H. R. *Hayden's Annual Cyclopedia of Insurance in the United States, 1900–1901.* Hartford, Conn.: Insurance Journal Co., 1901.

———. *Hayden's Annual Cyclopedia of Insurance in the United States 1901–02.* Hartford, Conn.: Insurance Journal Co., 1902.

———. *Hayden's Annual Cyclopedia of Insurance in the United States 1902–1903.* Hartford, *Conn.: Insurance Journal Co., 1903.*

———. *Hayden's Annual Cyclopedia of Insurance in the United States 1910-1911*. Hartford, Conn.: Insurance Journal Co., 1911.

———. *Annual Cyclopedia of Insurance in the United States 1915*. Hartford, Conn., 1915.

———, ed. *The Annual Cyclopedia of Insurance in the United States. 1890*. New York: Underwriter Printing and Publishing Co., 1891.

———. *The Annual Cyclopedia of Insurance in the United States, 1891*. New York: Underwriter Printing and Publishing Co., 1892.

Henry, Claude, and Marc Henry. "Formalization and applications of the Precautionary Principle." CECO-965. 2002. "hal-00243001."

Hidy, Ralph W., and Muriel E. Hidy. *Pioneering in Big Business History, 1882–1911: History of Standard Oil Company New Jersey*. New York: Harper & Brothers, 1955.

Hine, C. C. *A Book of Forms Containing Forms of Policies, Endorsements, Certificates, and other Valuable Matter, For the Use of Agents and Others*. Rev. ed. New York: Sackett & Mackay, 1866.

———. *Fire Insurance: A Book of Instructions for the Use of the Agents of the United States*. New York: Insurance Monitor, 1870.

Historical Review of South-East Texas and the Founders, Leaders and Representative Men of Its Commerce, Industry and Civic Affairs. Vol. 2. Ed. Dermot H. Hardy and Ingham S. Roberts. Chicago: Lewis Publishing Co., 1910.

Hohenemser, Christoph, and Jeanne X. Kasperson, eds. *Risk in the Technological Society*. Boulder, Colo.: Westview Press, 1982.

Holder, Jane, and Donald McGillivray, eds. *Locality and Identity: Environmental Issues in Law and Society*. Brookfield, Vt.: Ashgate, 1999.

Hoover, Stephen R. *Fire Protection for Industry*. New York: Van Nostrand Reinhold, 1992.

Horr, Norton T., and Alton A. Bemis. *A Treatise on the Power to Enact, Passage, Validity and Enforcement of Municipal Police Ordinances*. Cincinnati: Robert Clarke Co., 1895.

Howe, G. Melvyn. *People, Environment, Disease and Death: A Medical Geography of Britain Throughout the Ages*. 2nd ed. Cardiff: University of Wales Press, 1997.

Hughes, Thomas P. *American Genesis: A Century of Invention and Technological Enthusiasm, 1870–1970*. Chicago: University of Chicago Press, 2004.

Hunter, P. R., P. Payment, N. Ashbolt, and J. Bartram. "Assessment of Risk." In *Assessing Microbial Safety of Drinking Water: Improving Approaches and Methods*. Published on behalf of the WHO and the Organization for Economic Co-operation and Development. London: IWA Publishing, 2003.

ICC (International Code Council). *International Codes*. "State and Jurisdiction Code Adoption Data." Updated through 2012. www.iccsafe.org/content/state-and-jurisdiction-code-adoption-data/.

———. *IFC: International Fire Code 2009 Edition*. Country Club Hills, Ill.: International Code Council, 2009.

———. *International Fire Code 2012 Edition*. Country Club Hills, Ill.: International Code Council, 2012.

Illinois Central Railroad. The Commercial, Industrial and Financial Outlook for New Orleans. Cedar Rapids, Iowa: Republican Printing Co., 1894.

International City Managers' Association. Institute for Training in Municipal Administration, *Municipal Fire Administration*. 7th ed. Chicago: International City Managers' Association, 1967.

The Investor's Guardian and Insurance Almanac 1873. New York: J. H. and C. M. Goodsell, 1873.

IOM (Institute of Medicine). *Environmental Decisions in the Face of Uncertainty*. Washington, D.C.: National Academies Press, 2013.

Israel, Paul. *From Machine Shop to Industrial Laboratory: Telegraphy and the Changing Context of American Invention, 1830–1920*. Baltimore: Johns Hopkins University Press, 1992.

Jackson, Joy J. *New Orleans in the Gilded Age: Politics and Urban Progress, 1880–1896*. Baton Rouge: Louisiana State University Press, 1969.

Jackson, Kenneth T. *Crabgrass Frontier*. New York: Oxford University Press, 1985.

Jewell's Digest of the City Ordinances, Together With the Constitutional Provisions, Acts of the General Assembly and Decisions of the Courts Relative to the Government of the City of New Orleans. Comp. Edwin L. Jewell. Rev. ed. New Orleans: Edwin L. Jewell, 1882.

Jewell's Digest of the City Ordinances, Together With the Constitutional Provisions and Acts of the General Assembly, Relative to the Government of the City of New Orleans. Comp. Edwin L. Jewell. New Orleans: Edwin L. Jewell, 1887.

Joyce, Joseph Asbury. *A Treatise on Franchises Especially Those of Public Service Corporations*. New York: Banks Law Publishing Co., 1909.

Judicial and Statutory Definitions of Words and Phrases. 2nd ser., vol. 2: *Deposition–Lamp. St.* Paul, Minn.: West Publishing Co., 1914.

Kelman, Ari. *A River and Its City: The Nature of Landscape in New Orleans*. Berkeley: University of California Press, 2003.

Kerzner, Harold. *Project Management: A Systems Approach to Planning, Scheduling, and Controlling*. 10th ed. New York: John Wiley & Sons, Inc., 2009.

King, John O. *Joseph Stephen Cullinan: A Study of Leadership in the Texas Petroleum Industry, 1897–1937*. Nashville: Vanderbilt University Press, 1970.

Klinke, Andres, and Ortwin Renn. "Prometheus Unbound: A New Approach to Risk Evaluation." In *Precautionary Risk Appraisal and Management. An Orientation for Meeting the Precautionary Principle in the European Union*, ed. Ortwin Renn et al. Bremen: Europäischer Hockschulverlag GmbH & Co., 2009. Part 2: 120–71.

Klinoff, Robert W. *Introduction to Fire Protection*. 3rd ed. Clifton Park, New York: Thomson Delmar Learning, 2007.

Knowles, Scott Gabriel. *The Disaster Experts: Mastering Risk in Modern America*. Philadelphia: University of Pennsylvania Press, 2011.

Kwon, W. Jean, and Harold D. Skipper. *Risk Management and Insurance: Perspectives on a Global Economy*. Oxford. U.K.: Blackwell Publishing, 2007.

Labbé, Ronald M., and Jonathan Lurie. *The Slaughterhouse Cases: Regulation, Reconstruction, and the Fourteenth Amendment*. Lawrence: University Press of Kansas, 2003.

Landes, David S. *The Unbound Prometheus: Technological Change and Industrial Development in Western Europe from 1750 to the Present*. 2nd ed. New York: Cambridge University Press, 2003.

Larson, Erik. *Isaac's Storm: A Man, a Time, and the Deadliest Hurricane in History*. New York: Vintage Books, 2000.

The Laws and General Ordinances of the City of New Orleans, Together with the Acts of the Legislature, Decisions of the Supreme Court, and Constitutional Provisions, Relating to the City Government. Rev. and digested by Henry J. Leovy. New Orleans: E. C. Wharton, 1857.

The Laws and General Ordinances of the City of New Orleans, Together with the Acts of the Legislature, Decisions of the Supreme Court, and Constitutional Provisions Relating to the City Government. Rev. and digested by Henry J. Leovy and C. H. Luzenberg. New Orleans: Simmons & Co., 1870.

The Laws of Texas, 1822–1897. Comp. and arranged by H. P. N. Gammel. Introduction by C. W. Raines. Austin, Tex.: Gammel Book Co., 1898.

Lears, Jackson. *Rebirth of a Nation: The Making of Modern America, 1877–1920*. New York: HarperCollins, 2010.

———. *Something for Nothing: Luck in America*. New York: Viking Penguin, 2003.

Lectures on Fire Insurance: Being the Substance of Lectures given before the Evening Classes in Fire Insurance conducted by The Insurance Library Association of Boston during the Fall and Winter of Nineteen Hundred and Eleven and Twelve. Boston: Insurance Library Association of Boston, 1912.

Lee, Ji-Hyun, and Tian-Chiu Li. "Integrated Physically Based Manipulation and Decision-Making Tree for Navigation to Support Design Rationale." In *Human Interface and the Management of Information: Methods, Techniques and Tools in Information Design*. Symposium on Human Interface 2007. Proceedings, Part 1: 480–89. Ed. Michael J. Smith and Gavriel Salvendy. Berlin: Springer-Verlag, 2007.

Leonard, John W., ed. *The Book of Chicagoans: A Biographical Dictionary of Leading Living Men of the City of Chicago*. Chicago: A. N. Marquis & Co., 1905.

Lester, Paul. *The Great Galveston Disaster Containing a Full and Thrilling Account of the Most Appalling Calamity of Modern Times.* [Galveston], 1900.

Lewis, Pierce F. *New Orleans: The Making of an Urban Landscape.* Santa Fe, N.M.: Center for American Places, 2003.

Linebaugh, Donald W. *The Springfield Gas Machine: Illuminating Industry and Leisure, 1860s–1920s.* Knoxville: University of Tennessee Press, 2011.

Liverpool and London and Globe Insurance Company. *1848–1898: Fifty Years of Work in the United States of America, Being an Account of the Entry into the United States of the Company Now Known as The Liverpool and London and Globe In the Year 1848 and of Its Work There for Fifty Years; with Which is Included a Brief and General Operations of the Company.* New York: [J. C. & W. E. Powers Press, 1898.

Louisiana, State of. *Acts Passed by The General Assembly of the State of Louisiana at the Regular Session . . . May, 1896.* Baton Rouge: The Advocate, Official Journal of Louisiana, 1896.

———. Secretary of State. Business Services. "Charter of Gulf Refining Company of Louisiana. www.sos.la.gov/BusinessServices/Pages/default.aspx.

Lucier, Paul. *Scientists & Swindlers: Consulting on Coal and Oil in America, 1820–1890.* Baltimore: Johns Hopkins University Press, 2008.

Lupton, Deborah. *Risk.* London: Routledge, 1999. Rpt. eds. 2003, 2004.

———, ed. *Risk and Sociocultural Theory: New Directions and Perspectives.* New York: Cambridge University Press, 1999.

Map of the City of Mobile, Compiled from an Original Map Made by P. J. Pillans City Engineer, and Harry Pillans, Ass't Eng'r, With Additions to date, by D. M. N. Ross, City Engineer, 1901. Baltimore: Union Map and Atlas Co., 1901.

Massachusetts, State of. *Acts and Resolves Passed by the General Court of Massachusetts, in the Year 1866.* Boston: Secretary of the Commonwealth, 1866.

———. *Acts and Resolves Passed by the General Court of Massachusetts, in the Year 1867.* Boston: Secretary of the Commonwealth, 1867.

———. *The Public Statutes of the Commonwealth of Massachusetts, Enacted November 19, 1881; to Take Effect February 1, 1882.* Boston: Rand, Avery, & Co., 1882.

Mawhinney, Matthew H. *Practical Industrial Furnace Design.* New York: J. Wiley & Sons, 1928.

McClenachan, C.T., comp. *A Compilation of the Laws of the State of New York; Also, of the Ordinances, Resolutions, and Orders Established by the Mayor, Aldermen, and Commonalty of the City of New York, in Common Council Convened, Relating to the Fire Department of the City of New York, From 1812 to 1860.* New York: Edmund Jones & Co., 1859.

McComb, David G. *Galveston: A History.* Austin: University of Texas Press, 1986.

———. *Houston: The Bayou City.* Austin: University of Texas Press, 1969.

McKinney's Consolidated Laws of New York Annotated. Book 19: *General Business Law.* Northport, New York: Edward Thompson Co., 1917.

McKnight, Tom L., and Darrel Hess. *Physical Geography: A Landscape Appreciation. 7th ed.* Upper Saddle River, N.J.: Prentice-Hall, 2002.

Meiksins, Peter, and Chris Smith. *Engineering Labour: Technical Workers in Comparative Perspective.* New York: Verso, 1996.

Melville, George W. *Report of the U.S. Naval "Liquid Fuel Board" of Tests Conducted on the Hohenstein Water Tube Boiler.* Washington, D.C.: Government Printing Office, 1904.

Meyer, David R. *Networked Machinists: High-Technology Industries in Antebellum America.* Baltimore: Johns Hopkins University Press, 2006.

Mileti, Dennis S. *Disasters by Design: A Reassessment of Natural Hazards in the United States.* Washington, D.C.: John Henry Press, 1999.

Miller, Donald L. *City of the Century: The Epic of Chicago and the Making of America.* New York: Simon & Schuster, 1996.

Maloney, Thomas J. *Mobile City Directory 1903.* Vol. 38. Atlanta: Mutual Publishing Co., 1903.

The Mobile City Directory 1902. Mobile, Ala.: J. Wiggins & Co., 1902.

Mobile City Directory 1907. Mobile, Ala.: R. L. Polk & Co., 1907.

Mobile City Directory 1910. Mobile, Ala.: R. L. Polk & Co., 1910.

Monkkonen, Eric H. *America Becomes Urban: The Development of U.S. Cities & Towns, 1780–1980.* Berkeley: University of California Press, 1990.

Moore, Francis C. *Fire Insurance and How to Build. Combining Also a Guide to Insurance Agents Respecting Fire Prevention and Extinction, Special Features of Manufacturing Risks, Writing of Policies, Adjustment of Losses, Etc., Etc.* New York: Baker & Taylor Co., 1903.

Moore, Winfred B., Joseph H. Tripp, and Lyon G. Taylor Jr., eds. *Developing Dixie: Modernization in a Traditional Society.* New York: Greenwood Press, 1988.

Morgan, M. Granger, and Max Henrion. *Uncertainty: A Guide to Dealing with Uncertainty in Quantitative Risk and Policy Analysis.* Cambridge, U.K.: Cambridge University Press, 1990.

Morrison and Fourmy's General Directory of the City of Houston, 1902–1903. Galveston: Morrison & Fourmy, [1903].

Morrison, Andrew. *The Industries of New Orleans.* New Orleans: n.p., 1885.

———, comp. and ed. *New Orleans and the New South.* [New Orleans]: Metropolitan Publishing Co., [1888].

———, ed. *The Port of Galveston and the State of Texas.* Galveston: George W. Engelhardt & Co., 1890.

Morton, Henry, and William A. Anderson. *Electric Lighting, and the Underwriters' Standard Requirements in Reference Thereto, with Instructions for the Proper Inspec-*

tion of Electric Light Equipments, Illustrated, Prepared by Request of the National Board of Fire Underwriters, and the United Fire Underwriters in America. [New York]: National Board of Fire Underwriters and the United Fire Underwriters in America, 1882.

Moss, David. *When All Else Fails: Government as the Ultimate Risk Manager.* Cambridge, Mass.: Harvard University Press, 2002.

Nackenoff, Carol, and Julie Novkov, eds. *Statebuilding from the Margins Between Reconstruction and the New Deal.* Philadelphia: University of Pennsylvania Press, 2014.

Nash, Gerald D. *United States Oil Policy, 1890–1964: Business and Government in Twentieth Century America.* Pittsburgh: University of Pittsburgh Press, 1968.

National Academy of Science. National Research Council. Commission on Life Sciences. Board on Environmental Studies and Toxicology. Committee on Developmental Toxicology. *Scientific Frontiers in Developmental Toxicology and Risk Assessment. Washington, D.C.: National Academies Press, 2000.*

———. Commission on Life Sciences. Committee on the Institutional Means for Assessment of Risks to Public Health. *Risk Assessment in the Federal Government: Managing the Process.* Washington, D.C.: National Academies Press, 1983.

———. Committee on Disaster Research in the Social Sciences. *Facing Hazards and Disasters: Understanding Human Dimensions.* Washington, D.C.: National Academies Press, 2006.

National Association of Fire Engineers. *Proceedings of the Twenty-Second Annual Convention Held at Montreal, P.Q., August 14th, 15th, 16th and 17th, 1894.* Cincinnati: Keating & Co., 1894.

National Board of Fire Underwriters. *Annual Report of the Committee on Statistics of the National Board [of] Fire Underwriters. 1877–1878.* New York: Styles & Cash, 1878.

———. *Building Code Recommended by The National Board of Fire Underwriters.* New York: NBFU, 1905.

———. *Building Code Recommended by The National Board of Fire Underwriters.* New York: NBFU, 1909.

———. *Building Code Recommended by The National Board of Fire Underwriters.* New York: NBFU, 1915.

———. *Building Code Recommended by The National Board of Fire Underwriters.* 4th ed. New York: NBFU, 1922.

———. *Building Code Recommended by The National Board of Fire Underwriters: An Ordinance Providing for Fire Limits, and Regulations Governing the Construction, Alteration, Equipment, Repair, or Removal of Buildings or Structures.* 5th ed. Rpt. New York: NBFU, 1934.

———. *Fire Insurance: Its Importance; Its Relation to the Community.* New York: NBFU, 1899.

———. *Fire Prevention Code: A Code Prescribing Regulations Governing Conditions Hazardous to Life and Property from Fire;* also, *A Suggested Ordinance Adopting the Fire Prevention Code and Establishing a Bureau of Fire Prevention.* New York: [National Board of Fire Underwriters], 1953.

———. *Fire Prevention Code: A Code Prescribing Regulations Governing Conditions Hazardous to Life and Property from Fire;* also, *A Suggested Ordinance Adopting the Fire Prevention Code and Establishing a Bureau of Fire Prevention.* New York: [National Board of Fire Underwriters], 1956.

———. *Fire Prevention Code: A Code Prescribing Regulations Governing Conditions Hazardous to Life and Property from Fire;* also, *A Suggested Ordinance Adopting the Fire Prevention Code and Establishing a Bureau of Fire Prevention.* New York: [National Board of Fire Underwriters], 1960.

———. *Journal of the Proceedings of the First Annual Meeting of the National Board of Fire Underwriters, Held in New York, Feb. 20th, and 21st, 1867, Together with the Proceedings of the Executive Committee, February 19th, 1867, and the Names of Officers and Executive Committee, For 1867–68.* New York: [National Board of Fire Underwriters], 1867.

———. *Journal of the Proceedings of the Fourth Annual Meeting of the National Board of Fire Underwriters, Held in New York, April 20th, 1870, Together with the Proceedings of the Executive Committee, April 20th, 1870, and the Names of Officers and Executive Committee For 1870–71.* New York: [National Board of Fire Underwriters], 1870.

———. *Journal of the Proceedings of the Third Annual Meeting of the National Board of Fire Underwriters, Held in New York, April 21st, 22d and 23d, 1869, Together with the Proceedings of the Executive Committee, April 20th and 23d, 1869, and the Names of Officers and Executive Committee For 1869–70.* New York: [National Board of Fire Underwriters], 1869.

———. "Minutes of a Meeting of the Executive Committee of the National Board of Fire Underwriters, Held at Its Rooms 156 & 158 Broadway, New York, June 19th, 1873." In bound vol. Insurance Library Association, Boston.

———. "Minutes of a Meeting of the Executive Committee of the National Board of Fire Underwriters, Held at the Committee's Rooms, 156 and 158 Broadway, New York, February 24th and 25th, 1876." In bound vol. Insurance Library Association, Boston.

———. "Minutes of a Meeting of the Executive Committee of the National Board of Fire Underwriters, Held at the Committee's Rooms, 156 and 158 Broadway, New York, May 19th, 20th and 21st, 1875." In bound vol. Insurance Library Association, Boston.

———. "Minutes of a Meeting of the Executive Committee of the National Board of Fire Underwriters, Held at the Committee's Rooms, 156 Broadway, New York, May 11th and 12th, 1876." In bound vol. Insurance Library Association, Boston.

———. "Minutes of a Meeting of the Executive Committee of the National Board of Fire Underwriters, Held at the Committee's Rooms, 156 Broadway, New York. Thursday, January 4, 1877." In bound vol. Insurance Library Association, Boston.

———. Minutes of a Meeting of the Executive Committee of the National Board of Fire Underwriters, Held at the Franklin Fire Insurance Company, Philadelphia, Wednesday, June 14, 1876. In bound vol. Insurance Library Association, Boston.

———. "Minutes of a Meeting of the Executive Committee of the National Board of Fire Underwriters, Held at the Rooms of the Committee, Nos. 156 and 158 Broadway, New York, May 14th, 1873." In bound vol. Insurance Library Association, Boston.

———. *The National Building Code.* Golden Anniversary Ed. New York: National Board of Fire Underwriters, 1955.

———. *Pioneers of Progress: National Board of Fire Underwriters 1866–1941 (To Be Continued).* New York: NBFU, 1941.

———. *Proceedings of the Eighth Annual Meeting of the National Board of Fire Underwriters, Held in New York, April 22d and 23d, 1874.* New York: [National Board of Fire Underwriters], 1874. In bound vol. Insurance Library Association, Boston.

———. *Proceedings of the Executive Committee of the National Board of Fire Underwriters, at Meetings, Held In New York, April, 23d and 24th, May 14th and 15th, June 19th, and September 17th and 18th, In Philadelphia, November 12th and 13th, and In New York, December 11th and 12th, 1873.* New York: [National Board of Fire Underwriters], 1873.

———. *Proceedings of the Executive Committee of the National Board of Fire Underwriters, at Meetings, Held In New York, December 19th, 1873, January 29th & 30th, 1874, In Philadelphia, February 26th & 27th, 1874, and In New York, March 26th & 27th, 1874.* New York: [National Board of Fire Underwriters], 1874.

———. *Proceedings of the Executive Committee of the National Board of Fire Underwriters, at Meetings, Held In Philadelphia, January 28th and 29th, and In New York, February 25th and 26th, April 1st and 2d, and April, 21st and 22d, 1875.* New York: [National Board of Fire Underwriters], 1875.

———. *Proceedings of the Executive Committee of the National Board of Fire Underwriters at the Meeting Held in New York City, February 24th and 25th, 1869.* New York: [National Board of Fire Underwriters], 1869.

———. *Proceedings of the Executive Committee of the National Board of Fire Underwriters at the Meeting Held in New York City, December 8th and 9th, 1869.* New York: [National Board of Fire Underwriters], 1869.

———. *Proceedings of the Executive Committee of the National Board of Fire Underwriters at the Meetings of December 18th, 1872, and February 12th, 1873, Held at Underwriters' Hall, 158 Broadway, New York, and at that of March 27th, 1873, Held at the Rooms of the Philadelphia Association of Fire Underwriters 308 Walnut Street, Philadelphia.* New York: [National Board of Fire Underwriters], 1873.

———. *Proceedings of the Executive Committee of the National Board of Fire Underwriters at the Meetings of June 11th and 12th, and September 4th and 5th, 1872, Held at Underwriters' Hall, 158 Broadway, New York.* New York: [National Board of Fire Underwriters], 1872.

———. *Proceedings of the National Board of Fire Underwriters, and of the Executive Committee, From April, 1873, to March, 1874.* New York: N.Y. Economical Printing Co., 1874.

———. *Proceedings of the National Board of Fire Underwriters, and of the Executive Committee, From April 22, 1875, to April 25, 1876.* New York: [National Board of Fire Underwriters], 1876.

———. *Proceedings of the Seventh Annual Meeting of the National Board of Fire Underwriters, Held in New York April 23d and 24th, 1873. New York: [National Board of Fire Underwriters], 1873.* Bound in Pamphlets on Insurance 1861–1879, vol. 2. University of California at Berkeley.

———. *Proceedings of the Tenth Annual Meeting of the National Board of Fire Underwriters, Held in New York April 26, 27 & 28, 1876; Also Names of Officers & of Standing Committees For the Year.* New York: [National Board of Fire Underwriters], 1876.

———. *Proceedings of the Thirty-Fifth Annual Meeting of the National Board of Fire Underwriters Held at the Rooms of the New York Board of Fire Underwriters, New York, Thursday, May 9th, 1901.* New York: Styles & Cash, 1901.

———. *Proceedings of the Thirty-Fourth Annual Meeting of the National Board of Fire Underwriters, Held at the Rooms of the New York Board, 32 Nassau St., New York, Thursday, May 10, 1900.* New York: Styles & Cash, 1900.

———. *Proceedings of the Thirty-Third Annual Meeting of the National Board of Fire Underwriters, Held at the Rooms of the New York Board, 32 Nassau Street, New York, Thursday, May 11, 1899.* New York: Styles & Cash, 1899.

———. *Proceedings of the Twenty-Ninth Annual Meeting of the National Board of Fire Underwriters, Held at The Rooms of the New York Board, 32 Nassau St., New York, Thursday, May 9, 1895.* New York: Styles & Cash, 1895.

———. *Proceedings of the Twenty-Third Annual Meeting of the National Board of Fire Underwriters, Held at the Rooms of the New York Board of Fire Underwriters.* [New York: NBFU, 1889]. In bound vol. Insurance Library Association, Boston.

———. *Proposed Standard Schedule For Grading Cities and Towns of the United States With Reference To Their Fire Defenses and Physical Conditions.* New York: NBFU, 1915.

———. *Report of Proceedings of Session of Executive Committee, Held at Boston, Mass. Sept. 11 and 12, 1867.* New York: [National Board of Fire Underwriters], 1867.

———. *Report of the Eleventh Annual Meeting of the National Board of Fire Underwriters, Held in the City of New York, April 25th and 26th, 1877.* [1877]. In bound vol. Insurance Library Association, Boston.

———. *Report of the Ninth Annual Meeting of the National Board of Fire Underwriters, Held In Underwriters' Hall, 156 and 158 Broadway.* [1875].

———. *Report on the City of Galveston, Texas: National Board of Fire Underwriters, Committee on Fire Prevention.* New York: NBFU, 1910.

———. *Report on the City of Galveston, Texas: (superseding that of 1925): National Board of Fire Underwriters, Committee on Fire Prevention and Engineering Standards.* New York: NBFU, 1946.

———. *Reports of Thomas H. Montgomery, General Agent National Board of Fire Underwriters, September, 1872–April, 1877.* New York: Styles & Cash, 1877.

National Commission on the BP Deepwater Horizon Oil Spill and Offshore Drilling. *Deep Water: The Gulf Oil Disaster and the Future of Offshore Drilling.* Report to the President. 2011.

National Electric Light Association. *Proceedings of the National Electric Light Association at Its Eighth Convention, Semi-Annual Meeting, Held at New York City, August 29, 30 and 31, 1888.* Vol. 4. Boston: Press of Modern Light and Heat, 1888.

———. *Proceedings of the National Electric Light Association, at Its Eleventh Convention, Annual Meeting, Held at Kansas City, February 11, 12, 13 and 14, 1890.* Vol. 8. New York: NELA, 1890.

———. *Proceedings of the National Electric Light Association, At Its First Annual Convention, Feb., 1885, Held At Chicago, Ill., And At Its Semi-Annual Convention, August 1885, Held at New York City.* Vol. 1. Baltimore: Baltimore Publishing Co., 1886.

———. *Proceedings of the National Electric Light Association at Its Ninth Convention. Annual Meeting Held in the Exposition Building at Chicago, Ill., February 19, 20 and 21, 1889.* Vol. 5. Boston: Press of Modern Light and Heat, 1889.

———. *Proceedings of the National Electric Light Association, at Its Tenth Convention. Semi-Annual Meeting, Held in the Casino, at Niagara Falls, N.Y., August 6, 7 and 8, 1889.* Vol. 7. New York: NELA, 1890.

———. *Proceedings of the National Electric Light Association at the Third Semi-Annual Convention Held at Boston, Mass. August 9, 10 and 11, 1887. And at Its Fourth Annual Convention Held at Pittsburgh, Pa. February 21, 22 and 23, 1888.* Vol. 3. Boston: Press of Modern Light and Heat, 1888.

National Fire Protection Association. *Life Safety Code Handbook.* Ed. John A. Sherry. 1st ed. Based on 1976 ed. of Life Safety Code. Boston: NFPA, 1978.

———. *National Fire Codes: A Compilation of NFPA Codes, Standards, Recommended Practices, and Manuals.* Boston: National Fire Protection Association, [1967].

———. *Fire Prevention Code of the National Fire Protection Association* (NFPA No. 1). Boston: National Fire, 1975.

———. *National Fire Codes: A Compilation of NFPA Codes, Standards, Recommended Practices, and Manuals*. Vol. 1. Boston: National Fire Protections Association, 1977.

———. *National Fire Codes: A Compilation of NFPA Codes, Standards, Recommended Practices, and Manuals*. Vol. 3. Boston: National Fire Protections Association, 1977.

———. *National Fire Codes: A Compilation of NFPA Codes, Standards, Recommended Practices, and Manuals*. Quincy, MA: The Association, 1984.

———. *National Fire Codes: A Compilation of NFPA Codes, Standards, Recommended Practices, and Manuals*. Quincy, MA: National Fire Protection Association, [1991].

———. *National Fire Protection Association: A Partial Record of the Transactions at Third Annual Meeting Held in Boston June 13th, 14th and 15th, 1899*. Boston: Standard Publishing Co., 1899.

———. *National Fire Protection Association: Proceedings of Sixth Annual Meeting Held in Philadelphia May 13th, 14th and 15th, 1902*. New York: Press of Insurance Engineering, 1902.

———. *National Fire Protection Association: Proceedings of the Fifth Annual Meeting Held in Chicago June 11th, 12th and 13th, 1901*. Boston: Standard Publishing Co., 1901.

———. *National Fire Protection Association: Proceedings of the Fourth Annual Meeting Held in New York June 26th, 27th and 28th, 1900*. Boston: Standard Publishing, 1900.

———. *NFPA 1 Uniform Fire Code*. Quincy, MA: National Fire Protection Association, 2003.

———. *NFPA 1 Uniform Fire Code*. Quincy, MA: National Fire Protection Association, 2006.

———. *NFPA 101, Life Safety Code*. Quincy, MA: National Fire Protection Association, 1970.

———. *NFPA 101, Life Safety Code*. Quincy, MA: National Fire Protection Association, 1981.

———. *NFPA 101, Life Safety Code*. Quincy, Mass.: NFPA, 2000.

Nelson, Harold E. *"The Potential of Scientifically Based Fire Protection Engineering?" In Fire Safety: Science and Engineering: A Symposium Sponsored by ASTM Committee E-5 on Fire Standards and by the Society of Fire Protection Engineers. Denver, June 26–27, 1984*. Ed. T. Z. Harmathy. Philadelphia: ASTM, 1985. 3–7.

New England Insurance Exchange. *Tenth Anniversary of the New England Insurance Exchange Hotel Vendome, Boston, Mass., January 6, 1893*. Boston: Standard Publishing Co., 1893.

New York Board of Fire Underwriters. Special Committee on Gas Machines, Carburetters and Matters Connected Therewith. *Report of the Committee on Gas Machines, Carburetters, and the Handling and Using of the Lighter Products of Petroleum, Made to the New York Board of Fire Underwriters, and Adopted October, 29, 1869*. New York: Wm. H. Woglom & Reading, 1869.

New York, State of. Board of Health. Metropolitan Sanitary District. *Fourth Annual Report of the Metropolitan Board of Health of the State of New York. 1869.* New York: D. Appleton and Co., 1870.

———. Board of Health. *Report on the Methods and Apparatus for Testing Inflammable Oils: Based Upon Investigations, Ordered by the State Board of Health, for Establishing a Standard Safety-Test of Illuminating Oils. By Arthur H. Elliott.* Albany, N.Y.: Weed, Parsons and Co., 1882.

Novak, William J. *The People's Welfare: Law and Regulation in Nineteenth-Century America.* Chapel Hill: University of North Carolina Press, 1996.

Nystrom, Justin A. *New Orleans after the Civil War: Race, Politics, and a New Birth of Freedom.* Baltimore: Johns Hopkins University Press, 2010.

Oakley, Henry A. *Addresses as President of the National Board of Fire Underwriters of the United States, on Several Occasions, by Henry A. Oakley, 1871–1876.* New York: [National Board of Fire Underwriters], 1876.

———, and D. A. Heald. *Report to the Executive Committee of the National Board of Fire Underwriters on Gas Machines.* New York: [National Board of Fire Underwriters], 1876.

Oliphant, F. H. "Petroleum." In *The Mineral Industry, Its Statistics, Technology and Trade in the United States and Other Countries to the End of 1901*, ed. Joseph Struthers. 2nd ed. Vol. 10. New York: Engineering and Mining Journal, 1903. 503–16.

O'Rear, Mary Jo. *Storm Over the Bay: The People of Corpus Christi and Their Port.* College Station: Texas A&M University Press, 2009.

Orton, Edward. *Geological Survey of Ohio: Preliminary Report Upon Petroleum and Inflammable Gas.* Rpt. with Supplement. Columbus, Ohio: A. H. Smythe, 1887.

Ousley, Clarence, ed. *Galveston in Nineteen Hundred: The Authorized and Official Record of the Proud City of the Southwest as It Was Before and After the Hurricane of September 8, and a Logical Forecast of Its Future.* Atlanta: William C. Chase, 1900.

Owen, Thomas McAdory. *History of Alabama and Dictionary of Alabama Biography.* 4 vols. Chicago: S. J. Clarke Publishing Co., 1921.

Pascoe, Craig S., Karen Trahan Leathem, and Andy Ambrose, eds. *The American South in the Twentieth Century.* Athens: University of Georgia Press, 2005.

Pattison, William D. *Beginnings of the American Rectangular Land Survey System, 1784–1800.* Chicago: University of Chicago Press, 1964.

Pegram, Thomas R. *Partisans and Progressives: Private Interest and Public Policy in Illinois, 1870–1922.* Urbana: University of Illinois Press, 1992.

Pelling, Mark. *The Vulnerability of Cities: Natural Disasters and Social Resilience.* London: Earthscan Publications Ltd., 2003.

"Petroleum." In *The Mineral Industry during 1903*. Vol. 12. Ed. D. H. Newland. New York: Engineering and Mining Journal, 1904. 281–91.

Petroleum and Products. International Library of Technology. Scranton, [Pa.]: International Textbook Co., 1902.

Philadelphia Fire Underwriters' Association. Constitution, By-Laws, Agreement and Rules: Philadelphia Fire Underwriters' Association. [Philadelphia], 1906.

Phillips, William B., and S.H. Worrell. *The Fuels Used in Texas*. Bulletin of the University of Texas. No. 307. Austin: University of Texas, 1913.

Pinney, C. *Atlas of the City of Boston, surveyed, drawn, etc., expressly for the use of the New York and Boston Fire Insurance Cos. and Agents*. Boston: L. Prange, 1861.

Platt, Harold L. *City Building in the New South: The Growth of Public Services in Houston, Texas, 1830–1910*. Philadelphia: Temple University Press, 1983.

Powell, Evans. *A Five Years' Fight against Fire Waste and Its Possible Control in the United States by Fire Prevention and Protection: Comprising a Series of Addresses, Showing the Growth of an Idea on this Subject within Five Years—1908–12 Inclusive*. Philadelphia: n.p., 1912.

Pratt, Joseph A. "The Ascent of Oil: The Transition from Coal to Oil in Early Twentieth-Century America." In *Energy Transitions: Long-Term Perspectives*, ed. Lewis J. Perelman, August W. Giebelhaus, and Michael D. Yokell. Boulder, Colo.: Westview Press, 1981. 9–35.

Prentice, W. P. *Police Powers Arising Under The Law of Overruling Necessity. New York: Banks & Brothers, 1894.*

Quarantelli, E. L., ed. What Is a Disaster? Perspectives on the Question. London: Routledge, 1998.

Redpath, Lionel V., comp. *Petroleum in California A Concise and Reliable History of the Oil* Industry of the State. Los Angeles: Lionel V. Redpath, 1900.

Redwood, Boverton. *Cantor Lectures on Petroleum and Its Products. Delivered before the Society of Arts, March 8, 15, 22, 29, 1886*. London: Society for the Encouragement of Arts, Manufactures, and Commerce, 1886.

Redwood, Sir Boverton. Petroleum: A Treatise on the Geographical Distribution and Geological Occurrence of Petroleum and Natural Gas. 2nd ed. 2 vols. London: Charles Griffin and Co., 1906.

Rees, Alun R., and H. J. Purcell, eds. *Disease and the Environment. Proceedings of the Inaugural Conference of the Society for Environmental Therapy, Oxford, March 21–23, 1981*. New York: Wiley, 1982.

Renn, Ortwin, Pia-Johanna Schweizer, Ulrich Müller-Herold, and Andrew Stirling. *Precautionary Risk Appraisal and Management. An Orientation for Meeting the Precautionary Principle in the European Union. Bremen*. Germany: Europäischer Hochshulverlag GembH & Co., 2009.

Riegel, Robert, and H. J. Loman. *Insurance Principals and Practices. 2nd ed.* New York: Prentice-Hall, 1922.

Ring, Natalie J. *The Problem South: Religion, Empire, and the New Liberal State, 1880–1930*. Athens: University of Georgia Press, 2012.

Ripley, George, and Charles A. Dana, eds. *The American Cyclopædia: A Popular Dictionary of General Knowledge*. 16 vols. New York: D. Appleton and Co., 1873.

Ritter, Gretchen. *Goldbugs and Greenbacks: The Antimonopoly Tradition and the Politics of Finance in America, 1865–1896*. 1st paperback ed. Cambridge, U.K.: Cambridge University Press, 1999.

Rosen, Christine Meisner. *The Limits of Power: Great Fires and the Progress of City Growth in America. Cambridge, U.K.: Cambridge University Press, 1986.*

Ross, William Frazier. *The Estimation of Exposures, or External Hazards, in Fire Insurance. With Schedules and Formulas for Rating Risks, and Stenographic Formulas for Denoting Risks and Rates. Philadelphia: J. B. Lippincott & Co., 1874.*

Rubin, Harvey W. *Dictionary of Insurance Terms*. 6th ed. Hauppauge, New York: Barron's Education Series, Inc., 2013.

Russell, Andrew L. *Open Standards and the Digital Age. History, Ideology, and Networks*. New York: Cambridge University Press, 2014.

Russell, James M. *Atlanta, 1847–1890: City Building in the Old South and the New*. Baton Rouge: Louisiana State University Press, 1988.

Salisbury, Stephen. "The Emergence of an Early Large-Scale Technical System: The American Railroad Network." In *The Development of Large Technical Systems*, ed. Renate Mayntz and Thomas P. Hughes. Boulder, Colo.: Westview Press, 1988. 37–68.

Saward, Frederick E. *The Coal Trade: A Compendium of Valuable Information Relative to Coal Production, Prices, Transportation, Etc.* 22nd ed. [New York]: n.p., 1895.

———. *The Coal Trade. A Compendium of Valuable Information Relative to Coal Production, Prices, Transportation, Etc.* 32nd ed. [New York]: n.p., 1905.

Schurr, Sam H., and Bruce C. Netschert. Energy in the American Economy, 1850–1975: an Economic Study of Its History and Prospects. Baltimore: Johns Hopkins University Press, 1960.

Scott, W. Richard. Organizations: Rational, Natural and Open Systems. Upper Saddle River, N.J.: Prentice-Hall, 2002.

Shapiro, Sidney A., and Robert L. Glicksman. *Risk Regulation at Risk: Restoring a Pragmatic Approach*. Stanford, Calif.: Stanford University Press, 2003.

Shenhav, Yhouda. *Manufacturing Rationality: The Engineering Foundations of the Managerial Revolution*. New York: Oxford University Press, 2002.

Sibley, Marilyn McAdams. *The Port of Houston: A History*. Austin: University of Texas Press, 1968.

Sinclair, Bruce. *Philadelphia's Philosopher Mechanics: A History of the Franklin Institute 1824–1865*. Baltimore: Johns Hopkins University Press, 1974.

Singer, Jonathan W. *Broken Trusts: The Texas Attorney General Versus the Oil Industry, 1889–1909*. College Station: Texas A&M University Press, 2002.

Sledge, John S. *The Mobile River.* Columbia: University of South Carolina Press, 2015.

Slovic, Paul. *The Perception of Risk: Risk, Society and Policy.* Rpt. 2001. London: Earthscan Publications, Ltd., 2000.

Smelser, Neil J., and Paul B. Baltes, eds. *International Encyclopedia of the Social & Behavioral Sciences.* Vol. 20. Amsterdam: Elsevier Science, 2001.

Smil, Vaclav. *Energies: An Illustrated Guide to the Biosphere and Civilization*. Cambridge, Mass.: MIT Press, 1999.

———. *Energy: A Beginner's Guide*. Oxford, U.K.: Oneworld Publications, 2006.

———. *Energy in Nature and Society: General Energetics of Complex Systems*. Cambridge, Mass.: MIT Press, 2008.

South-Eastern Tariff Association. *Book of Policy Forms, January 1, 1903*. Atlanta: South-Eastern Tariff Association, 1903.

———. *Proceedings of the Twenty-First Annual Meeting of the South-Eastern Tariff Association . . . May 21-22, 1902*. Atlanta, Ga.: South-Eastern Tariff Association Print, 1902.

———. *Quarter-Centennial History of The South-Eastern Tariff Association: 1882–1907*. Digital Library of Georgia. http://dlg.galileo.usg.edu/georgiabooks/do-pdf:gb5094

Southern Building Code Congress. *Southern Standard Fire Prevention Code.* Birmingham, Alabama: Southern Building Code Congress, 1974.

The Spectator Company. *Fire Prevention and Protection*. 3rd ed. Rev. A. C. Hutson. New York: Spectator Co., 1916.

———. *The Insurance Year Book, 1892–93 [Fire and Marine*]. Twentieth Annual Issue. New York: Spectator Co., 1892.

———. *The Insurance Year Book, 1900–1901 [Fire and Marine*]. Twenty-Eighth Annual Issue. New York: Spectator Co., 1900.

———. *The Insurance Year Book, 1902–1903 [Fire and Marine*]. 30th Annual Issue. New York: Spectator Co., 1902.

Stanonis, Anthony J. *The Big Easy: New Orleans and the Emergence of Modern Tourism, 1918–1945*. Athens: University of Georgia Press, 2006.

Steckel, Richard H., and Roderick Floud, eds. *Health and Welfare during Industrialization*. Chicago: University of Chicago Press, 1997.

Steeb, George Velten. *Agents' and Inspectors Pocket-Book of Fire Protection.* New York: Spectator Co., 1902.

Steele, Jenny. *Risks and Legal Theory.* Legal Theory Today. Portland, Ore.: Hart Publishing, 2004.

Stephens, Hugh W. *The Texas City Disaster, 1947.* Austin: University of Texas Press, 1997.

Stern, Paul C., and Harvey V. Fineberg, eds. *Understanding Risk: Informing Decisions in a Democratic Society.* Committee on Risk Characterization. Commission on Behavioral and Social Sciences and Education. National Research Council. Washington, D.C.: National Academies Press, 1996.

Stradling, David. *Smokestacks and Progressives: Environmentalists, Engineers, and Air Quality in America, 1881–1951.* Baltimore: Johns Hopkins University Press, 1999.

Strahler, Alan, and Arthur Strahler. *Physical Geography: Science and Systems of the Human Environment.* 2nd ed. New York: John Wiley & Sons, 2002.

Tabak, John. *Probability and Statistics: The Science of Uncertainty.* New York: Checkmark Books, 2005.

Teaford, Jon C. *The Unheralded Triumph: City Government in America, 1870–1900.* Baltimore: Johns Hopkins University Press, 1984.

Tebeau, Mark. *Eating Smoke: Fire in Urban America, 1800–1950.* Baltimore: Johns Hopkins University Press, 2003.

Texas, State of. Department of Agriculture. *Twenty-Fifth Annual Report of the Commissioner of Agriculture, Insurance, Statistics, and History for the Year 1899.* Austin, Tex.: Department of Agriculture, 1900.

———. *General Laws of the State of Texas Passed at the Regular Session of the Twenty-Sixth Legislature.* Austin: Boeckmann, Moore & Schultz, 1899.

———. Secretary of State. Corporation Section. "Charter of the Mississippi-Texas Oil Co." Record #10224, filed May, 1901. www.sos.state.tx.us/corp/sosda/index.shtml.

———. *Twenty-Ninth Annual Report of the Commissioner of Agriculture, Insurance, Statistics, and History for the Year 1903.* Austin, Tex.: Department of Agriculture, 1904.

Texas Water Development Board. *100 Years of Rule of Capture: from East to Groundwater Management. Texas Water Development Board Report,* vol. 361. Ed. W. F. Mullican and Suzanne Schwartz. [Austin]: Texas Water Development Board, 2004.

Thompson, Craig. *Since Spindletop: A Human Story of Gulf's First Half-Century.* [Pittsburgh], [Gulf Oil Corporation], [ca. 1951].

Thomson, J. H., and Sir Boverton Redwood. *Handbook on Petroleum for Inspectors Under the Petroleum Acts.* 2nd ed. London: Charles Griffin and Co., 1906.

Thomson, Ross. *Structures of Change in the Mechanical Age: Technological Innovation in the United States, 1790–1865.* Baltimore: Johns Hopkins University Press, 2009.

Tiedeman, Christopher G. *A Treatise on the Limitations of Police Power in the United States.* 1886. Rpt. New York: Da Capo Press, 1971.

Tiffany, H. S. *Tiffany's Book of Policy Forms for Fire Insurance Agents.* 4th ed. Chicago: H. S. Tiffany & Co., 1890.

———, comp. *Tariffs and Schedules, for Rating Risks. Also Mandatory Forms, Rates and Rules Promulgated by the Various Organizations of Fire Underwriters.* Chicago: H. S. Tiffany & Co., 1891.

———, comp. *Tiffany's Special Agents' and Adjusters' Assistant Containing Numerous Citations to the Law of Insurance and Other Valuable Statistics and References.* 10th ed. Chicago: H. S. Tiffany & Co., 1891.

Todd, Alden L. *A Spark Lighted in Portland: the Record of the National Board of Fire Underwriters.* New York: McGraw-Hill Book Co., 1966.

Tolbert, Lisa C. *Connecting Townscapes: Space and Society in Antebellum Tennessee.* Chapel Hill: University of North Carolina Press, 1999.

Toulmin, Stephen. *Cosmopolis: The Hidden Agenda of Modernity.* Chicago: University of Chicago Press, 1990.

Trinks, Charles L. Willibald. *Industrial Furnaces.* 2nd ed. 2 vols. New York: Wiley, 1923–25.

Underwriters' Laboratories. *The Organization, Purpose and Methods of Underwriters' Laboratories.* [Chicago]: Underwriters' Laboratories, 1916.

United Nations Educational, Scientific and Cultural Organization. *Global Trends in Water-Related Disasters: An Insight for Policy Makers.* World Water Assessment Programme. By Yoganath Adikari and Junichi Yoshitani. Paris: UNESCO, 2009.

United States Department of Commerce. *Fourteenth Census of the United States.* Vol. 9: *Manufacturers 1919.* Washington, D.C.: Government Printing Office, 1923.

———. *Thirteenth Census of the United States Taken in the Year 1910.* Vol. 3: *Population 1910. Nebraska-Wyoming, Alaska, Hawaii, and Porto Rico.* Washington, D.C.: Government Printing Office, 1913.

———. *National Bureau of Standards. Standardization Activities in the United States: A Descriptive Directory.* By Sherman F. Booth. NBS Misc. Pub. 230. Washington D.C.: Government Printing Office, 1960.

———. *Standardization Activities of National Technical and Trade Organizations.* By Robert A. Martino. NBS Misc. Pub. M169. Washington, D.C.: Government Printing Office, 1941.

United States Department of Commerce and Labor. *Thirteenth Census of the United States Taken in the Year 1910: Statistics for Alabama.* Washington, D.C.: Government Printing Office, 1913.

———. *Thirteenth Census of the United States Taken in the Year 1910: Statistics for Louisiana.* Washington, D.C.: Government Printing Office, 1913.

———. National Oceanic and Atmosphere Administration. Committee on Environment and Natural Resources (CENR). *Ecological Risk Assessment in the Federal Government.* May 1999. CENR/5–99/001.

United States. Department of State. *Consular Reports: Commerce, Manufactures, Etc.* Vol. 54. Report no. 203. Washington, D.C.: Government Printing Office, 1897.

United States Department of the Interior. Bureau of Mines. The Flash Point of Oils: Methods and Apparatus for Its Determination. Technical Paper 49. By Irving C. Allen and A. S. Crossfield. Washington, D.C.: Government Printing Office, 1914.

———. *Report on the Production, Technology, and Uses of Petroleum and Its Products.* By S. F. Peckham. Washington, D.C.: Government Printing Office, 1885.

———. U.S. Geological Survey. Division of Mining and Mineral Resources. *Twenty-First Annual Report of the United States Geological Survey to the Secretary of the Interior 1899–1900.* Part 6: *Mineral Resources of the United States, 1899 Metallic Products, Coal, and Coke.* Washington, D.C.: Government Printing Office, 1901.

———. U.S. Geological Survey. *Mineral Resources of the United States, 1900.* By David T. Day. Washington, D.C.: U.S. Government Printing Office, 1901.

———. U.S. Geological Survey. *Mineral Resources of the United States, 1901.* By David T. Day. Washington, D.C.: U.S. Government Printing Office, 1902.

———. U.S. Geological Survey. *Mineral Resources of the United States, 1902.* By David T. Day. Washington, D.C.: U.S. Government Printing Office, 1904.

———. U.S. Geological Survey. *The Production of Petroleum in 1894.* By Joseph D. Weeks (1895). From the *Sixteenth Annual Report of the Director, 1894–95,* Part 4: *Mineral Resources of the United States, Calendar Year 1894.* Washington, D.C.: Government Printing Office, 1895. 315–19.

United States. Department of the Navy. *Report of the Secretary of the Navy, with An Appendix, Containing Bureau Reports, Etc.* Washington, D.C.: Government Printing Office, December 1867.

United States. Environmental Protection Agency. *An Examination of EPA Risk Assessment Principles and Practices.* Staff Paper, Risk Assessment Task Force. EPA/100/B-04/001. Washington, D.C.: U.S. Environmental Protection Agency, 2004.

———. Environmental Protection Agency. Risk Assessment Forum. *Framework for Cumulative Risk Assessment.* EPA/630/P-02/001F. Washington, D.C.: U.S. Environmental Protection Agency, 2003.

———. National Weather Service. National Hurricane Center, Miami. *The Deadliest, Costliest, and Most Intense United States Tropical Cyclones from 1851 to 2006.* By Eric S. Blake, Edward N. Rappaport, and Christopher W. Landsea. NOAA Technical Memorandum NWS TPC-5. Miami: National Hurricane Center, 2007.

———. Nuclear Regulatory Agency. Glossary. "Performance-based Regulation." www.nrc.gov/reading-rm/basic-ref/glossary.html.

United States. House of Representatives. *Annual Reports of the War Department for the Fiscal Year Ended June 30, 1898.* "Report of the Secretary of War." *Miscellaneous Reports,* 55th Congress, 3d sess., 1898, doc. no. 2. Washington, D.C.: Government Printing Office, 1901.

———. *Annual Reports of the War Department for the Fiscal Year Ended June 30, 1898. Report of the Chief of Engineers, Part 1.* Washington: Government Printing Office, 1898.

———. *Annual Reports of the War Department for the Fiscal Year Ended June 30, 1899. Report of the Chief of Engineers, Part 1*. Washington: Government Printing Office, 1899.

———. *Annual Reports of the War Department for the Fiscal Year Ended June 30, 1901. Report of the Chief of Engineers, Part 1*. Washington: Government Printing Office, 1901.

U.S. Chemical Safety and Hazard Investigation Board. *Investigation Report: Refinery Explosion and Fire, BP Texas City, Texas, March 23, 2005*. Report No. 2005-04-I-TX. www.CSB.gov/assets/ document/CSBFinalReportBP.pdf.

Vermaas, Peter E., Peter Kroes, Ibo van de Poel, Maarten Franssen, and Wybo Houkes. *A Philosophy of Technology from Technical Artefacts to Sociotechnical Systems.* Williston, Vt.: Morgan & Claypool Publishers, 2011.

Walford, Cornelius. *The Insurance Cyclopaedia.* 6 vols. London: Charles and Edwin Layton, 1878.

Wallis, Michael. *Oil Man: The Story of Frank Phillips and the Birth of Phillips Petroleum.* Garden City, N.Y.: Doubleday, 1988.

Waring, Alan E., and A. Ian Glendon. *Managing Risk: Critical Issues for Survival and Success in the 21st Century.* London: International Thomson Business Press, 1998.

Watson, Alan. *Wilmington, North Carolina, to 1861.* Jefferson, N.C.: McFarland & Co., 2003.

Weems, John Edward. *A Weekend in September.* College Station: Texas A&M Press, 2000.

Wells, David A., ed. *Annual of Scientific Discovery; or, Year-Book of Facts in Science and Art For 1865.* Boston: Gould and Lincoln, 1865.

Wermiel, Sarah E. *The Fireproof Building: Technology and Public Safety in the Nineteenth-Century American City.* Baltimore: Johns Hopkins University Press, 2000.

Whiting, John E. *A Schedule of the Buildings and Their Occupancy on the Principal Streets and Wharves in the City of Boston: Surveyed and Published Expressly for the Use of Insurance Companies.* Boston: W. L. Deland, 1877.

Wiebe, Robert H. *The Search for Order, 1877–1920.* New York: Hill and Wang, 1967.

Wilcox, Delos F. *Municipal Franchises: A Description of the Terms and Conditions upon which Private Corporations Enjoy Special Privileges in the Streets of American Cities.* 2 vols. Rochester, New York: Gervaise Press, 1910–11.

Williamson, Harold F., and Arnold R. Daum. *The American Petroleum Industry. The Age of Illumination 1859–1899.* Evanston, Ill.: Northwestern University Press, 1959.

Williamson, Harold F., Ralph L. Andreano, Arnold R. Daum, and Gilbert C. Klose. *The American Petroleum Industry: The Age of Energy 1899–1959.* Evanston, Ill.: Northwestern University Press, 1963.

Wills, Jocelyn. *Boosters, Hustlers, and Speculators: Entrepreneurial Culture and the Rise of Minneapolis and St. Paul, 1849–1883.* St. Paul: Minnesota Historical Society, 2005.

Winner, Langdon. *Autonomous Technology: Technics-Out-of-Control as a Theme in Political Thought.* Cambridge, Mass.: MIT Press, 1977.

Woodbury, C. J. H. *The Engineer as an Economist: An Address Given at the Commencement Exercises of the Thomas S. Clarkson Memorial School of Technology June 9, 1905.* Potsdam, N.Y.: n.p., 1905.

———. *The Fire Protection of Mills; and Construction of Mill Floors: Containing Tests of Full Size Wood Mill Columns.* New York: John Wiley & Sons, 1882.

———. *History of the National Electrical Code: An Address Before The National Electrical Contractors' Association At Its Sixth Annual Convention Cleveland, Ohio July 19, 1906.* [Cleveland, Ohio: National Electrical Contractors' Association, 1906].

———. *The Influence of the Underwriters' Rules on Electrical Development: A Paper Read before the New York Electrical Society, March 27, 1907.* [New York: New York Electrical Society, 1907].

Woodward, C. Vann. *Origins of the New South, 1877–1913.* Baton Rouge: Louisiana State University Press, 1971.

Wright, Gavin. *Old South New South: Revolutions in the Southern Economy Since the Civil War.* New York: Basic Books, 1986.

The Writers' Press Association. *The Successful American.* Vol. 5. New York: The Writers' Press Association, 1902.

Yorke, Dane. *Able Men of Boston: The Remarkable Story of the First 100 Years of the Boston Manufacturers Mutual Fire Insurance Company: Its Men, Its Times, and Its Growth—in Insurance Protection, in Fire Prevention Know-how, in ever Vital Service and Security to North American Industry, 1850–1950.* Boston: Boston Manufacturers Mutual Fire Insurance Co., 1950.

Young, Earle B. *The Tracks to the Sea: Galveston and Western Railroad Development, 1866–1900.* College Station: Texas A&M University Press, 1999.

Zacharie, James S. *New Orleans Guide, With Descriptions of the Routes to New Orleans, Sights of the City Arranged Alphabetically, and Other Information Useful to Travellers; Also Outlines of the History of Louisiana.* New Orleans: New Orleans News Co., 1885.

Zunz, Olivier. *Why the American Century?* Chicago: University of Chicago Press, 1998.

ARTICLES

Adger, W. Neil. "Vulnerability." *Global Environmental Change* 16, no. 3 (2006): 268–81.

Allen, Zachariah. "Explosibility of Coal Oils." In *Annual Report of the Board of Regents of the Smithsonian Institution, Showing the Operations, Expenditures, and Condition of the Institution for the Year 1861.* H.R. Misc. Doc. No. 77. Washington, D.C.: Government Printing Office, 1862. 330–42.

Atkinson, Edward. "Preliminary Report Upon the Lubrication of Heavy Bearings." In *Boston Manufacturers Mutual Fire Insurance Companies, Special Report,* No. 14: *Report Upon Coverings for Steam Pipes By Prof. John M. Ordway, of the Massachusetts Institute of Technology; Also a Preliminary Report on the Lubrication of Heavy Bearings.* Boston: BMMFIC, 1883.

Aven, Terje. "Interpretations of Alternative Uncertainty Representations in a Reliability Risk Analysis Context." *RE&SS* 96, no. 3 (March 2011): 353–60.

———. "On How to Define, Understand and Describe Risk." *RE&SS* 95, no. 6 (June 2010): 623–31.

———. "On the Meaning of a Black Swan in a Risk Context." *Safety Science* 57 (August 2013): 44–51.

———. "Perspectives on Risk: Review and Discussion of the Basis for Establishing a Unified and Holistic Approach." *RE&SS* 90, no. 1 (October 2005): 1–14.

———. "Practical Implications of the New Risk Perspectives." *RE&SS* 115 (July 2013): 136–45.

———. "A Risk Concept Applicable for Both Probabilistic and Non-Probabilistic Perspectives." *Safety Science* 49 (2011): 1080–86.

———. "The Risk Concept—Historical and Recent Development Trends." *RE&SS* 99 (2012): 33–44.

———. "Selective Critique of Risk Assessment with Recommendations for Improving Methodology and Practice." *RE&SS* 96, no. 5 (May 2011): 509–14.

Aven, Terje, and Bodil S. Krohn. "A New Perspective on How to Understand, Assess and Manage Risk and the Unforeseen." *RE&SS* 121 (January 2014): 1–10.

Aven, Terje, Ortwin Renn, and Eugene A. Rosa. "On the Ontological Status of the Concept of Risk." *Safety Science* 49 (2011): 1074–79.

Aven, Terje, and Seth Guikema. "Whose Uncertainty Assessments (Probability Distributions) Does a Risk Assessment Report: the Analysts' or the Experts'?" *RE&SS* 96, no. 10 (October 2011): 1257–62.

Bankoff, Greg. "Constructing Vulnerability: The Historical, Natural and Social Generation of Flooding in Metropolitan Manila." *Disasters* 27, no. 3 (September 2003): 223–38.

Baranoff, Dalit. "A Policy of Cooperation: the Cartelisation of American Fire Insurance, 1873–1906." *Financial History Review* 10, no. 2 (2003): 119–36.

Bénézech, Danièle, Gilles Lambert, Blandine Lanoux, Christophe Lerch, and Joceylne Loos-Baroin. "Completion of Knowledge Codification: An Illustration through the ISO 9000 Standards Implementation Process." *Research Policy* 30, no. 9 (December 2001): 1395–1407.

Ben-Haim, Yakov. "Why Risk Analysis Is Difficult, and Some Thoughts on How to Proceed." *Risk Analysis* 32, no. 10 (October 2012): 1638–46.

Berkes, Fikret. "Understanding Uncertainty and Reducing Vulnerability: Lessons from Resilience Thinking." *Natural Hazards* 41 (2007): 283–95.

Björkman, Torsten. "The Rationalisation Movement in Perspective and Some Ergonomic Implications." *Applied Ergonomics* 27, no. 2 (1996): 111–17.

Bolton, Ronan, and Timothy J. Foxon. "Infrastructure Transformation as a Socio-Technical Process—Implications for the Governance of Energy Distribution Networks in the UK." *Technological Forecasting & Social Change* 90, part B (January 2015): 538–50.

Boudia, Soraya, and Nathalie Jas. "Introduction: Risk and 'Risk Society' in Historical Perspective." *History and Technology* 23, no. 4 (December 2007): 317–31.

Brophy, William. "Report of the N.E.L.A. Committee on Safe Wiring." *Electrical Engineer* 23, no. 476 (June 16, 1897): 672–73.

Budgen, Eurling Paul J. "The Engineer at Risk: A Proposal to Use Hazard and Risk Assessment as a General Management Tool." *Engineering Management Journal* 9, no. 6 (December 1999): 265–72.

Christensen, Frans Møller, Ole Andersen, Nijs Jan Duijm, and Poul Harremoës. "Risk Terminology—a Platform for Common Understanding and Better Communication." *Journal of Hazardous Materials* A103, no. 3 (2003): 181–203.

Chwastiak, Michele. "Rationality, Performance Measures and Representations of Reality: Planning, Programming and Budgeting and the Vietnam War." *Critical Perspectives on Accounting* 17, no. 1 (January 2006): 29–55.

Clark, John G. "New Orleans and the River: A Study in Attitudes and Responses." *Louisiana History* 8, no. 2 (Spring 1967): 117–35.

Clarke, Sally H. "Unmanageable Risks: MacPherson v. Buick and the Emergence of a Mass Consumer Market." *Law & History Review* 23, no. 1 (Spring 2005): 1–52.

Cobb, James C. "Beyond Planters and Industrialists: A New Perspective on the New South." *Journal of Southern History* 54, no. 1 (February 1988): 45–68.

Cohen, Bernard L. "Probabilistic Risk Analysis for a High-Level Radioactive Waste Repository." *Risk Analysis* 23, no. 5 (October 2003): 909–15.

Cohendet, Patrick, and Frieder Meyer-Krahmer. "The Theoretical and Policy Implications of Knowledge Codification." *Research Policy* 30, no. 9 (December 2001): 1563–91.

Colyvan, Mark. "Is Probability the Only Coherent Approach to Uncertainty?" *Risk Analysis* 28, no. 3 (June 2008): 645–52.

Constant II, Edward W. "Reliable Knowledge and Unreliable Stuff: On the Practical Role of Rational Beliefs." *Technology and Culture* 40, no. 2 (April 1999): 324–57.

Covello, Vincent T., and Jeryl Mumpower. "Risk Analysis and Risk Management: An Historical Perspective." *Risk Analysis* 5, no. 2 (June 1985): 103–20.

Crowe, Robert M., and Ronald C. Horn. "Curricular Concepts in Risk and Insurance." *Journal of Risk and Insurance* 34, no. 3 (September 1967): 459–74.

Cumming, Robert B. "Is Risk Assessment A Science?" *Risk Analysis* 1, no. 1 (March 1981): 1–3.

Day, Thomas H. "Concentric Wiring." *Electrical Review and Western Electrician* 68, no. 9 (February 26, 1916): 383–87.

D'Ippolito, Beatrice, Marcela Miozzo and Davide Consoli. "Knowledge Systematisation, Reconfiguration and the Organisation of Firms and Industry: The Case of Design." *Research Policy* 43, no. 8 (October 2014): 1334–52.

Eaton, Gary H., and Donald E. Little, "Risk Assessing & Mitigating to Deliver Sustainable Safety Performance." *Professional Safety* 56, no. 7 (July 2011): 35–41.

"Electrical Hazards: Annual Meeting of the Electrical Committee of the Underwriters' National Electric Association—Some History." *Insurance Press* 5, no. 119 (December 8, 1897): 2.

"Electric Hazards: Last Week's Meeting of the Electric Committee of the Underwriters' National Electric Association—Full Report of the Proceedings." *Insurance Press* 1, no. 15 (December 18, 1895): 1–3.

"Electrical Inspection and the National Electric Code." *Quarterly of the National Fire Protection Association* 7, no. 1 (July 1913): 53–58.

Ersdal, Gerhard, and Terje Aven. "Risk Informed Decision-Making and Its Ethical Basis." *RE&SS* 93, no. 2 (February 2008): 197–205.

Fouquet, Roger, and Peter J. G. Pearson. "Past and Prospective Energy Transitions: Insights from History." *Energy Policy* 50 (November 2012): 1–7.

Fressoz, Jean-Baptiste. "Beck Back in the 19th Century: Towards a Genealogy of Risk Society." *History and Technology* 23, no. 4 (December 2007): 333–50.

Fuenfschilling, Lea, and Bernhard Truffer. "The Structuration of Socio-Technical Regimes—Conceptual Foundations from Institutional Theory." *Research Policy* 43, no. 4 (May 2014): 772–91.

Geels, Frank W. "Transformations of Large Technical Systems: A Multilevel Analysis of the Dutch Highway System (1950–2000)." *Science, Technology, & Human Values* 32, no. 2 (March 2007): 123–49.

Godschalk, David R. "Urban Hazard Mitigation: Creating Resilient Cities." *Natural Hazards Review* 4, no. 3 (August 1, 2003): 139–40.

Gorner, Tomas, and Michal Simon. "Using the Theory of Technical Systems to Describe the Interaction between Human and Technical Systems with the Ergonomic System." *Procedia Engineering* 100 (2015): 592–601.

Graham, John D. "Historical Perspective on Risk Assessment in the Federal Government." *Toxicology* 102, no. 1–2 (September 1995): 29–52.

Grubler, Arnulf. "Energy Transitions Research: Insights and Cautionary Tales." *Energy Policy* 50 (November 2012): 8–16.

Hexamer, Charles A. "Fire Insurance—Rates and Schedule Rating." *Annals of the American Academy of Political and Social Science* 26 (September 1905): 211–23.

Heyvaert, Veerle. "Reconceptualizing Risk Assessment." *Review of European Community and International Environmental Law* 8, no. 2 (1999): 135–43.

Humphrey, Henry H. "On the Use of Beaumont Oil as Fuel." *Journal of the Association of Engineering Societies* 30, no. 3 (March 1903): 95–103.

Hunt, A.M. "California Crude Oil and Its Use as Fuel." *"Industry": A Magazine Devoted to Science, Engineering, and the Mechanic Arts, Especially on the Pacific Coast,* no. 77 (December 1894): 731–44.

Jackson, Joy. "Bosses and Businessmen in Gilded Age New Orleans Politics." *Louisiana History* 5, no. 4 (Autumn 1964): 387–400.

Jannadi, Osama Ahmed Jannadi, and Salman Almishari. "Risk Assessment in Construction." *Journal of Construction Engineering and Management* 129, no. 5 (October 1, 2003): 492–500.

"J. H. Hill on Embargoes." *Traffic World* 15, no. 17 (April 24, 1915): 911–14.

Joskow, Paul L. "Cartels, Competition and Regulation in the Property-Liability Insurance Industry." *Bell Journal of Economics and Management Science* 4, no. 2 (Autumn 1973): 375–427.

Kates, Robert W., and Jeanne X. Kasperson. "Comparative Risk Analysis of Technological Hazards (A Review)." *Proceedings of the National Academy of Sciences of the United States of America* 80, no. 22 (November 1983): 7027–38.

Keller, William, and Mohammad Modarres. "A Historical Overview of Probabilistic Risk Assessment Development and Its Use in the Nuclear Power Industry: A Tribute to the Late Professor Norman Carl Rasmussen." *RE&SS* 89, no. 3 (September 2005): 271–85.

Klein, Jonathan H., and Robin B. Cork. "An Approach to Technical Risk Assessment." International Journal of Project Management 16, no. 6 (December 1998): 345–51.

Kletz, T. A. "The Origins and History of Loss Prevention." *Transactions of the Institution of Chemical Engineers* 77, part B (May 1999): 109–16.

Klinke, Andres, and Ortwin Renn. "Precautionary Principle and Discursive Strategies: Classifying and Managing Risks." *Journal of Risk Research* 4, no. 2 (2001): 159–173.

Knowles, Scott Gabriel. "Lessons in the Rubble: the World Trade Center and the History of Disaster Investigations in the United States." *History and Technology* 19, no. 1 (March 2003): 9–28.

Kottow, Michael H. "Vulnerability: What Kind of Principle Is It?" *Medicine, Health Care and Philosophy* 7 (2004): 281–87.

Kristensen, V., T. Aven, and D. Ford. "A New Perspective on Renn and Klinke's Approach to Risk Evaluation and Management." *RE&SS* 91, no. 4 (April 2006): 421–32.

Laperche, Blandine, Gilliane Lefebvre, and Denis Langlet, "Innovation Strategies of Industrial Groups in the Global Crisis: Rationalization and New Paths." *Technological Forecasting & Social Change* 78, no. 8 (October 2011): 1319–31.

Leveson, Nancy. "A Systems Approach to Risk Management through Leading Safety Indicators." *RE&SS* 136 (April 2015): 17–34.

Li, Francis G. N., Evelina Trutnevyte, and Neil Strachan, "A Review of Socio-Technical Energy Transition (STET) Models." *Technological Forecasting & Social Change* 100 (November 2015): 290–305.

Linebaugh, Donald W. "Illuminating a Country Retreat: Gas Lighting at President Lincoln's Cottage." *APT Bulletin* 43, no. 2–3 (2012): 59–69.

Lord, J. W. "The Fire Hazard of Fuel Oil." *American Concrete Institute: Proceedings of the Fifteenth Annual Convention Held at Atlantic City, N.J., June 27 and 28, 1919,* vol. 15 (1919): 179–85.

"The Louisiana Sugar Planters' Association." *Louisiana Planter and Sugar Manufacturer* 26, no. 15 (April 13, 1901): 225.

Lowrey, Walter M. "The Engineers and the Mississippi." *Louisiana History* 5, no. 3 (Summer 1964): 233–56.

Lubken, Uwe, and Christof Mauch. "Uncertain Environments: Natural Hazards, Risk and Insurance in Historical Perspective." *Environment and History* 17, no. 1 (February 2011): 1–12.

MacLean, Douglas. "Quantification, Regulation, and Risk Assessment." *PSA: Proceedings of the Biennial Meeting of the Philosophy of Science Association* 2 (1982): 243–60.

Marhavilas, P. K, D. Koulouriotis, and V. Gemeni. "Risk Analysis and Assessment Methodologies in the Work Sites: On a Review, Classification and Comparative Study of the Scientific Literature of the Period 2000–2009." *Journal of Loss Prevention in the Process Industries* 24, no. 5 (September 2011): 477–523.

Marszal, Edward M. "Tolerable Risk Guidelines." *ISA Transactions* 40, no. 4 (September 2001): 391–99.

Matthews, Derek. "The Technical Transformation of the Late Nineteenth-Century Gas Industry." *Journal of Economic History* 47, no. 4 (December 1987): 967–80.

McSwain, James B. "Fire Hazards and Protection of Property: Municipal Regulation of the Storage and Supply of Fuel Oil in Mobile, Alabama, 1894–1910." *Journal of Urban History* 28, no. 5 (July 2002): 599–628.

———. "Urban Government and Environmental Policies: Regulating the Storage and Distribution of Fuel Oil in Houston, Texas, 1901–1915." *Journal of Southern History* 71, no. 2 (May 2005): 279–320.

Meyer, David R. "The Industrial Retardation of Southern Cities, 1860–1880." Explorations in *Economic History* 25, no. 4 (October 1988): 366–86.

———. "Midwestern Industrialization and the American Manufacturing Belt in the Nineteenth Century." *Journal of Economic History* 49, no. 4 (December 1989): 921–37.

Mitello, Lucia, and Fabrizio Rufo, "Individual Risk and Collective Fears: Towards a Governance of Medical Uncertainty." *Topoi* 23, no. 2 (September 2004): 221–27.

Murphy, Colleen, and Paolo Gardoni, "The Acceptability and the Tolerability of Societal Risks: A Capabilities-based Approach." *Science and Engineering Ethics* 14 (2008): 77–92.

"National Electrical Code History: Exhaustive Review Made by F. Elliot Cabot of Boston Board Before Electrical Engineers in Providence, R.I." *The Standard* 91, no. 18 (October 28, 1922): 682–84.

Oerlemans, Leon A. G., Marius T. H. Meeus, and Frans W. M. Boekeman. "Innovation and Space: Theoretical Perspectives." *Eindhoven Centre for Innovation Studies, Working Paper* 99, no. 3 (February 1999).

O'Laughlin, Jay. "Policies for Risk Assessment in Federal Land and Resource Management Decisions." *Forest Ecology and Management* 211, no. 1–2 (June 2005): 15–27.

Ormerod, R. J. "Critical Rationalism in Practice: Strategies to Manage Subjectivity in OR Investigations." *European Journal of Operational Research* 235, no. 3 (June 16, 2014): 784–97.

Paté-Cornell, Elisabeth. "On 'Black Swans' and 'Perfect Storms': Risk Analysis and Management When Statistics Are Not Enough." *Risk Analysis* 32, no. 11 (November 2012): 1823–33.

Patwardhan, Anand, and Upasna Sharma. "Improving the Methodology for Assessing Natural Hazard Impacts." *Global and Planetary Change* 47 (2005): 253–65.

Peabody, Ernest H. "Oil Fuel." *Transactions of the International Engineering Congress* 10 (1915): *Naval Architecture and Marine Engineering,* Paper No. 214: 466–596.

Pease, Jane H., and William H. Pease. "The Blood-Thirsty Tiger: Charleston and the Psychology of Fire." *South Carolina Historical Magazine* 79, no. 4 (October 1978): 281–95.

Platt, Harold L. "Energy and Urban Growth: A Comparison of Houston and Chicago." *Southwestern Historical Quarterly* 91, no. 1 (July 1987): 1–18.

Pratt, Joseph A. "8F and Many More: Business and Civic Leadership in Modern Houston." *Houston History Magazine* 1, no. 2 (Summer 2004): 2–7, 31–37, 40–42, 44.

———. "Growth or a Clean Environment? Responses to Petroleum-related Pollution in the Gulf Coast Refining Region." *Business History Review* 52 (Spring 1978): 1–29.

———. "Letter from the Editor: The Houston Ship Channel and Its Port." *Houston History* 12, no. 1 (Fall 2014).

———. "Letting the Grandchildren Do It: Environmental Planning During the Ascent of Oil as a Major Energy Source." *Public Historian: A Journal of Public History* 2, no. 4 (Summer 1980): 28–61.

The Press Biographical Company. *The Successful American (Illustrated)*. Vol. 1: *1900*. New York: Press Biographical Co., 1900.

Prince, K. Stephen. "A Rebel Yell for Yankee Doodle: Selling the New South at the 1881 Atlanta International Cotton Exposition." *Georgia Historical Quarterly* 92, no. 3 (Fall 2008): 340–71.

Rasmussen, Jens. "Risk Management in a Dynamic Society: A Modelling Problem." *Safety Science* 27, no. 2–3 (November–December 1997): 183–213.

Redmill, Felix. "Human Factors in Risk Analysis." *Engineering Management Journal* 12, no. 4 (August 2002): 171–76.

———. "Risk Analysis—a Subjective Process." *Engineering Management Journal* 12, no. 2 (April 2002): 91–96.

Renn, Ortwin. "The Role of Risk Perception for Risk Management." *RE&SS* 59, no. 1 (January 1998): 49–62.

Righi, Angela Weber, and Tarcisio Abreu Saurin. "Complex Socio-Technical Systems: Characterization and Management Guidelines." *Applied Ergonomics* 50 (September 2015): 19–30.

Ristow, Walter W. "United States Fire Insurance and Underwriters Maps, 1852–1968." *Quarterly Journal of the Library of Congress* 25, no. 3 (July 1968): 194–218.

Ross, Michael A. "Resisting the New South: Commercial Crisis and Decline in New Orleans, 1865–85." *American Nineteenth Century History* 4, no. 1 (Spring 2003): 59–76.

Rutter, Paul, and James Keirstead. "A Brief History and the Possible Future of Urban Energy Systems." *Energy Policy* 50 (November 2012): 72–80.

Saleh, J.H., and K. Marais. "Highlights from the Early (and Pre-) History of Reliability Engineering." *RE&SS* 91, no. 2 (February 2006): 249–56.

Salter, John. "Risk Management in a Disaster Management Context." *Journal of Contingencies and Crisis Management* 5, no. 1 (March 1997): 60–65.

Schöttl, Florian, and Udo Lindemann. "Quantifying the Complexity of Socio-Technical Systems: A Generic, Interdisciplinary Approach." *Procedia Computer Science* 44 (2015): 1–10.

Schyfter, Pablo. "The Bootstrapped Artefact: a Collectivist Account of Technological Ontology, Functions, and Normativity." *Studies in History and Philosophy of Science* 40, no. 1 (March 2009): 102–11.

Sherman, Mimi. "A Look at Nineteenth-Century Lighting: Lighting Devices from the Merchant's House Museum." *APT Bulletin* (Association for Preservation Technology International), vol. 31, no. 1 (2000): 37–43.

Shin, Dong-Hee, Hongbum Kim, and Junseok Hwang. "Standardization Revisited: A Critical Literature Review on Standards and Innovation." *Computer Standards & Interfaces* 38 (February 2015): 152–57.

Slovic, Paul, Melissa L. Finucane, Ellen Peters, and Donald G. MacGregor. "Risk as Analysis and Risk as Feelings: Some Thoughts about Affect, Reason, Risk, and Rationality." *Risk Analysis* 24, no. 2 (April 2004): 311–22.

Stanislawski, Dan. "The Origin and Spread of the Grid-Pattern Town." *Geographical Review* 36, no. 1 (January 1946): 105–20.

Stapleton, Larry, David Smith, and Fiona Murphy. "Systems Engineering Methodologies, Tacit Knowledge and Communities of Practice." *AI and Society* 19, no. 2 (April 2005): 159–80.

Sweetland, Ralph. "Electrical Fire Hazards . . . Lecture 2: Early Recognition of Hazards of Electricity." In *Lectures on Fire Insurance,* 10–12.

Tainter, Joseph A. "Energy, Complexity, and Sustainability: A Historical Perspective." *Environmental Innovation and Societal Transitions* 1 (2011): 89–95.

Teaford, Jon C. "New Life for an Old Subject: Investigating the Structure of Urban Rule." *American Quarterly* 37, no. 3 (1985): 346–56.

"Texas Oil Notes." *Age of Steel* 90, no. 8 (August 24, 1901): 22.

Tolbert, George D. "Residual Risk Reduction: Systematically Deciding What Is 'Safe.'" *Professional Safety* 50, no. 11 (November 2005): 25–33.

Vanem, Erick. "Ethics and Fundamental Principles of Risk Acceptance Criteria." *Safety Science* 50, no. 4 (April 2012): 958–67.

Veland, H., and T. Aven. "Risk Communication in the Light of Different Risk Perspectives." *Reliability Engineering and System Safety* 110 (February 2013): 34–40.

Vermaas, Pieter E. "The Physical Connection: Engineering Function Ascriptions to Technical Artefacts and Their Components." *Studies in History and Philosophy of Science* 37, no. 1 (March 2006): 62–75.

Westgaard, R. H., and J. Winkel, "Occupational Musculoskeletal and Mental Health: Significance of Rationalization and Opportunities to Create Sustainable Production Systems—A Systematic Review." *Applied Ergonomics* 42, no. 2 (January 2011): 261–96.

Wilson, Charlie. "Up-Scaling, Formative Phases, and Learning in the Historical Diffusion of Energy Technologies." *Energy Policy* 50 (November 2012): 81–94.

Wright, Helena. "Insurance Mapping and Industrial Archeology." *IA: The Journal of the Society for Industrial Archeology* 9, no. 1 (1983): 1–18.

Wu, Paul Pao-Yen, Clinton Fookes, Jegar Pitchforth, and Kerrie Mengersen. "A Framework for Model Integration and Holistic Modelling of Socio-Technical Systems." *Decision Support Systems* 71 (March 2015): 14–27.

Zio, Enrico, and Terje Aven. "Industrial Disasters: Extreme Events, Extremely Rare. Some Reflections on the Treatment of Uncertainties in the Assessment of Associated Risks." *Process Safety and Environmental Protection* 91, no. 1–2 (January–March 2013): 31–45.

THESES AND DISSERTATIONS

Alsobrook, David E. "Alabama's Port City: Mobile during the Progressive Era, 1896–1917." PhD diss., Auburn University, 1983.

Greenwood, Richard Eric. "'Scientific Engineering and Useful Improvements': The Manufacturing Career of Zachariah Allen, 1822–1872." PhD diss., Brown University, 1996. UMI Microform 9704036.

INDEX

Page numbers in italics refer to illustrations.